T0324930

Decision Support Systems for Risk-Based
Management of Contaminated Sites

Antonio Marcomini · Glenn W. Suter II ·
Andrea Critto
Editors

Decision Support Systems for Risk-Based Management of Contaminated Sites

 Springer

Editors

Antonio Marcomini
University Ca' Foscari of Venice
Italy
marcom@unive.it

Glenn W. Suter II
U.S. Environmental Protection Agency
Cincinnati, OH
USA
suter.glenn@epa.gov

Andrea Critto
University Ca' Foscari of Venice
Italy
critto@unive.it

ISBN: 978-0-387-09721-3   e-ISBN: 978-0-387-09722-0
DOI 10.1007/978-0-387-09722-0

Library of Congress Control Number: 2008931334

Printed on acid-free paper

springer.com

Preface

Decision making in environmental risk management is a very complex process that must integrate multiple and often conflicting objectives, knowledge and expertise from different disciplines and the views of multiple parties. In fact, the complexity of the process is caused by the need to look at the conservation of the environment along with the improvement of human well being in a more cost-effective way. Environmental protection and cleanup objectives are increasingly connected to the aspirations of involved populations and to the economic profits derived by the redevelopment of contaminated or abandoned areas.

Recognition of this complexity has led to the development of several frameworks and methods for establishing and rationalizing the management processes. These activities include significant attempts to codify specialist expertise into decision support tools, to facilitate reproducible and transparent decision–making and support the decision makers in defining possible options of intervention to solve a problem.

Every day, decision makers must perform a complex process of defining environmental quality objectives for contaminated river basins, coastal lagoons and other ecosystems and selecting remedial options that will achieve those goals. Although each environmental management problem is unique and should be supported by a site-specific analysis, many of the key decisions are similar in structure and objectives. This consideration has encouraged the definition of standard management frameworks and approaches, sometimes adopted in specific national or international environmental regulations, and the development of dedicated decision support systems. In fact, there is a wide multiplicity of Decision Support Systems that concern specific areas of management, such as the financial or environmental ones, but few of them tackle the assessment and management of pollutants released in the environment, and their effects. Moreover, few of them can respond to the abovementioned problem complexity with a corresponding complete but concurrently user-friendly organization.

The book, after an introduction about the environmental and socio-economic relevance of contaminated sites in Europe and the United States and about the risk based environmental management concepts and principles, provides an analysis of the main steps and tools for the development of decision support systems: environmental risk assessment, multi-criteria analysis, spatial analysis and geographic information system, indices and indicators definition.

Finally, specific chapters are dedicated to the review of decision support systems for contaminated land, river basins and coastal lagoons, including the discussion of management problem formulation and the description of the application of specific decision support systems. The case studies are selected to discuss these themes in a more illustrative and tangible way.

The decision support systems presented in this volume encompass a range of types of support and decisions to be supported. In some cases, they provide purely technical support concerning a particular decision such as, are wastes the cause of observed environmental impairments or health effects? At the other extreme, they integrate value judgments with technical information to help identify the optimum management action.

This book is addressed primarily to environmental risk managers and to decision makers involved in a sustainable management process for contaminated sites, including contaminated lands, river basins and coastal lagoons. Decision support systems are intended to meet their needs. They must be aware of what decision support systems and capabilities are available so that they can decide which to use and what new systems should be developed. A secondary audience is the environmental scientists who gather data and perform assessments to support decisions. They provide input to the decision support systems and often implement the systems for the managers. Other audiences include developers of decision support systems, students of environmental science and members of the public who wish to understand the assessment science that supports remedial decisions.

Cincinnati, USA   Glenn W. Suter II
Venice, Italy   Antonio Marcomini
                                                                                        Andrea Critto

Contents

Contributors

Paola Agostini   IDEAS (Interdepartmental Centre for Dynamic Interactions between Economy, Environment and Society), University Ca' Foscari, of Venice San Giobbe 873, I-30121, Venice, Italy; Consorzio Venezia Ricerche, via della Libertà 5-12, I-30175, Venice, Italy, e-mail: paola.agostini@unive.it

Roger Argus   Tetra Tech EMI, 1230 Columbia Street, 10th Floor, Suite 1000, San Diego, CA 92101, USA, e-mail: roger.argus@ttemi.com

Kelly Black   Neptune and Company, Inc. 8550 West 14th Ave., Suite 100, Lakewood, CO 80215, USA, e-mail: kblack@neptuneinc.org

Paul Black   Neptune and Company, Inc. 8550 West 14th Ave., Suite 100, Lakewood, CO 80215, USA, e-mail: www.neptune.com

Pieter Booth   Exponent, 15375 SE 30th Place, Suite 250, Bellevue, WA 98007, USA, e-mail: boothp@exponent.com

Adriana Brancia   Consorzio Venezia Ricerche, via della Libertà 12, Venice, Italy, e-mail: giuppelia@libero.it

Claudio Carlon   European Commission Joint Research Centre, Institute for Environment and Sustainability, Rural Water and Ecosystem Resources Unit, Via E. Fermi 1, I-21020 Ispra (VA), Italy, e-mail: Claudio.CARLON@echa.europa.eu

Marco Casini   Centro per lo Studio dei Sistemi Complessi, Università di Siena, Via Tommaso Pendola 37, 53100 Siena, Italy, e-mail: casini@dii.unisi.it

José-Manuel Zaldívar Comenges   Institute for Environment and Sustainability, Joint Research Center, European Commission, Via Enrico Fermi 1, TP 272, 21020 Ispra (VA), Italy, e-mail: jose.zaldivar-comenges@jrc.it

Susan M. Cormier   National Center for Environmental Assessment, U.S. Environmental Protection Agency, 26 West Martin Luther King, Cincinnati, OH, 45268 USA, e-mail: cormier.susan@epa.gov

Andrea Critto   Department of Environmental Sciences and Centre IDEAS, University Ca' Foscari of Venice, Calle Larga S. Marta 2137, I-30123 Venice, Italy, e-mail: critto@unive.it

Paul B. Duda AQUA TERRA Consultants, 150 E. Ponce de Leon Ave, Suite 355, Decatur, GA 30030, USA, e-mail: pbduda@aquaterra.com

Annette M. Gatchett Acting Deputy Director, National Risk Management Research Laboratory, US Environmental Protection Agency, 26 West Martin Luther King, Cincinnati, OH 45268 USA, e-mail: gatchett.annette@epa.gov

Gianmarco Giordani Dipartimento di Scienze Ambientali, Università di Parma, Viale Usberti, 33/A, 43100 Parma, Italy, e-mail: giordani@nemo.unipr.it

Silvio Giove Department of Applied Mathematics, University Ca' Foscari of Venice, Dozsoduzo 3825/E, 30123 Venice, Italy, e-mail: sgiove@unive.it

Elisa Giubilato IDEAS, Interdepartmental Centre for the Dynamic Interactions between Economy, Environment and Society, University Ca' Foscari of Venice, San Giobbe 873, Venice, Italy, e-mail: giubilato@unive.it

Stefania Gottardo Consorzio Venezia Ricerche, via della Libertà 12, 30175 Marghera, Venice, Italy; IDEAS, Interdepartmental Centre for the Dynamic Interactions between Economy, Environment and Society, University Ca' Foscari of Venice, San Giobbe 873, 30121 Venice, Italy, e-mail: stefania.gottardo@unive.it

Alexandre Grebenkov Joint Institute for Power and Nuclear Research – Sosny, Minsk, Belarus, e-mail: greb@sosny.bas-net

Bruce K. Hope Air Quality Division, Oregon Department of Environmental Quality, 811 SW Sixth Avenue, Portland, OR 97204-1390, USA, e-mail: hope.bruce@deq.state.or.us

Mark S. Johnson U.S. Army Center for Health Promotion and Preventive Medicine, Health Effects Research Program, 5158 Blackhawk Rd. Aberdeen Proving Ground, MD 21010-5403, USA, e-mail: mark.s.johnson@us.army.mil

Russell S. Kinerson Retired U. S. Environmental Protection Agency, 6527 El Reno Lane, Joplin, MO 64804, USA, e-mail: rskinerson@gotsky.com

John L. Kittle, Jr., AQUA TERRA Consultants, 150 E. Ponce de Leon Ave, Suite 355, Decatur, GA 30030, USA, e-mail: jlkittle@aquaterra.com

Marek Korcz Department of Environmental Management, Institute for Ecology of Industrial Areas, Kossutha 40-844, Poland, e-mail: korcz@ietu.katowice.pl

Sheryl A. Law Exponent, 15375 SE 30th Place, Suite 250, Bellevue, WA 98007, USA, e-mail: slaw@exponent.com

Igor Linkov US Army Engineer Research and Development Centre, Concord, MA, USA, e-mail: ilinkov@intertox.com

Antonio Marcomini Department of Environmental Sciences and Centre IDEAS, University of Ca' Foscari, Calle Larga S. Marta 2137, I-30123, Venice, Italy, e-mail: marcom@unive.it

Charles A. Menzie Exponent, 1800 Diagonal Road, Suite 300, Alexandria, VA 22314, USA, e-mail: camenzie@exponent.com

Christian Micheletti Department of Environmental Sciences, University Ca' Foscari of Venice, Calle Larga S. Marta, 30123, Venice, Italy e-mail: c.micheletti@unive.it

Chiara Mocenni Centro per lo Studio dei Sistemi Complessi, Università di Siena, Via Tommaso Pendola 37, 53100 Siena, Italy, e-mail: mocenni@dii.unisi.it

Susan B. Norton National Center for Environmental Assessment, U.S. Environmental Protection Agency, 1200 PA Ave., NW, Washington, DC 20460 USA, e-mail: norton.susan@epa.gov

Simone Paoletti Centro per lo Studio dei Sistemi Complessi, Università di Siena, Via Tommaso Pendola 37, 53100 Siena, Italy, e-mail: paoletti@dii.unisi.it

Lisa Pizzol Consorzio Venezia Ricerche Via della Libertà 5-12, I-30175 Marghera, Venice, Italy, e-mail: lisa.pizzol@unive.it

S. Thomas Purucker U.S. Environmental Protection Agency, 960 College Station Road, Athens, GA 30605, USA, e-mail: Purucker.Tom@epa.gov

Francesca Quercia APAT, Agency for Environmental Protection and Technical Services, via V. Brancati, 48 00144 Roma, Italy, e-mail: quercia@apat.it

Michiel Rutgers National Institute for Public Health and the Environment, RIVM, P.O. BOX , 3720 BA Bilthoven, The Netherlands, e-mail: Michiel.Rutgers@rivm.nl

F. Kyle Satterstrom Harvard School of Engineering and Applied Sciences, Cambridge, MA, USA, e-mail: satterst@fas.harvard.edu

Kate Schofield National Center for Environmental Assessment, U.S. Environmental Protection Agency, 1200 PA Ave., NW, Washington, DC 20460, USA, e-mail: schofield.kate@epa.gov

Elena Semenzin IDEAS (Interdepartmental Centre for Dynamic Interactions between Economy, Environment and Society), University Ca' Foscari, San Giobbe 873, I-30121, Venice, Italy; Consorzio Venezia Ricerche, Via della Libertà 2-12, Marghera-Venice, Italy, e-mail: semenzin@unive.it

Patricia Shaw-Allen National Center for Environmental Assessment, U.S. Environmental Protection Agency, 26 West Martin Luther King, Cincinnati, OH 45268 USA, e-mail: shaw-allen.patricia@epa.gov

Katherine von Stackelberg Exponent, 8 Winchester Place, Suite 303, Winchester, MA 01890; Harvard Center for Risk Analysis and Exponent, Inc., Winchester, MA, 15677, USA, e-mail: kvon@igc.org

Robert N. Stewart Department of Ecology and Evolutionary Biology, University of Tennessee, Knoxville, TN 37996, USA, e-mail: stewart@utk.edu

Neil Stiber US Environmental Protection Agency Headquarters, Ariel Rios Building, Mail Code: 8105R, 1200 Pennsylvania Ave., NW, Washington, DC 20460, USA, e-mail: stiber.neil@epa.gov

Tom Stockton Neptune and Company, Inc. 1505 15th Street, Suite B, Los Alamos, NM 87544, USA, e-mail: www.neptune.com

Terry Sullivan Brookhaven National Laboratory, Upton, NY, USA, e-mail: tsullivan@bnl.gov

Glenn W. Suter II National Center for Environmental Assessment, U.S. Environmental Protection Agency, 26 West Martin Luther King, Cincinnati, OH 45268 USA, e-mail: suter.glenn@epamail.epa.gov

Silvia Torresan Consorzio Venezia Ricerche, via della Libertà 12, 30175 Marghera, Venice, Italy, e-mail: torresan@unive.it

Ann Vega US Environmental Protection Agency, 26 W. Martin Luther King Dr., Cincinnati, OH 45268, USA, e-mail: vega.ann@epa.gov

Pierluigi Viaroli Dipartimento di Scienze Ambientali, Università di Parma, Viale Usberti, 33/A, 43100 Parma, Italy, e-mail: pierluigi.viaroli@unipr.it

Chris J.E. Welsh Department of Ecology and Evolutionary Biology, University of Tennessee, Knoxville, TN 37996, USA, e-mail: cwelsh@utk.edu

Boris Yatsalo Obninsk State Technical University of Nuclear Power Engineering, Obninsk, Russia, e-mail: yatsalo@prana.obninsk.org

Lester Yuan National Center for Environmental Assessment, U.S. Environmental Protection Agency, 1200 PA Ave., NW, Washington, DC 20460, USA, e-mail: yuan.lester@epa.gov

Alex Zabeo University Ca' Foscari of Venice, Dpt. of Informatics, Via Torino 155, 30170 Venezia Mestre, Italy, e-mail: zabeo@dsi.unive.it

C. Richard Ziegler National Center for Environmental Assessment, U.S. Environmental Protection Agency, Washington DC, USA, e-mail: ziegler.rick@epa.gov

Introduction

Annette M. Gatchett, Antonio Marcomini and Glenn W. Suter II

A book on Decision Support Systems for Risk-based Management of contaminated sites is appealing for two reasons. First, it addresses the problem of contaminated sites, which has worldwide importance. Second, it presents Decision Support Systems (DSSs), which are powerful computer-based tools for assessment and management in complex interdisciplinary decision-making processes. In this Introduction, the two aspects will be presented to explain the complexity of contaminated site assessment and management, the diversity of current policy and practise, and the helpful support provided by DSSs. These themes will be discussed in more detail in the following chapters of this book, which provide reviews of both methods and applications.

Importance of Contaminated Sites

Contaminated sites often raise greater public concern than other environmental issues, because the occurrence of wastes at a local site is vivid and may be threatening to human health. This consideration by itself is sufficient to elevate the stature of assessing and managing contaminated sites among the most critical objectives of policies and practises. Moreover, environmental protection regulations and policies give priority to the rehabilitation of contaminated resources, such as soil or water, rather than exploiting unspoiled natural resources. As a result, the rehabilitation of contaminated or potentially contaminated sites, referred to as, is preferred to the development of agricultural or natural spaces known as greenfields. This is especially true in countries, such as many European States, where green, open spaces are scarce, due to intensive land use.

Statistics support the importance of redeveloping contaminated sites. A recent European Environment Agency (EEA, 2007) report estimates that

A.M. Gatchett (✉)
Acting Deputy Director, National Risk Management Research Laboratory, US Environmental Protection Agency, Cincinnati, OH, USA

potentially polluting activities have occurred at nearly 3 million sites in Europe, of which approximately 250,000 are judged to require clean up. By contrast, more than 80,000 sites have been cleaned up in the last 30 years in European countries where data on remediation are available. European countries are not alone in having numerous contaminated sites. It is estimated that the US alone has more than 450,000 brownfield sites requiring some cleanup. Over the past 25 years the US has cleaned up more than 966 or 62% of Superfund sites (CERCLA, 1980), which are generally considered the most contaminated brownfield sites. Work continues on more than 400 Superfund sites.

Contamination affects surface water, groundwater, sediments and soils. Currently, the sources of aqueous contamination are primarily non-point, because industrial and urban effluents and other point sources are well regulated and treated. Soils have been contaminated primarily by industrial activities that involve uncontrolled disposal or leakage from inadequate storage. However, accidental spills continue to be important sources of contamination.

Another critical concern is the cost of remediation. EEA (2007) reports that annual national expenditures for the management of contaminated sites (including only soil contamination) are on average about 12 Euro per capita, which corresponds to nearly a thousandth of the average national Gross Domestic Product. Of these costs, 60% are employed for remediation measures and 40% for site investigation activities.

The importance of contaminated site management is also underlined by the fact that the previous data provided by the EEA are included in the Core Set of Indicators (CSI) produced by the Agency for the whole European community. The CSI collects those indicators that have high relevance for the EU policies' priority issues and represent the most important information for assessment of environmental topics and definition of management measures. Therefore, the inclusion of an indicator such as the "Progress in management of contaminated sites (CSI 015)" is a clear sign of the attention paid to the problem by the European Union.

Because Brownfield contaminated sites are mostly a community issue, public authorities usually play the leading role in their assessment and management, particularly when human health is at stake. However, industrial organizations and other private institutions also have an interest in the restoration of contaminated sites and should participate in remedial actions.

An example of collaboration between public institutions, regulators, private parties and industries is the establishment of scientific networks dealing with contamination topics, such as the European CLARINET (Contaminated Land Rehabilitation Network for Environmental Technologies in Europe) and NICOLE (Network for Industrially Contaminated Land in Europe) and in the US through state organizations such as the Interstate Technology and Regulatory Council (ITRC). For example, during the last 10 years CLARINET has defined strategies and approaches for addressing contaminated site issues.

It should not be forgotten that the contaminated sites problem is not only a subject of environmental debate but is also often an economic and social concern (already proved by the rehabilitation costs previously reported). Rehabilitation of contaminated sites has a socio-economic impact when important industrial activities are involved (e.g., the industrial ports of Rotterdam in the Netherlands or of Porto Marghera in Italy where pollution problems affect critical economic activities), when human communities are concerned (e.g., in urban areas, where contaminated sites are usually characterized by economic and social degradation and abandonment), and when the economic and social revenues of the remediation process, in terms of job creation, sustainable land use and social well being, are considered.

The contaminated sites issue is also addressed in several laws and regulations. In Europe, recent EU legislations have addressed contaminated sites. In the Water Framework Directive (EU Directive 2000/60/EC), the contamination of inland surface waters, transitional waters, coastal waters and groundwater is tackled in an integrated and holistic approach to water resources management (see more details about this Directive and the contamination issue in Chapter 14). Equally, another European Directive, under final preparation, regards soil and its management, where also risk assessment and management are expected to have a significant role. Moreover, liability issues are addressed by the Directive 2004/35/CE of the European Parliament and of the Council on environmental liability with regard to the prevention and remedying of environmental damage, which supports the "polluter pays" principle. While this principle remains true in the case of Superfund, the US is moving toward a more collaborative approach to encourage cleanup without strict regulatory enforcement. This was made apparent by the passage of the Small Business Liability Relief and Brownfields Revitalization Act (Brownfields, 2002).

Decision Processes for Contaminated Sites

Acknowledging the significance of the contaminated sites problem does not by itself facilitate finding its solution. The decision process for assessing and managing contaminated sites is controversial and difficult, in part because of its diverse aspects (economic interest, environmental restoration, social acceptance, technological application, land planning, and other influences). For these reasons, during recent years, many studies and regulatory applications have emphasized defining the decision process for the assessment and management of contaminated sites.

For example, the risk-based land management approach developed by CLARINET (2002) highlights three main goals: fitness for use (proper land use which is accepted by concerned people), protection for the environment, and long-term care (taking into account intergenerational effects of present choices in light of sustainability).

The various approaches used by different countries can be generalized as including three main stages: inventory; characterization and risk assessment; and remediation or management.

- The inventory stage identifies contaminated sites, performs a preliminary investigation and produces a priority list.
- The characterization and risk assessment stage collects relevant information and estimates risks to human health and the environment for use in developing management objectives.
- The last stage, management, includes selecting and implementing the remedial actions on the site that lead ultimately to its remediation, reuse and monitoring the outcomes.

Each stage calls for a decision and an assessment to provide supporting information: (1) what sites should be listed and in what priority, (2) are the risks at a site sufficient to justify remedial actions, and (3) what is the best action given the risks and other relevant information?

From this general overview, the process may be perceived as quite structured and unproblematic. In practice, the process is complex since no two sites are exactly the same. Therefore many issues and questions arise at all three stages.

For example, in the risk assessment, the set of potential remedial actions must be identified and then each must be assessed to determine its risks, costs, benefits and compliance with regulatory frameworks. The range of technologies must include plausible and practical technologies. There are rare cases in which no action is selected. In the US Superfund program, a no action alternative is assessed to determine whether remediation is needed and then for comparison to proposed remedial actions before a decision is made. Equally, in the remediation stage and more generally in the management stage, different redevelopment solutions for the site should be compared, and the concerns and preferences of stakeholders should be duly considered.

The information from the assessment stage, including risk features, economic and technological aspects and stakeholders' perceptions, should be integrated, processed and evaluated in the management stage. Therefore, decision making for contaminated sites remains a complex process, requiring considerable administrative, economic and technical efforts.

Decision Support Systems Role

A Decision Support System (DSS) can be defined as a computer-based tool used to support complex decision-making and problem solving (Shim et al., 2002). Nothing could be more fitting to the complexity of risk-based contaminated sites management, as described in the previous section.

Specifically when public authorities have to manage complex contamination issues, tools that facilitate their challenging task in a framework that efficiently provides ideas, best practices and searchable resources are of great benefit. The help that a computer-based system can offer in data summarization, in logical and quantitative analyses and in communicating analytical results and basic information is widely recognized, and is indeed the topic of this book. The many examples of DSSs detailed herein indicate the effectiveness of these instruments.

Decision Support Systems may help to answer different management questions. Examples include the following. What is the level of risk? What are the remedial technology options? What are the costs? Will the regulatory targets be achieved?

DSSs also allow the integration of different types of information. They can include integrative methodologies, such as cost-benefit analyses, that evaluate site management alternatives. DSSs can also provide powerful functionalities for analysis, visualization, simulation and information storage that are essential to complex decision processes. Information and options can be presented in an ordered structure, visualized in a space-time perspective, elaborated in simulated scenarios and therefore more easily discussed among the interested parties to reach a common rehabilitation objective.

Moreover, DSSs can facilitate one of the most important aspects of contaminated sites management, which is communication. The use of a DSS allows the parties to openly discuss potential decisions and their implications under different scenarios and assumptions of risks, benefits and costs. This fosters a dialogue and consensus building process among decision-makers, experts and stakeholders. It provides greater transparency in the decision making process for communication with the wider public and higher authorities.

References

CERCLA (1980) Comprehensive Environmental Response, Compensation and Liability Act 42 U.S.C. §9601–9675.

CLARINET (2002) Sustainable Management of Contaminated Land: an overview. A report from the Contaminated Land Rehabilitation Network for environmental technologies, p. 128.

EEA (European Environment Agency) (2007) Progress in the management of contaminated sites (CSI 015) – assessment published in Aug 2007.

Shim J P, Warkentin M, Courtney J F, Power D J, Sharda R and Carlsson C. (2002) Past present and future of decision support technology. Decision Support Systems 33: 111–126.

Small Business Liability Relief and Brownfields Revitalization Act (2002) Pub. L. No. 107-118, 115 stat. 2356.

Chapter 1
Basic Steps for the Development of Decision Support Systems

Paul Black and Tom Stockton

Abstract There is a growing desire to develop effective and efficient computational methods and tools that facilitate environmental analysis, evaluation and problem solving. Environmental problems of interest may include concerns as apparently dissimilar as revitalization of contaminated land, and effective management of inland and coastal waters. The approach to effective problem solving in both of these examples can involve the development of what are commonly called Decision Support Systems (DSSs).

Standard DSSs might be characterized as computational systems that provide access to a wealth of information pertaining to a specific problem. The types of information that might be available include information content, maps, and data. This information can be contained in databases and geographic information systems (GIS). Access is often provided through interfaces to queries that ease the task of sifting through the often large amounts of information available. These DSSs facilitate some numerical analysis (e.g., overlays of data on GIS images, rudimentary statistical analysis of data), but usually only indirectly affect evaluation and problem solving. Currently, DSSs of this form are the most common. However, an option exists to incorporate evaluation and problem solving directly into a DSS by using statistical decision tools such as sensitivity analysis and multi-criteria decision analysis. These systems may be thought of as decision analysis (MCDA) support systems.

Development of a DSS requires consideration of both the problem to be solved and the computational tools that are appropriate or needed. In terms of the problem, important components include: definition of objectives; links to the legislative or regulatory context; model structuring including identification of, and relationships between, parameters; cost factors; and value judgments. These should encompass environmental, economic and socio-political concerns. This is the standard approach to performing decision analysis using MCDA tailored specifically to environmental problem solving. A further

P. Black (✉)
Neptune and Company, Inc., Denver Colorado and Los Alamos, New Mexico, US
e-mail: www.neptune.com

A. Marcomini et al. (eds.), *Decision Support Systems for Risk-Based Management of Contaminated Sites*, DOI 10.1007/978-0-387-09722-0_1,
© Springer Science+Business Media, LLC 2009

consideration is how to gather and present case studies, once the DSS is developed. Computational issues that are faced include: database management (e.g., information, data, GIS); analysis tools (statistics, fate and transport modeling, risk assessment, MCDA); visualization of the problem; presentation of results; document production; feedback mechanisms; help; and advice. The user interface to each of these components, the navigation through these components, and degree of openness of each component of the DSS must also be considered. Openness, including communication and stakeholder involvement, is very important for maintaining transparency and defensibility in all aspects of the DSS.

1.1 Introduction

More and more people are becoming aware of the seriousness of environmental problems. Consequently, they often look to scientists to provide solutions. One way that scientists are attempting to meet this challenge is to develop efficient computational methods and tools that facilitate environmental analysis and problem solving. Environmental problems of interest may include concerns as apparently dissimilar as revitalization of contaminated land, evaluation of the impacts of ecological risk, and effective management of inland and coastal waters. Approaches to effective problem solving for these types of problems can involve the development of Decision Support Systems (DSSs).

A DSS is a system for helping to choose among alternative actions. Although a full taxonomy of DSSs could include non-computer applications (paper or back-of-the-envelope calculations, checklists, books, encyclopedias, for example), the main focus of this chapter is how to build a DSS in a computer-based environment. There are many DSS application areas, possible approaches to making decisions, and levels at which decisions can be supported. A DSS might support decision-making for a specific problem or type of problems. If the problem becomes remotely complex, perhaps involving more than a few pages of information, or a handful of quantified factors, then the DSS is usually computer assisted. If quantitative analysis is involved, then computer models are usually needed.

Before building a DSS, a designer must consider application areas, functionality, technical complexity, structural development, stakeholder involvement in both development and use, and approaches to hardware and software implementation. Further consideration must be given to the level at which the DSS should function. A DSS can be specific to an application, generic to an application area (an application framework), or generic to all possible applications (generic framework). Given these considerations, Fig. 1.1 provides an overview of some options that are available for implementing a DSS. This set of options can be implemented in a wide variety of combinations depending upon the technical level at which decisions need to be supported, the technical approach to decision-making, and the type of application. The following subsections

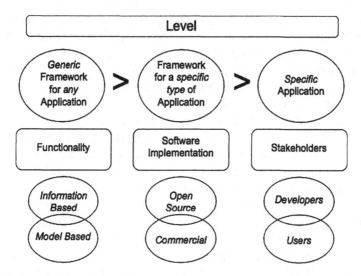

Fig. 1.1 Overview of options for decision support systems

describe these options for DSSs, and how the options affect functionality and construction of, and stakeholder interaction with, a computer-based DSS application.

1.2 Types of Decision Support System

The range of "systems" that might be called a DSS is large. The least complex are information retrieval systems that provide access to information with the goal of using the information to help make a decision. The most complex are frameworks that allow any decision to be modeled and solved using a formal decision analysis approach. A DSS can be designed that is:

- very specific to a particular decision or component of a particular decision (e.g., a watershed nutrient loading model built for a specific watershed, a Brownfield revitalization model built for a specific industrial site),
- a framework that allows a particular type of application to be modeled (e.g., watershed management, site revitalization, sustainable land reuse), or,
- a generic framework for modeling any type of decision (e.g., Analytica (lumina.com), GoldSim, (goldsim.com))

Most example DSSs presented in the succeeding chapters of this book are framework programs for specific applications. For example, DESYRE and SMARTe (smarte.org) provide frameworks for solving site contamination and revitalization problems. Both of these DSSs allow the user to evaluate a specific site. Similarly, CADDIS (epa.gov/caddis) and MODELKEY (modelk-ey.ufz.de) are application framework DSSs with a focus instead on evaluation of aquatic ecosystems. Again, the user can apply these DSSs to a specific site or location (watershed, stream system, etc.). Specific application DSSs can be built by using these framework programs, or they can be built as stand-alone applications. For example, a stand-alone DSS has been developed for the probability of volcanic hazard assessment at Yucca Mountain, a proposed nuclear waste disposal facility (CRWMS, 1996).

A potential benefit of application framework DSSs is that they provide a common format for users, so that similar problems can be approached in the same way. However, a potential benefit of building a stand-alone DSS is the greater flexibility that can be achieved by customized programming from start to finish. That is, the constraints or limitations of an application framework are removed. This depends on the amount of flexibility that is built into the framework DSS, but often it is not possible to accommodate every nuance of all the potential specific applications in a framework program.

One advantage of generic framework DSSs is that they are available for immediate use. Analytica and GoldSim are examples of commercial general framework DSSs. Analytica offers a graphical user interface that can be used to create, analyze, and communicate quantitative decision models. GoldSim is aimed instead at constructing environmental models to support decision making, although decision analysis options can be built into this visual framework program.

The focus of Analytica is on solving numerical decision problems. However, DSSs do not have to have that focus. A DSS, instead, might simply provide access to information content, as opposed to data. This distinguishes model-based and information-based DSSs. Table 1.1 uses some specific examples of DSSs to provide a summary of the interaction between types of DSSs and this level of functionality.

The World Wide Web (Web) is a prime example of an information-based DSS. Some sites on the Web are also examples of information-based DSSs (e.g., Wikipedia[1], Google). The Web holds an enormous amount of information that people use every day to help make better decisions. Consequently, the Web can be thought of as a DSS. There are still constraints for this high-level framework

Table 1.1 Examples of types of decision support systems

Functionality	Types of DSS		
	Generic framework	Specific framework	Specific application
Information-based	WWW Encyclopedia	www.epa.gov wedMD	Radon (in epa.gov) asthma (in webMD)
Model-based	Analytica GoldSim	SMARTe DESYRE	Greenville (SMARTe) Yucca Mountain PVHA

---

[1] Wikipedia provides a web page on Decision Support Systems, with many related references and examples of DSSs.

DSS, under which almost any type of information-based application can be run. The constraints are based on the information content and quality. A specific application DSS might be constructed, instead, to provide access to more information.

Thinking of the Web as a DSS opens up many more possibilities for the implementation of a DSS. DSSs can be built as stand-alone desktop programs for an individual's personal computer (PC), or they can be built as shared resources on a server. The Web is an ideal location for serving or sharing of information, and, hence offers a forum for housing DSSs. This can be true for specific applications that can be shared with a project team, for application framework DSSs that can be used to collectively build a specific application DSS, and for generic framework DSSs that can be used to build a specific application DSS. Although the Web is not a requirement for sharing such a resource, it can facilitate sharing with a broad audience. The main disadvantage of a web-based application is efficiency. That is, a DSS will usually run faster as a desktop application than as a web-based application.

The distinctions between framework and application are not always obvious. A matrix of two examples for each type of DSS is shown in Table 1.1. This table is not meant to be definitive, so much as providing some insight into the different levels of DSS and the different ways they might be assembled. Of the information-based DSS types, the World Wide Web accommodates any form of web-based development and hence can be thought of as a generic framework for DSS. An encyclopedia is another example, although it does not require a computer application. WebMD (webmd.com) could be considered a specific framework for medical information, and EPA's website acts as a specific framework for environmental information. Specific applications then follow. On the model-based DSS types, generic framework programs such as Analytica and GoldSim allow the user to build decision models for any type of application. Specific framework programs, including most DSSs presented in this book, allow the user to build specific applications within a topical area (e.g., environmental, or, more specifically, brownfields revitalization, stressor identification in aquatic systems). Consequently, specific applications include those developed using an application framework (e.g., Greenville is an application of SMARTe), or they can be stand-alone (e.g., the PVHA [probability of volcanic hazard] at Yucca Mountain was programmed using FORTRAN).

1.3 Functionality

Despite the application or framework designation of a specific DSS, the issue of functionality depends on technical objectives. Figure 1.2 shows a general technical architecture for a DSS, including the basic elements that should be considered for DSS development. The main technical components are a knowledge base, analysis tools and inference engine. Other components, such as a

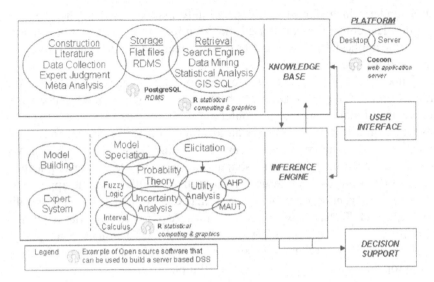

Fig. 1.2 General technical architecture for decision support systems

user interface, stakeholder interaction and quality assurance or feedback are important for effectiveness of the functionality, but do not directly impact the technical components. Figure 1.2 also highlights open source software (OSS, opensource.org) options that currently exist for implementation of the various elements. This is done to provide examples of some tools that are available for DSS construction, and to recognize that DSS construction does not require use of commercial products.

The capabilities built into a DSS depend on the role the DSS will play in meeting the application decision objectives. The role of the DSS may be to provide access to a knowledge base that contains only text, tables, and graphics, in which case it might manage, retrieve, and search information. Examples of framework DSSs of this type include the Web, Wikipedia, and web-served search engines. Specific application DSSs usually involve accessing data or summaries of data, as well as non-numerical information. When data are involved, data retrieval might occur in a variety of ways. For example, searching, sorting, subsetting, querying, plotting, data mining, and statistical analysis tools can be developed to access the data. Geographic information systems (GIS) can also be accommodated in these types of DSSs. Collectively, these might be called information-based DSSs. Information management is an essential part of building the DSS so that information retrieval is made possible. Databases might be maintained in flat files, however, in the case of large, complex datasets, data base management will often include relational databases.

Information-driven DSSs support decision-making simply by providing access to information and data. There is no attempt to fully and numerically integrate the information or data into the decision process to support quantitative inference, prediction, or decision analysis, although data summaries might be provided. Processing information to support quantitative analysis requires what is often called an inference engine. The inference engine propagates the information or data through a model that is structured to provide numerical solutions to decision support. Model development then requires structuring a model by defining relationships between factors, parameters, or variables, and then specifying values for those factors, parameters, or variables.

The approach to decision support depends on the type of DSS, and the type of quantitative analysis that is performed for a model-driven DSS. In the remainder of this section, development of information-based DSSs is described first, including static analysis tools that might be used to display data or information. Development of model-based DSSs is then described, including various options for inference engines. This is followed by a discussion of other aspects that are important to DSS development, such as developer and user interaction, quality assurance, and some philosophical comments on the building and utility of the DSSs.

1.3.1 Information-Based DSS

An information-based DSS includes information upon which decisions are supported. The simplest form provides access to textual information, possibly including static tables, pictures, and graphics. In this form, the information is used qualitatively to support decision-making. For example, Web-searching is often performed to provide information that will better inform a decision. Checklists might also be included, which allow the user to select an option from a list. This is the simplest form of DSS. It involves no numerical information or data, and, hence involves no explicit quantitative analysis. Decision-making is supported by examining and possibly qualitatively evaluating the available information content.

The next level of information-based DSS includes numerical data in addition to information. The numerical data might be used in summary form to support decision-making, but no formal propagation of the numerical information is attempted. These data are not used for the purposes of inference, prediction, or decision analysis, but are used to provide the *ad hoc* decision support that is inherent in an information-based DSS. That is, the data are used primarily as another source of information. Analytical tools can be incorporated into the DSS to view and explore data; however decision models and inference engines are not contemplated.

1.3.1.1 Structures in an Information-Based DSS

Construction of the simple information-based DSS involves database access to textual information. This is usually achievable with flat database structures that are populated with information content that might be supplied by developers, contributors or users. Key words may need to be identified in the construction process to enhance search functionality of these simple DSSs. Search engines and key word identification might lead to more complex database structures. User contributions can be set up so that they are input directly online in a Web-based DSS. Quality assurance includes checking and debugging the software, reviewing the content before it is included in the DSS, and evaluating user supplied feedback that is effective in continually monitoring the DSS and which might result in its improvement or refinement.

In the case of numerical data, flat or relational database structures can be used. GIS systems can integrate map data and data retrieval and summary. Numerical data can be structured using data that are collected specifically to populate the DSS, or data from literature reviews, meta-analysis, or expert elicitation. Sometimes, data formed from literature reviews and meta-analysis are termed secondary data. However, it is not clear that this is a useful distinction. It is more important to acknowledge the source of the data, associated quality in relation to other information sources, and how the data are assimilated or processed prior to entry into the database. This might be more important when the data are used in an inference engine or model-based DSS. The quality of the data usually becomes clear during analysis, but also might be evaluated based on user feedback. That is, if the data are not regarded as useful, it is unlikely that they will be used. This directly impacts the usefulness of the DSS.

Literature review data are not, in general, ideally suited to immediate use because the reason for their collection might not match their intended use in another application. They are usually collected and summarized, manipulated or otherwise processed prior to inclusion in a DSS database unless the literature is referenced verbatim. It is possible to include these data pre-processing steps in the DSS, so that data processing is traceable to its source through the DSS. In an open source DSS this might also be considered preferable. Options include assembling a dataset based on literature reviews, or assessing some summary statistics that imply a distribution or an important value related to the factor, parameter or variable of interest. Meta-analysis and expert opinion are likely to play much smaller roles in information-based DSS, but, if needed, user interfaces can be created to facilitate their inclusion.

Information-based DSSs can involve analysis capabilities, even if there is no intent to quantify inference, prediction or decision-making. The first step in analysis is retrieval of information. Pure content DSSs can be analyzed using search engines that can retrieve any part of the potentially massive, database. Following retrieval functions, another simple form of analysis tool is search and selection. For example, checklists can be made available so that the user can choose an option from the list. The user could, in general, select any subset of the information that is available. If numerical data are involved, then there are many more options to consider for analysis capabilities. A numerical database can be queried, statistical analysis can be performed, and, in the case of GIS data, geographic images or maps can be presented. Querying provides access to the data and allows subsets of the data to be obtained for presentation or for further analysis.

Functionally, analysis tools can be provided within an information-based DSS to analyze the supporting database. These tools span querying data to performing complex statistical analysis, including GIS interfaces. If numerical data are collected on a temporal or repetitive basis, then tools can be developed to update the database automatically, for example, using Bayesian updating. Other forms of updating can also be automated by building user interfaces for addition of data. Some analysis tools carry more weight when they are engaged in a model-based framework DSS, in which they can be used to not only analyze the data, but also to build specific DSS applications.

The structure of analysis tools in DSSs starts with providing an interface to facilitate their implementation. For information and data queries, interfaces can be built that simplify the process so that a querying language does not have to be used directly. This can be seen in examples from the Web for which queries and searches are often implemented with a simple user interface. In general, the user-interface and navigation scheme is designed to efficiently access the database by setting up some form of dialog box that translates the user preferences into a querying language command. This allows subsets of the data to be obtained for presentation or for further analysis.

At a basic level, all the analysis tools are constructed with a user interface that entails objects similar to dialog boxes. For statistical analysis tools an option exists to interface directly with a statistical programming language. Consequently, statistical analysis tools are built using dialog boxes that interface with the statistical programming language. The same basic approach of constructing dialog boxes that interface with the appropriate programming language can be taken for construction of GIS tools, expert elicitation tools, automation tools, and any other technical tools that support analysis in a DSS.

1.3.1.2 Software Implementation of an Information-Based DSS

There are many programs that can be used for software implementation of a simple information-based DSS. Here, the focus is on how DSSs can be implemented using custom Web-based programming tools and open source software, and based on World Wide Web Consortium (W3C, w3c.org) Web specifications. These tools facilitate Web-based DSS development but they can also be used for stand-alone systems. However, there are clear advantages to serving an information-based DSS on the Web given the external information that can be accessed and the large audience that can potentially benefit.

For simple information-based DSSs the information content and user interface can be developed using the eXtensible Markup Language (XML, w3c.org). XML is platform independent; for example, Windows or Linux can be used. Also, XML is software independent; the content can be edited with any text editor. XML content can be developed directly through the Open Office suite of programs. Software compatible presentation can be handled using the eXtensible Stylesheet Language (XSL, w3c.org) and Cascading Style Sheets (CSS, w3c.org). This separation facilitates providing accessibility to the broadest possible audience using the W3C Web Accessibility Initiative specifications. By using XML as a content management tool, content is separated from, and thus is independent of, presentation style, technique, and technology. This separation allows great flexibility and variety in the choice of presentation. Information content, for example, can be seamlessly transformed into HTML (w3c.org) for Web browsing and into Adobe Acrobat PDF for reporting.

For numerical data there are many database management programs and tools available, both commercial and open source. Open source options for database management include PostgreSQL (postgresql.org) and, for simple databases, XML. PostgreSQL is a highly-scalable, Structured Query Language (SQL, ansi.org) compliant, open source, object-relational database management system. User queries of PostgreSQL databases are performed using SQL based on user interaction with a database search user interface. The user interface is presented in the user's browser as an HTML form. Form parameters can be translated into an SQL query using Java and Javascript (AJAX). User queries of XML-based databases can also be based upon user interaction with an HTML form. However, for XML-based databases, form parameters can be used directly in Java or Javascript to search the database.

Many of the same tools play important roles in the development of analysis tools. Dialog boxes can be built using Web programming languages such as XML, Java, or Javascript. Of import here is how these tools can be interfaced with other programming languages that perform the analysis. Database queries for use in technical calculations can be performed using pre-defined SQL queries from R or Java. R is an open source statistical programming language (r-project.org). R is well-suited for building statistical tools for DSSs. Dialog boxes can be constructed in XML or html so that both queries and R commands are developed seamlessly. For the user this is important. For a contributing developer, an option exists to build the interface for content using XML and for analysis using R in Open Office. Since Open Office is a suite of office programs, this can ease development for most contributors, removing the need for the contributing developer to understand XML or R.

User choices in the dialog boxes are translated into the appropriate language for technical analysis. In the case of statistical tools, R is an option for the language. In the case of GIS there are open source options such as PostGIS (postgis.refractions.net), MapServer (mapserver.gis.umn.edu), and Grass (grass.itc.it). R can also be used with GIS databases to present spatial data. R can interface with the GIS software through shape files that can be provided by the user. R provides many statistical options for overlaying spatial data and statistics on these shape files, forming a nexus of geographic and topical data presentation. There are many DSSs that are aimed primarily at sharing geographic information in this form. Raw data are presented on the GIS images, and no statistical summarization is otherwise performed. However, GIS programs also offer some statistical presentations of data in a geographic presentation, but GIS programs do not offer the same power or flexibility as statistical programming languages.

There are many other software options for development of a DSS, including other open source and commercial programming tools. In addition, changes and enhancements in technology are happening constantly thus requiring constant tracking to make sure that the greatest advantage is gained from the most recently developed tools. This is perhaps particularly true for open source software, which is obviously in a continuous state of innovative development.

1.3.2 Model-Based DSSs

The main distinction between information-based DSSs and model-based DSSs is the way in which decision-making is supported. For information-based DSSs, decision-making is supported indirectly by providing access to information that is relevant to the decision at hand. Although numerical analysis tools might be involved, the decision component is qualitative. For model-based DSSs, a further step is taken in order to quantitatively support decision-making. Generally, this step might be described as building a decision model and using an inference engine to process the model. Basic construction includes building a model structure, specifying the model, and processing the model.

Model-driven DSSs provide numerical solutions that support decision-making. These solutions do not necessarily involve decision analysis, and might instead, only involve inference or prediction. The difference is in the objectives. For example, modeling of events may be sufficient for inference or prediction, or it may be necessary to include costs and value judgments into a complete decision analysis. Incorporating valuation and problem solving directly into a DSS using decision analysis tools produces a system that could be thought of as a decision analysis support system (DASS). This is accomplished by using statistical decision tools such as decision analysis, sensitivity analysis, and value of information analysis.

A decision is a choice between alternatives based on estimates of the values of those alternatives. Supporting a decision means helping people working alone or in a group to gather intelligence, generate alternatives, and make choices. Comparing alternatives in a decision analysis requires inputs that capture uncertainty in possible events, and costs or value judgments of outcomes. If costs and value judgments are not included, then the decision support in a model-based DSS stops at inference or prediction. If cost and value judgments are included then decision analysis can be performed. A hierarchy can be described within model-based DSS that ends with decision analysis.

Development of a model-based DSS requires consideration of both the problem to be solved and the computational tools that are appropriate or needed. In terms of an environmental problem, important components include:

definition of objectives; links to the legislative or regulatory context; model structuring including identification of, and relationships between, parameters; uncertainty specification; and, specification of cost factors and value judgments. These should encompass environmental, economic, and socio-political concerns. Computational issues that are faced include: database management (e.g., information, data, GIS); analysis tools (e.g., statistical analysis, fate and transport modeling, risk assessment, multi-criteria decision analysis); visualization of the problem; presentation of results; document production; feedback mechanisms; help; advice; and training. The user interface to each of these components, the navigation through these components, and degree of openness of each component of the DSS must also be considered. Openness, including communication and stakeholder involvement, is very important for maintaining transparency and defensibility in all aspects of the DSS (Henrion, 2006).

1.3.2.1 Structures in a Model-Based DSS

The structure and resulting complexity of a model-based DSS is dependent on the type of decision support envisioned. Potential components of a model-based structure are presented below in a hierarchical fashion from relatively simple to more complex. The move to more complex approaches is in large part a move towards a more a holistic framework based on decision analytic approaches aimed at adaptive decision-making in which uncertainties, decision consequences, and stakeholder values are accounted for. Information sources (e.g., hyper-links to web pages containing relevant information), presentation capabilities (e.g., GIS), and database accessibility to obtain data that might be used in statistical summaries or data analysis can, and should, also be made available in a model-based DSS.

DSS models are aimed at solving a specific question. Therefore, the initial structuring steps include specifying the decision to be made and identifying the decision options. DSS model structuring beyond identification of decision options includes components for model building and model specification. Models consist of variables and parameters, and factors and relationships between variables. Often variables are characterized as nodes, and relationships as rules between nodes. Relationships between variables can involve empirical models or mechanistic ones, also termed functional models. Complex hierarchical models can be built using rule-based systems, including expert systems and related model structures. Decisions are usually made in the face of uncertainty (Morgan and Henrion, 1990), in which case an uncertainty calculus should, arguably, be a component of an expert system or a model-based DSS. Probability theory offers one option, but there are others such as fuzzy logic (Zadeh, 1983), random sets (Molchanov, 2005), belief functions (Dempster, 1967; Shafer, 1976), and lower probability theory (Walley, 1991). Specification of cost and value functions can be performed with a wide variety of methods that attempt to characterize the utility of attributes of the decision problem (Keeney and Raiffa, 1993). Some methods of model building and model specification are described below.

1.3.2.2 Expert Systems

An expert system can be described as a tool that incorporates concepts derived from experts in a field and that structures those concepts in a way that facilitates problem solving (Giarratano and Riley, 2005). In this sense, an expert system is an integral part of a model-based DSS. Issues to consider for the modeling components include structuring and specification. Structuring can involve setting up rule-based systems or more general forms of relationships between variables (e.g., probabilistic relationships or functional relationships). Examples of rule-based systems include decision trees and production systems; whereas, influence diagrams, probabilistic networks, Bayesian belief networks, and neural networks are examples of the more general form.

Specification of an expert system, or a model, can be *deterministic* or *probabilistic* (Crowell et al., 1999). Relationships can be logical rules, probabilistic or uncertainty rules, or functional rules. The term deterministic is used to mean that constant values specify each parameter in the model. The use of probability is qualified here because there are many other theories of uncertainty that could be used instead. Functional rules can also be expressed with uncertainty. Typically, an *expert system* consists of three parts.

Expert System = Knowledge Base + Inference Engine + User Interface (1.1)

The *knowledge base* consists of the relevant domain-specific information available for a particular decision. The *inference engine* consists of the algorithms for processing the information in the knowledge base. The user interface facilitates model structuring, model specification, and navigation through the model.

The knowledge base needs to be processed by an engine that can provide useable, synthesized information. This engine is referred to as the *inference engine* and can include tools such as a user interface that guides one to relevant pieces of the knowledge base, logic rules, statistical analysis, modeling tools, and decision analysis. There is a fine line between what is considered part of the knowledge base versus the inference engine. For example, a GIS can have a relational database management component with SQL querying that may be considered as part of either the knowledge base or the inference engine. Distinct separation between the knowledge base and the inference engine increases the modularity and adaptability of the system.

Diagnostic decision trees and taxonomic identification keys are forms of expert system in which an investigator responds to a structured sequence of questions. Early uses of diagnostic decision trees are seen in medical diagnosis. In general, diagnostic decision trees are straightforward to understand, are easy to visualize, and can be implemented with or without a computer. These types of decision trees can also be automatically constructed using tree-based classification estimated by *statistical analysis* of the knowledge base (Ripley, 1996). The disadvantages of decision trees can include (Crowell et al., 1999):

- One unexpected response can lead to a misdiagnosis or incorrect conclusion
- Missing information is not handled well
- The tree is not easily updated with new information.

Crowell et al., (1999) suggest these drawbacks stem from the lack of separation between the knowledge base and the inference engine. The knowledge base is the foundation of a reliable expert system. A clean separation allows the knowledge base to be updated through learning from mistakes made in using the inference engine.

Production systems represent another form of rule-based system that attempts to perform symbolic reasoning using logic rules (Giarratano and Riley, 2005). Production systems consist of three components: working memory, the rule base, and the interpreter. The working memory contains the information the system has currently elicited. The production rules are formulated as IF-THEN statements that provide a degree of explanation for a particular case under consideration. The interpreter executes the rules based on the current working memory. The number of rules can grow quickly and it can be difficult to ensure consistency. Production systems differ from decision trees in that they need not be applied in a sequential manner.

1.3.2.3 Uncertainty

The choice of which uncertainty-based calculus to use is important. Uncertainty can be represented using, for example Bayesian probability theory (Berger, 1985), fuzzy logic (Zadeh, 1983), random sets (Molchanov, 2005), belief functions (Shafer, 1976) and lower probability theory (Walley, 1991). It has been demonstrated that the theories of fuzzy logic and random sets are identical in terms of both their static representations of uncertainty and in their dynamic combination rules (Goodman, 1994). Research has also shown that fuzzy logic is consistent and coherent only if redefined probabilistically and only for decision trees (Heckerman, 1986). Belief functions are as rich statically, but they use different combination rules. Fuzzy logic, random sets, and belief functions use an interval calculus, meaning that the measure of an event is bounded by both lower and upper values. Arguably, the interval basis means that two measures are available to characterize both uncertainty and ignorance, corresponding to the location and width of interval (Black, 1996b). Probability theory, instead, uses a single-valued calculus, although distributions can be assigned to any parameter or variable. For fuzzy logic, random sets, and belief functions, problems arise because of computational complexity, because the static representations are not sufficiently rich, and the dynamic combination rules lead to lack of coherence or rationality as defined by Bayesian probability theory. In particular regarding dynamic combination, models that are built using fuzzy logic or belief functions suffer from not being able to specify joint distributions through conditioning. Lower probability theory overcomes some of these deficiencies, but computational problems persist. Lower probability theory also uses an interval-based calculus, and lower probability theory is the most general extension of Bayesian probability theory. However, computational complexity is still an issue, and there are some foundational issues in higher dimensional spaces (Papamarcou and Fine, 1986). Bayesian probability theory conforms to normative notions of coherence, and thus does not suffer from the computational complexity issues that can plague paradigms that are based on an interval calculus.

To support decision analysis, Bayesian methods are the only ones that provide a normative, rational response (Savage, 1954). That does not mean that other methods could not be used. For example, Seidenfeld recognized that classical statistics will support the correct decision most of the time (Seidenfeld, 1992). The same applies to fuzzy logic and belief functions (Black, 1996a). Nevertheless, the remainder of this chapter will focus on Bayesian probability theory to represent uncertainty.

Probability networks are graphical models that depict the nature of relationships among a number of variables (Crowell et al., 1999). These graphical models consist of two components, one a qualitative representation of system dependencies, typically in the form of a graph or influence diagram, and the other a quantitative description of the system dependencies in terms of probability distributions. The initial step of graphical modeling allows experts to qualitatively structure the problem prior to quantitatively specifying uncertainties. A *graph*, loosely defined, is a set of nodes (vertices) connected by edges (relationships between nodes). Bayes' Theorem is the algorithmic foundation for making inferences in probabilistic networks.

The relationships in a probability network represent a blend of knowledge of the physical processes and experience, and expert opinion where knowledge of the physical mechanism is lacking. This approach is Bayesian in nature in that the focus is on probability as a belief about the system rather than a physical characteristic of the system. The approach is very flexible in that all types and sources of information can be incorporated with the input probability distributions. The outputs of a probability network are then probability distributions, rather than single values. This is a real advantage of probabilistic networks over the traditional expert system because distributions provide a more complete level of explanation than just classification. Furthermore, probability network models support adaptive decision-making under uncertainty with a built-in mechanism, Bayes' Theorem, for updating with new information, which can be implemented with Markov Chain Monte Carlo methods if necessary. A commonly cited disadvantage of probability networks models is the difficulties involved in specifying all the conditional probabilities in a large, complex network. This need not be seen as a limitation since sensitivity and decision analysis can be used in the initial stages of model development to prioritize variable selection and data collection.

An *influence diagram* (Howard and Matheson, 1981), sometimes called a decision network, is a compact graphical and mathematical representation of a decision problem. It is a generalization of a probability network where not only probabilistic inference problems but also decision problems can be modeled and solved. Consequently, influence diagrams include nodes that represent values and possible decision outcomes as well as uncertainty. General framework DSS programs such as Analytica facilitate development of influence diagrams, and provide the mathematical machinery necessary for their solution.

1.3.2.4 Utility or Value Judgments

Value judgment measurements usually encompass economic, political, and social costs. For environmental problems, values of human and ecological health are also important. Utility functions are the basis for characterizing values in formal decision analysis (Bernado and Smith, 1994). Utility functions are used to measure the relative preference of each of a set of possible decision outcomes. For example, a person might choose to take an umbrella on a walk simply because that person does not like to get wet. That person's utility for not getting wet is high. Clearly utilities are personal. Another individual might not take an umbrella on a walk because his preferences are not to carry an object compared with the consequence of getting wet. Utility functions can become complex if many attributes of the decision outcomes need to be considered. For example, when deciding whether to carry an umbrella, consideration might be given to attributes such as: preference to stay as dry as possible, need to be unencumbered to do work, desire to be seen in public with an umbrella, and cost of buying the umbrella.

For complex environmental problems the economic, political, and social consequences are often diverse and far-reaching. Possible concerns include tangible costs such as sampling costs, remediation costs, redevelopment costs, and insurance costs, as well as intangible costs related to community or ecological benefits and quality of life issues. To measure total utility, all attributes must be placed on the same scale. The scale can be monetary, or more abstract scales can be used based on scores, ranks, and weights. Whatever common scale is used, relationships between the attributes must also be formed. These can be simple, and measured, for example, in terms of scoring and weighting systems, or they can be complex, requiring specification of full joint utility functions using conditioning and independence, much the same way as a joint probability distribution can be formed.

Just as there are several competing theories of uncertainty, there are also competing theories for specifying utility or value judgments. One option is to specify values in terms of dollar costs, although this can sometimes be difficult for some attributes. Use of monetary units usually implies performing a cost-benefit analysis (Morgan and Henrion, 1990). If a comparative analysis between decision options is needed, and actual cost is relatively unimportant or is considered difficult to specify for some factors, then utility can potentially be specified instead using, for example, multi-attribute utility theory (MAUT) (Keeney and Raiffa, 1993), Analytical Hierarchy Process (AHP) (Saaty, 1980), and *outranking* methods (Rogers and Bruen, 1998). However, if, in any of these approaches, a single attribute is specified in monetary terms, then, in a complete model, the value of all other attributes could be translated into monetary value.

MAUT involves scoring to compare preferences across options. Scores are established for each attribute, and are aggregated, often using weighting mechanisms. The different attributes are combined into one common scale. AHP achieves the same end result, but does so through pairwise comparisons of decision criteria instead of utility and weighting. This is advantageous in that relative judgments are made instead of absolute judgments. However, there are disadvantages in that AHP can lead to contradictions in the preference order-ing. Outranking focuses instead on finding a decision option that has a degree of dominance over the other options. In some ways, this is a further relaxation on specifying values or preferences. There are many other methods available for specifying or ranking preferences (see, for example, Kiker et al., 2005 for environmental applications), most of which are aimed at scoring, ranking, or weighting of attributes. MAUT is discussed in greater detail in Chapter 3.

Decisions can also be made without regard for utility or expected utility. For example, *rights-based criteria* can be used in which decisions are mandated according to some rules (Morgan and Henrion, 1990). Environmental examples include complete elimination of environmental risk regardless of the cost or benefit, or setting a target risk threshold that must be satisfied regardless of cost or benefit. This type of analysis is common in environmental restoration work.

Cost-benefit analysis and MAUT serve the process of maximizing expected utility, or searching for a globally optimal decision. These methods can be considered normative, or rational, or coherent. Other methods use more relaxed sets of assumptions, which often ease the elicitation burden, but can result in inconsistencies. This trade-off is an important consideration when deciding which approach to use.

Multi-criteria decision-making refers to desires and preferences across many attributes of the decision problem (Keeney and Raiffa, 1993). Human health and ecological risk assessment deal with the science or technical side of envir-onmental problems, an aspect that is well supported by probability theory. The utility function aspects of multi-criteria decision-making are more aligned with environmental risk management activities, providing a paradigm by which environmental risk management can be quantified. A complete decision analy-sis requires both components, as described below.

1.3.2.5 Bayesian Statistical Decision Analysis

A complete decision support system will contain a knowledge base that pro-vides different types of information, some of which might be directly used in a decision analysis model, and an inference engine that allows decisions to be assessed quantitatively. To complete the quantitative components, the decision analysis model must include information and an inference engine pertaining to both the probability or science-based components and the utility or preference based components. The paradigm that can accommodate both aspects and achieve coherence is termed Bayesian statistical decision analysis (Berger. 1985). Under the full Bayesian paradigm, decision analysis is performed by maximizing expected utility, and updating of information is performed using Bayes' Theorem. Probability networks are extended with multi-criteria decision-making models to achieve a complete quantitative Bayesian decision support system. Other metrics for addressing uncertainty or decision analysis are possible, including fuzzy logic, random sets, and belief functions; however, as mentioned earlier, none of these metrics satisfy the coherence axioms of Bayesian decision analysis (Black, 1996a).

A complete quantitative decision support system takes probabilistic networks a step further in a decision analysis framework by involving utility functions associated with specific decision outcomes. Under Bayesian decision analysis the "best" decision is the decision that maximizes expected utility (Morgan and Henrion, 1990). The basic steps of a Bayesian statistical decision analysis can be summarized as follows:

- Identify decision options.
- Develop a decision analysis model structure that will support selection of the optimal decision action while accounting for uncertainty and cost and value judgments.
- Elicit utility, loss, and cost functions for decision consequences.
- Develop prior distributions and likelihood functions based on available information.
- Couple utility, prior distributions, and data likelihoods to identify optimal decision, that is, maximize expected utility.
- Determine the need for further information through sensitivity and uncertainty analysis, and use Bayes' rule to update the system if further information is collected.

Implemented in this way, Bayesian decision analysis is fully aligned with the basic tenets of the scientific method (Berry, 1996). That is, identify the question to be asked or the problem to be solved, form a decision model, collect available information, choose the best decision, and determine if the decision is supported well enough or if more data or information are needed to reduce uncertainty. Arguably, methods that deviate from Bayesian norms do not as obviously follow this basic approach to problem solving.

1.3.3 Software Implementation of a Model-Based DSS

As with information-based DSSs, both Web-based and desktop inference engine interfaces can be developed using a wide variety of powerful open source development tools including Java, C++, Visual Basic, Perl, HTML and XML for hyperlinking, and Web search engines. Several Java class proposals for implementing production expert systems are already under development. Tools can also be developed using higher-level languages such as R. Knowledge base interface tools would include relational databases (RDMS), structure query language (SQL), geographic information systems (GIS), and statistics. Statistical approaches that can be used to synthesize the knowledge base include data mining and knowledge discovery in databases (KDD), using neural networks and recursive partitioning, and meta-analysis approaches for combining information from different sources. This expert system could also largely be implemented with free open source computing tools. Building this tool under an open source paradigm expands both the availability to a wide variety of users as well as expanding the base of potential contributors. Also, using open source software guarantees that future users and contributors will not be restricted to using proprietary software in order to update or modify the tool.

From a software perspective, *model-based DDSs* differ from *information-based DDSs* largely in the reliance on a computational and probabilistic simulation engine. This engine potentially needs to be able to manage, query, and store data, integrate and control calls to external programs, and integrate elicited user input as required by the particular decision framework. In the open source community, R is a statistical programming language that can be used for the inference engine, providing capabilities for:

- programming any necessary numerical calculations,
- conducting probabilistic simulations (e.g., both Monte Carlo and Markov Chain Monte Carlo approaches),
- connecting and interacting with RDMS's via SQL (e.g., JDBC connection to a PostgreSQL database),
- contacting external programs within the probabilistic simulation (e.g., ecological process model, groundwater model),
- integrating probabilistic simulation results into the decision model,
- generating statistical and graphical interpretation of the decision simulation,
- conducting sensitivity analysis of the simulation results.

Apache Cocoon (cocoon.apache.org) is an open source Web application framework that can provide the glue that connects the user input (via a web browser) to a decision engine built in R.

1.4 Stakeholder Involvement

Stakeholders in a DSS have varied roles. Stakeholders potentially include DSS developers, contributors, and users. For simple information-based DSSs, user interaction involves developing content and using the DSS to gather information. For model-based and framework DSSs, contributions can be technical as well as content oriented, and these DSSs can be used to both gather and analyze information.

A DSS is constructed by a development team that can be responsible for project management, planning, development, maintenance, and testing or quality assurance. For environmental DSSs, environmental practitioners should be involved at all levels of development. That is, the development should be performed as a team that collectively understands the goals of the environmental problems being addressed by the DSS.

Some DSSs might also accommodate development support from contributors. Contribution is usually made externally to the DSS by submitting material to the development team. Contributors can provide content in text files or office products that can be inserted directly into the DSS, although some translation might be needed. For example for open source DSSs, the content might be translated into XML for inclusion in the DSS. Translation into XML can be facilitated using Open Office products.

Technical analysis tools or functions can also be contributed to a DSS, but external contribution might be constrained by the programming language used to build the DSS. That is, the contribution either needs to be in the same programming language or languages used by the DSS, or it will need to be translated, which often means re-programming. For commercial DSS products, the opportunities to contribute are usually limited, and direct external contribution is probably impossible. However, for open source systems, all the programming is available for any stakeholder or potential contributor. This potentially makes contribution easier, although it also raises quality assurance and testing issues.

A further option with open source DSS applications is that another party can obtain the software and tailor it to their specific needs. This party becomes a developer in its own right. This could significantly increase the usefulness and flexibility of a DSS. Sharing of open source software is strongly encouraged if not required by the license agreements. In this case, a DSS built for a specific problem could be modified directly for a similar problem by changing the program code. Whereas there are advantages to this approach, another option is for a framework DSS to capture and store specific applications for re-use. Essentially, examples can be saved and used as templates for similar situations. This can potentially result in considerable efficiencies when building a new specific application DSS.

Another option is to provide an interface within the DSS for contributions. This approach still requires some level of QA review, so it seems doubtful that contributions could immediately be loaded to the DSS. However, the interface could constrain the user input so that it more immediately matches the programming needs of the DSS. Contributions should always be acknowledged. If the contributions are submitted directly through a DSS user interface, then they can be easier to track. Receiving contributions in this way is also an organizational issue for the development team. The benefits are the potential for faster development of large DSSs. The disadvantage is managing or organizing many contributing developers. For example the statistical programming language

R has many contributors, and the contributions are managed by the R project team (r-project.org).

A user interface can also be set up to receive feedback on the DSS. The mechanism can be similar to one used for contributions, but the focus instead is on receiving comments. This might include identification of problems, suggestions for improvement or refinements, and any other feedback of interest. The feedback comments can be stored in a database, prioritized, and implemented to improve the functionality of the DSS. Once implemented, responses can also be directed to the commenter. It is important for a DSS to find ways to receive feedback. Automated electronic options exist for DSSs and are better for handling a large comment database. An open source example is Bugzilla, a product of Mozilla (bugzilla.org). Bugzilla was originally a bug-tracking tool, but can be used far more extensively as a comment-tracking tool. Commercial software such as Analytica and GoldSim provide similar feedback options.

Depending on the type of DSS, users can play different roles in building a DSS. For a specific application DSS, the role of the user is probably only to interpret the results because the DSS has been built, and information content and data are available. However, the user may also use analysis tools to process the information, and may then interpret the results. This is the level at which decision support applies to a specific application DSS.

For a framework DSS, however, the user options are more varied. The user can now build an application using the framework. In some sense, the distinction between user and developer becomes blurred, as the user runs the framework program to develop a specific application. Focusing on the framework program, the user specifies information and data that are used to build the specific application. This can involve providing content or specifying parameter values or distributions. Specification of parameter values or distributions can involve access to databases or to elicitation tools. That is, the user can request the DSS to access the internal databases directly to produce summary information and to use information in a specific application. Alternatively, the user might access external information, summarize the information externally, and then input the summaries into the application specific DSS. In this latter case, literature review data, meta-analysis, and expert elicitation might be involved. User interfaces are provided to complete the specific application from the framework program.

Within a framework program it is also desirable to have the capability to store the specific application. This serves two purposes. One is that a project team can have simultaneous access to the specific application at all times, and the other is that the specific application is saved or archived. This is a natural component of framework programs such as Analytica and GoldSim. These are desktop programs, so applications can be saved locally. For Web-based framework DSSs, for which applications are saved on the server side, database management of the user input is required. The benefit of a web-based framework DSS is that the user is provided with options to save information, to share the project, and to work in teams. Desktop framework DSSs are more difficult to share in real-time.

For environmental applications the stakeholder user groups are varied because environmental work tends to be multi-disciplinary, encompassing technical, management, regulatory, and political components. The list of environmental DSS users can include regulators, community groups, environmental engineers and consultants, property owners, developers, financial institutions, insurers, and lawyers. The roles that these users or stakeholders play in the development of a DSS depends on the nature of the DSS (application or framework), and on the functionality of the DSS. However, in general, they will play the role of user for a specific application DSS, and, in the case of a framework DSS they will play the role of user of the framework and developer of a specific application. In an open source system, the users will also play a role in quality assurance, both through feedback or comment mechanisms if available, and through general use of the DSS. That is, if the DSS is functional and useful, then the user groups will use it. If not, then the user groups will provide feedback or perhaps search for another alternative. This approach to quality assurance is different from tradition. However, it is more powerful in the sense that the product must prove its overall value before it will be widely used. The user community determines value, and, hence plays a critical role in quality assurance.

1.5 Environmental DSS Vision and Philosophy

Although there are many types of environmental problems that need solutions, the same decision strategy is often needed for each type of problem. For example, watershed management is similar in watersheds or stream systems around the World, land revitalization is similar from site to site, or environmental restoration at Superfund sites in the U.S. follows the same regulations and guidance. There are potential advantages in consistency of approach that can be realized by forming framework DSSs for some different types of problems. At another level, all environmental problems are similar in that they require consideration of the same basic economic, environmental, and socio-political factors. Consequently, framework DSSs that address different types of environmental problems can also share functionality and approaches to problem solving.

As described in the previous sections, DSSs span the gamut from information-based DSS to model-based DSS, the most holistic example of which might be called a decision analysis support system (DASS). Arguably, for the World to effectively value environmental or ecological systems, it is important that valuations of these systems are included in environmental-related decisions. It is important that they become a part of the economic equation. A framework DASS can be constructed to provide a holistic decision analysis system that integrates all aspects of environmental problem solving while facilitating communication and discussion among all stakeholders through presentation and document production capabilities. Is this absolutely necessary? The answer is probably no; there are probably cases of decision support that do not require this level of technical analysis (von Winterfeldt and Edwards, 1986). However, when decisions need to be made that affect environmental, financial, and community outcomes, then a DASS can effectively provide analytical capabilities to support decision making.

The output from a DSS often must be shared broadly. Therefore, when designing a DSS it is important to consider how to communicate and share the results. Transparency, and reproducibility of the process builds trust in the output and can begin with the programming itself. Openness in computer applications is a concept that is relatively new. It is embraced by the open source community, but does not have to be so limited in its domain (Henrion, 2006). Open source program code is publicly available for anyone to download, review, run, and modify. Arguably, the same should be required for any computer program or DSS that supports environmental decision-making because of the policy implications of many environmental decisions. Benefits of open source approaches to programming a DSS include greater transparency, reproducibility, and defensibility of the decisions that are supported. In addition, development of open source DSS is a collaborative process with a potentially wide range of contributors. The success of the open source movement is seen in products such as the Linux operating system, the Apache server system, the PostgreSQL database management program, the Firefox Web browser, the Web-based information system Wikipedia, and the statistical software program R. The same basic approach to openness can equally benefit DSS that are used for environmental problem solving. Some of the open source products are Web-based. There are clear advantages to presenting a DSS or DASS on the Web. The power of the Web can be combined with analysis and presentation tools in a comprehensive model-based environmental DSS or DASS.

Use of open source architectures helps create an environment of sharing, transparency and reproducibility, but the same can be achieved with proprietary software if the developers are willing to make all their code available. That is, open source is not a requirement for openness. What is important for DSS is that openness, transparency and reproducibility of information, and technical analysis are inherent components of the systems. This allows all potential users or stakeholders to access all the relevant information and make decisions that are as informed as possible. This means that an independent reanalysis would arrive at essentially the same results. This also means that the DSS is subject to review from all users and stakeholders, which provides much greater defensibility than is possible in a closed system.

DSS and DASS can combine knowledge bases, expert system technology, database and GIS access, analysis tools (e.g., environmental modeling, risk assessment, statistics, economic modeling, and decision analysis), and documentation and presentation capabilities, to provide a truly interactive analysis of environmental decision problems. However, the technical approach taken to solving environmental decision problems is also important. As described previously, there are many options for technical paradigms for decision support. These include Bayesian decision theory, and some interval-based calculi such as fuzzy logic, random sets, belief functions, and lower probability theory. These were discussed and compared briefly in Section 1.3.1. Arguably, the technical philosophy behind an environmental DASS should follow the intent of the scientific method and use the technical framework of Bayesian decision analysis. Within a Bayesian analysis, uncertainty is captured with probability distributions, and with costs and value judgments that are provided explicitly as loss or utility functions. As noted above, to complete a proper environmental DASS, it is critical that uncertainty and cost/value cover the full range of environmental, economic, and socio-political components of environmental decision support.

An environmental DSS can be described as providing a complete project management system for environmental decision problems that is comprised of the following components:

- Guidance for each aspect and function of the DSS, including interpretation of results and explanation of technical terms and methods.
- Access to and integration of project-specific knowledge bases with further access to the wealth of information available on the Web.
- Database management, including SQL queries and GIS access.
- Environmental modeling capability including fate and transport, risk assessment, and statistical and decision analysis tools.
- Expert system components that help the user navigate the technical choices available within the DSS analysis tools (e.g., risk assessment, financial, and social options, or statistical and decision analysis options).
- A presentation system that can be tailored to the specific needs of the users.
- A document production system that can be tailored to any form of computational output (e.g., Web-based, PDF, Office products).
- Quality assurance (QA) that is continuously measured and evaluated through user supplied feedback as well as more traditional QA techniques.
- Interactive training in each aspect of the DSS.

For a DASS, an inference engine based on Bayesian decision theory can operate within the system to directly support numerical insights into decision problems. The philosophy also embraces open source concepts including sharing, transparency, traceability, defensibility, continual development in response to feedback from the user community, and quality assurance (QA) through internal testing and user participation. The challenge of building such a DASS is how to capitalize on available resources, expertise, and knowledge, and effectively share and transfer that information to the organizations and individuals responsible for making decisions and implementing revitalization.

1.6 Conclusion

Decision support systems for contaminated sites help to organize information that enables decision making. DDSs can be categorized according to the level of application and the approach taken for technical decision support, but the ultimate test will always be: "Did the output of the process using the DSS lead to a scientifically informed decision that resulted in actions that met the environmental objectives?"

Perhaps this book should be viewed as a DSS for selecting a DSS for your needs. Therefore, we have further classified and characterized DSSs as application specific, as an application oriented framework program, or as a general framework program. A DSS can be information-based, providing access to information content and data, or model-based so that an inference engine is involved for inference, prediction, or decision analysis. Analytical tools can be part of either an information-based or model-based DSS. A model-based DSS can support an inference without actually performing decision analysis. In which case, it seems reasonable to define a DASS as the final extension that provides numerical decision analysis support. En toto, this chapter introduces the range of systems that you will encounter throughout this book and may help you to understand some of the basic assumptions thus enabling you to select DSSs based on your needs and their likely performance rather than on technical novelty or glitzy appeal.

For specific types of environmental problems, it seems reasonable to construct a framework DSS or DASS, provide access to a wide range of information through the Web or otherwise provide analysis tools, and ultimately support an inference engine that can propagate information through a model. This describes many of the example framework DSS that are included in later chapters of this book. Some models stop short of performing decision analysis; but formal decision analysis provides a method for fully integrating the value of ecological systems and community benefits in an environmental analysis, which seems like a worthwhile goal. The largest obstacle at this time is probably the challenges associated with valuing human health, ecosystems, or community benefits. However, methods are now being developed to address this problem. It should not be long before an environmental DASS can fully support reasonable decision analysis that involves comprehensive value judgments as well as uncertainty management. In the meantime, we encourage you to use or construct a DSS that works for you.

Acknowledgments The value of reviews by Dr. Paul Bardos (The University of Reading) and Dr. Neil Stiber (EPA)

References

Berger J (1985) Statistical Decision Theory and Bayesian Analysis. Springer-Verlag, New York
Bernado JM, Smith AFM (1994) Bayesian Theory. John Wiley and Sons, Inc., Chichester, England

Berry D (1996) Statistics: A Bayesian Perspective. Duxbury Press

Black PK (1996a) An Examination of Belief Functions and Other Monotone Capacities, Ph.D. dissertation, Department of Statistics, Carnegie Mellon University

Black PK (1996b) Geometric Structures of Lower Probability, in Applications and Theory of Random Sets, August 22–24. Institute for Mathematics and Its Applications (IMA), University of Minnesota, Minneapolis, Minnesota USA

Crowell RG, Dawid AP, Lauritzen SL, Spiegelhalter DJ (1999) Probabilistic Networks and Expert Systems. Springer-Verlag, New York

CRWMS M&O (Civilian Radioactive Waste Management System, Management and Operating Contractor) (1996) Probabilistic Volcanic Hazard Analysis for Yucca Mountain, Nevada, BA0000000-1717-2200-00082, Rev. 0, June 1996

Dempster AP (1967) Upper and lower probabilities induced by a multivalued mapping. The Annals of Mathematical Statistics, 38:325–339

Giarratano JC, Riley G (2005) Expert Systems Principles and Programming. PWS Publishing Company, Boston MA

GoldSim Technology Group (accessed 2008) GoldSim. <http://www.goldsim.com>

Goodman IR (1994) A new characterization of fuzzy logic operators producing homomorphic-like relations with one-point coverages of random sets. In: Wang P (eds) Advances in Fuzzy Theory and Technology Vol. 2. Duke University, Durham NC, pp. 133–160

Heckerman D (1986) Probabilistic interpretations for MYCIN's certainty factors. In: Kanal LN, Lemmer JF (eds) Uncertainty in Artificial Intelligence. North-Holland, Amsterdam, The Netherlands, pp. 167–196

Henrion M (2006) Open-Source Policy Modeling. Lumina Decision Systems, Inc. <http://www.lumina.com/>

Howard RA, Matheson JE (1981) Influence diagrams. In: Howard RA, Matheson JE (eds) (1984) Readings on the Principles and Applications of Decision Analysis Vol. 2. Menlo Park, CA, pp. 719–762

Keeney R, Raiffa H (1993) Decisions with Multiple Objectives. Cambridge University Press, Cambridge, UK

Kiker G, Bridges T, Varghese AS, Seager TP, Linkov I (2005) Application of multi-criteria decision analysis in environmental management. Integrated Environmental Assessment and Management 1(2):49–58

Lumina Decision Systems, Inc. (accessed 2008) Analytica. < http://www.lumina.com/ >

Modelkey Partnership (accessed 2008) Models for Assessing and Forecasting the Impact of Environmental Key Pollutants on Marine and Freshwater Ecosystems and Biodiversity. <http://www.modelkey.ufz.de/>

Molchanov I (2005) The Theory of Random Sets (Probability and its Applications). Springer, New York

Morgan MG, Henrion M (1990) Uncertainty: A Guide to Dealing with Uncertainty in Quantitative Risk and Policy Analysis. Cambridge University Press, Cambridge, UK

Open Source Geospatial Foundation (accessed 2008) Geographic Resources Analysis Support System. <http://grass.itc.it/>

Open Source Initiative (accessed 2008) Open Source Initiative. <http://www.opensource.org>

Papamarcou A, Fine TL (1986) A Note on undominated lower probabilities. Annals of Probability 14(2):710–723

PostgreSQL Global Development Group (accessed 2008) PostgreSQL. <http://www.postgresql.org>

Refractions Research (accessed 2008) PostGIS. <http://postgis.refractions.net>

Regents of the University of Minnesota (accessed 2008) MapServer. <http://mapserver.gis.umn.edu>

Ripley, BD (1996) Pattern Recognition and Neural Networks. Cambridge University Press, Cambridge, UK.

Rogers M, Bruen M (1998) A new system for weighting environmental criteria for use within ELECTRE III. European Journal of Operational Research 107:552–563

Saaty TL (1980) The Analytic Hierarchy Process. McGraw Hill, New York

Savage LJ (1954) The Foundations of Statistics. Dover Publications, New York

Seidenfeld T (1992) Fisher's fiducial argument and Bayes' Theorem. Statistical Science 7:358–368

Shafer G (1976) A Mathematical Theory of Evidence. Princeton University Press, Princeton, New Jersey

The Apache Software Foundation (accessed 2008) Cocoon. <http://cocoon.apache.org>

The Mozilla Organization (accessed 2008) Bugzilla. <http://www.bugzilla.org>

The R Foundation for Statistical Computing (accessed 2008) The R Project for Statistical Computing. <http://www.r-project.org>

U.S. Environmental Protection Agency (2007) Causal Analysis/Diagnosis Decision Information System. <http://cfpub.epa.gov/caddis/>

U.S. Environmental Protection Agency, German Federal Ministry of Education and Research, Interstate Technology & Regulatory Council (accessed 2008) Sustainable Management Approaches and Revitalization Tools – electronic (SMARTe). <http://www.smarte.org>

Von Winterfeldt D, Edwards W (1986) Decision Analysis and Behavioral Research. Cambridge University Press, Cambridge, UK

Walley P (1991) Statistical Reasoning with Imprecise Probabilities. Chapman-Hall

WebMD LLC (accessed 2008) WebMD. <http://webmd.com>

World Wide Web Consortium (accessed 2008) World Wide Web Consortium. <http://www.w3c.org>

Zadeh LA (1983) The role of fuzzy logic in the management of uncertainty in expert systems. Fuzzy Sets and Systems, 11:199–228

Chapter 2
Environmental Risk Assessment

Andrea Critto and Glenn W. Suter II

Abstract Environmental risk assessment is the process of evaluating the likelihood that adverse human health and ecological effects may occur or are occurring as a result of exposure to one or more agents. At contaminated sites, environmental risk assessment estimates risks of effects if no remedial action is taken and if each of the proposed alternatives were implemented. The purpose is to provide relevant and useful information to inform the remedial decision. The environmental risk assessment process begins with a planning phase that defines the scope and goals or the assessment. Next an analytical phase estimates exposure levels and exposure-response relationships for the contaminants of concern and the endpoint entities. Then a synthesis phase brings the analytical results together to estimate risks and associated uncertainties. The risks must then be communicated to the decision makers and stakeholders in a useful form. If a decision analysis is used to support the decision process, the risk assessment results should provide the needed input. Environmental risk assessments may be simple comparisons of point estimates of exposure to toxicological threshold values (e.g., no observed adverse effect levels). At the other extreme, they may include spatial analysis, probabilistic modeling, weighing of evidence, and other advanced techniques. Decision support systems may make environmental risk assessments quicker, easier, and more consistent and provide access to advanced analytical techniques.

2.1 Introduction

Risk may be defined as "the combination of the probability, or frequency, of occurrence of a defined hazard and the magnitude of the consequences of the occurrence" (US-National Research Council 1983; Royal Society 1992). It should be differentiated from hazard, which is commonly defined as "a property or situation that in particular circumstances could lead to harm" (Royal Society

A. Critto (✉)
Department of Environmental Sciences, University Ca' Foscari, Venice, Italy

A. Marcomini et al. (eds.), *Decision Support Systems for Risk-Based Management of Contaminated Sites*, DOI 10.1007/978-0-387-09722-0_2,
© Springer Science+Business Media, LLC 2009

1992). In fact, it is the likelihood of harm as a result of exposure to hazards which distinguishes risk from hazard.

Accordingly, Risk Assessment is the procedure in which the risks posed by hazards associated with processes or situations are estimated either quantitatively or qualitatively. Specifically, Environmental Risk Assessment is the examination of risks resulting from hazards in the environment that threaten ecosystems, plants, animals and people. It includes human health risk assessment and ecological risk assessment. Within environmental risk assessment, Ecological Risk Assessment is a process for organising and analysing data, information, assumptions, and uncertainties to evaluate the likelihood of adverse ecological effects (US-EPA 1998). This definition emphasizes the role and benefit of risk analysis as a methodology for systematically gathering, structuring and analysing relatively large bodies of complex information.

In the context of this book, risk based management of contaminated sites, environmental risk assessment informs remedial decisions by estimating the likelihood of adverse effects on human health and the environment. Typically, this is done in two stages. First, the risks associated with the baseline condition are assessed to determine whether risks from the unremediated site are acceptable. Second, if baseline risks are unacceptable, the risks associated with alternative remedial actions (e.g., capping, removal, or land use restrictions) are assessed. These remedial assessments consider whether sufficient risk reduction would be achieved and whether significant risks are associated with the remedial process itself.

Subsequent to risk assessment, Risk Management is the decision-making process for identifying, evaluating, selecting and implementing actions to prevent, reduce or control risks to human health and the environment (CRARM 1997). The process involves comparing the risks of taking no action with the risks associated with each remedial alternative, while taking into account social, cultural, ethical, economic, political, and legal considerations. It is often performed informally and subjectively by the decision-maker, but it may be informed by a formal management assessment employing cost-benefit analysis, net benefit analysis, decision analysis, or another technique. It should result in risks being reduced to an "acceptable" level within the constraints of the available resources.

In the process of assessing and managing risk, risk perception is a major determinant in whether a risk is deemed to be "acceptable" and whether the risk management measures undertaken are seen to resolve the problem (Fairman et al. 1998). Risk perception involves people's beliefs, attitudes, judgements and feelings, as well as the wider social or cultural values that people adopt toward hazards and the benefits of technology. It is closely linked to risk communication, which is an increasingly important area of risk management concerned with the way in which information relating to risks is communicated and includes the exchange of information about health and environmental risks among risk assessors, risk managers, the public, media, interested groups and others.

2.2 Historical Development

Risk assessment has its roots in gambling and insurance, but it became formalized as a method of addressing health concerns in the early 1970s (Ross 1995; Barnard 1994).

The primary impetus for risk assessment of contaminated sites was the promulgation in the United States of the Comprehensive Environmental Response, Compensation and Liability Act (CERCLA, or "Superfund") in 1980. It established a national program for remediating sites that were contaminated by releases of hazardous substances. The overarching mandate of the Superfund program is to protect human health and the environment from potential threats posed by environmental contamination. To meet this mandate, the United States Environmental Protection Agency (US-EPA) developed a human health and ecological evaluation process as part of its remedial response program. This process of gathering and assessing human health risk information was carried out by adapting chemical risk assessment principles and procedures first proposed by the National Academy of Sciences (NRC/NAS 1983).

The first formal human health risk assessment guidance manual was known as the Superfund Public Health Evaluation Manual (US-EPA 1986). After several years of Superfund program experience conducting risk assessment at hazardous waste sites, US-EPA updated the Public Health Evaluation Manual with the Human Health Evaluation Manual (US-EPA 1989). That guidance is currently used for the evaluation of hazardous waste and other sites in many states around the USA and internationally (CARACAS 1998, 1999).

During the 1990s, an increased attention was paid to developing ecological risk assessment and probabilistic risk assessment procedures.

In 1992, the US-EPA's Framework for Ecological Risk Assessment (US-EPA 1992) proposed a structure, principles and terminology for the ecological risk assessment process. The Agency then developed Guidelines for Ecological Risk Assessment that explain how to implement the framework (US-EPA 1998). Specific guidance for ecological risk assessments for Superfund is provided by the U.S. EPA (1997).

The concept of risk implies uncertainty and probabilistic analyses. The US-EPA has provided a "Process for Conducting Probabilistic Risk Assessment" at contaminated sites (US-EPA 1999).

Risk assessment approaches are becoming widely used in Europe to support the implementation of the recent European environmental policies concerning, for instance, the production and use of chemicals, the remediation of contaminated sites, the management of water quality and the protection of human health.

Risk assessment is the key tool for setting acceptable practices and uses of chemicals, as well as to set restrictions and risk reduction needs (Tarazona 2002). Currently, different requirements and protocols are established for several groups of chemicals. Industrial chemicals are covered by two categories: "existing" and "notified" (new) substances, to which a common risk assessment protocol is applied: the Technical Guidance Document (EC 2003). Other regulations and technical notes for guidance cover the risk assessment of pesticides, biocides, pharmaceuticals, feed additives, etc. However, a revision of the European chemicals policy is on-going, as result of the discussion on implementation of the Commission's White Paper issued in February 2001 (EC 2001) and the introduction of the REACH (Registration, Evaluation, Authorization and restriction of Chemicals) legislation. Two basic ideas are focusing the discussion. First, scientifically sound risk assessment methodologies should be used in the decision process whenever possible. Second, the "Precautionary Principle" should be applied when a scientifically sound risk assessment cannot be conducted. Moreover, in the context of the REACH legislation, the focus has shifted from risk assessment to risk management and from the principle of the authorities identifying and regulating the risks to industry taking responsibility for doing the assessments and for implementing the necessary measures to adequately control the risks (i.e. the *reversal of the burden of proof*).

For contaminated sites, the risk assessment approaches applied over the EU have been thoroughly explored by CARACAS, CLARINET and NICOLE networks. Results of the studies (CARACAS 1998, 1999; CLARINET 2003) have shown that, even if the rationale is very similar, several differences exist among Member States in terms of approaches. Moreover, to assist in the convergence of thinking and the development of solutions for the problems presented by contaminated land in Europe, the concept of risk based land management (RBLM) was proposed by CLARINET (2003) and included in the recommendation contained in the European Thematic Strategy for Soil Protection.

While the idea of a unique risk assessment model for soil contamination is generally not supported, there is a consensus for developing a common framework and a tool box, comprising a set of models and common data bases. Recent initiatives for building an Exposure Factors Sourcebook for Europe are in line with this approach and similar projects could be addressed to physico-chemical and toxicological data bases. Moreover, much of the work done to develop and harmonize risk assessment methods in related fields (e.g., for the risk assessment of new and existing substances coordinated by the European Chemical Bureau), can contribute to the improvement of risk assessment practices for contaminated sites.

2.3 The Role of Risk Assessment in Environmental Management

Environmental risk assessment has became a fundamental tool for the environmental decision making process, especially for chemical risk control.

Several complementary factors led to the definition of this fundamental role. The most important of these was increased public concern about pollution and environmental risks, which increased demand for prevention and protection.

As a consequence, the development of environmental regulations and policies were accelerated, in order to define stringent environmental benchmarks (i.e., environmental quality standards) and innovative assessment approaches to support environmental management processes.

As a result, risk assessment and management techniques are used more and more as decision-making tools (Fairman et al. 1998) for: (a) designing regulations (e.g., the EU legislation relating to new and existing hazardous substances); (b) providing a basis for site-specific decisions (contaminated land sites are an example where risk-based regulation is being used in Europe); (c) ranking environmental risks (e.g., prioritisation of chemicals); and (d) comparing risks.

Environmental risk assessment has become useful for planning and managing land use and for defining environmental monitoring plans.

Because environmental risk assessment has been used by regulators, it is also increasingly used by industry. In fact, companies use environmental risk assessment to determine the levels of risk associated with certain processes or plants and for industrial financial planning (Salgueiro et al. 2001). Finally, risk assessment and management can be important decision-making tools in the prioritisation and evaluation of industrial risk reduction measures.

2.4 Overview of Risk Assessment Frameworks and Approaches

The procedure for human health risk assessment proposed by the National Academy of Sciences, consisting of (a) hazard identification, (b) exposure assessment, (c) dose-response assessment, and (d) risk characterization, has been widely used and accepted (NRC/NAS 1983). It has also been elaborated and adapted for other organizations. For example, Covello and Merkhofer (1993), proposed a procedure that includes: (a) problem formulation; (b) hazard identification; (c) release assessment; (d) exposure assessment; (e) consequence assessment; (f) risk estimation; and (g) risk evaluation (see Fig. 2.1).

Although the human health and ecological risk assessment processes are conceptually similar (ecological risk assessment having been developed from human health risk assessment), they developed differently (Fairman et al. 1998).

The US-EPA (1992, 1998) pioneered the development of ecological risk assessment, by developing a framework and guidelines. They include three major steps, as shown in Fig. 2.2.

Problem formulation is the planning and scoping process that converts the goals and constraints provided by the risk manager into an operational plan for performing the risk assessment. The main expected results from the problem formulation step are: (a) the selection of assessment endpoints, which are the ecosystem's components or attributes of concern (e.g., the abundance of a fish population or the number of soil invertebrate taxa); (b) a conceptual model that represents the hypothesized pathways by which human activities induce effects on the assessment endpoints (Fig. 2.3); and (c) an analysis plan that designs the analysis program, including identification of the data needs, means for data

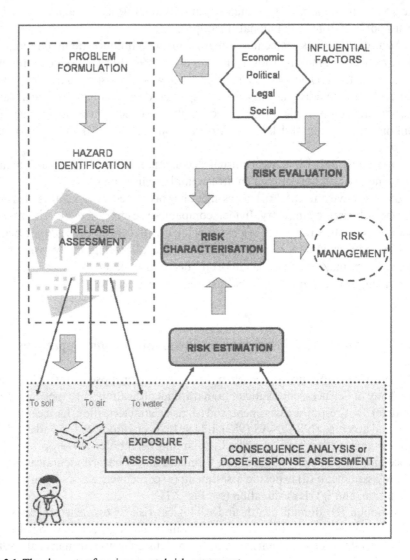

Fig. 2.1 The elements of environmental risk assessment

generation, and methods for conducting the subsequent steps of the risk assessment. The *analysis phase* contains the characterizations of exposure and of ecological effects. In the characterization of exposure, the assessors quantify the release, migration and fate of contaminants, and characterise the exposure of the receptors. In the characterization of effects, the assessors evaluate the evidence of any cause and effects relationships and defines the exposure-response relationships. *Risk characterisation*, the final step, provides risk estimates, through integration of the results of the exposure and effects characterisations, and characterizations of uncertainties.

Fig. 2.2 The framework for
ecological risk assessment

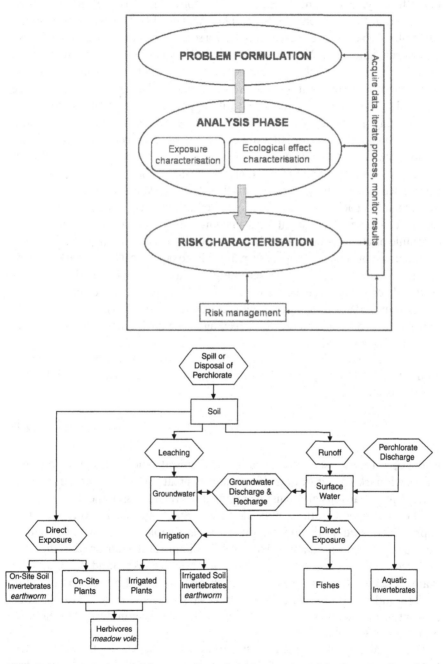

Fig. 2.3 A conceptual model for an ecological risk assessment of lands contaminated by perchlorate

2.4.1 The Tiered Approach: Screening and Definitive Risk Assessment

For efficiency, both human health and ecological risk assessments are often performed in tiers of increasingly complex analyses. In assessments of contaminated sites, two tiers are typically employed, a screening assessment and a definitive assessment (Suter et al. 2000; US-EPA 2001). Screening assessments are intended to narrow the scope of subsequent assessments by screening out chemicals, media, or routes of exposure that are not credible hazards. Definitive assessments are intended to provide risk estimates that will support a management decision.

Screening assessments are like the screening of soil to remove rocks and roots. At contaminated sites, they are used primarily to screen out chemicals that do not occur at sufficient concentrations, over a sufficient area or with sufficient frequency to be contaminants of concern. Usually, sufficient concentration is determined by comparing a conservatively-estimated exposure concentration to a toxicological benchmark such as a soil screening level (US-EPA 2005). The exposure concentration might be a maximum observed concentration, a high percentile of the distribution of observed concentrations, or a modelled concentration based on a high-exposure scenario. The toxicological benchmarks may be values developed specifically for screening such as the human health and ecological soil screening levels (US-EPA 1996, 2005), the aquatic screening benchmarks (Suter 1996) or values developed ad hoc. If the conservative exposure concentration does not exceed the screening benchmark concentration, the chemical may be screened out.

The combined toxicity of multiple chemicals may also be screened. The only model that is routinely used for that purpose is the hazard index (HI). The assessor calculates:

$$HI = \sum(C_{ei}/C_{bi}) \qquad (2.1)$$

where C_{ei} is the exposure concentration of chemical i and C_{bi} is the corresponding benchmark concentration. If the sum is greater than one, the mixture is potentially hazardous and must be retained for further assessment.

Chemical concentrations may also be screened against background. This approach is based on the idea that it is not reasonable to remediate concentrations below background levels. However, it is important to ensure that the form of the contaminant is not different from the form in background materials. For example, concentrations of hexavalent chromium should not be compared to background chromium which is nearly always trivalent. Also, naturally high concentrations, such as soils in areas with metal ores, should not be treated as part of the regional background.

Commonly, two screening assessments are performed at contaminated sites. First, a screening assessment of pre-existing data is used to inform the problem formulation. Then, when a preliminary sampling and analysis program has been completed, another screening assessment may be performed to revise the complete sampling and analysis program that will inform the definitive assessment.

After screening assessments and data generation are complete, definitive assessments are performed to inform the remedial decision. They tend to differ from screening assessments in several important respects.

They are more focused on the priorities of the decision makers and stakeholders. The results of definitive assessments must, as far as possible, address the concerns of stakeholders and provide the information needed to select a remedial action that adequately addresses environmental risks.

They are focused on those contaminants, media and receptors that passed the screening assessment. Because screening assessments typically eliminate most contaminants of potential concern and even some media and receptors, definitive assessments can generate and analyze information that is specific to a relatively few issues.

They are more site specific. Definitive assessments should replace generic assumptions and models with site-specific information and models that represent the distribution and movements of contaminants and receptors on the site.

They are more spatial. Because remedial actions must be performed in specified areas, definitive risk assessments must associate risks with specific areas of land, reaches of streams, or other contaminated areas. In addition, the risks are dependent on the area contaminated. Small contaminated areas typically provide small risks to wide-ranging animals. The exceptions are area independent routes of exposure, such as a highly contaminated waste pond that provides drinking water.

They are more specific to the endpoint receptors. For example, while a screening assessment may use the highest observed concentration and the lowest toxicity value, a definitive assessment may use data for important site species and limit concentration data to those seasons in which sensitive life stages are present on the site.

They must incorporate more lines of evidence. Definitive assessments may use data from biological surveys of the site, toxicity tests performed with site media, laboratory toxicity tests, and biomarkers and body burdens of contaminants in organisms from the site. As discussed below, these lines of evidence must be weighed and integrated to estimate risks.

They use more quantitative analyses of uncertainty. In place of the conservative assumptions that are used in screening assessments, definitive assessments may use uncertainty analysis to express the implications of missing or inaccurate information.

They consider the risks and benefits of alternative actions. Different remedial actions including no action (also called natural attenuation) have a mixture of benefits from removal or isolation of contaminants and risks from dredging, earth moving, disposal, incineration, etc.

Decision support systems play different roles in screening and definitive assessments. Screening assessments are relatively simple and the entire assessment can be readily converted to an algorithm and implemented in software.

Definitive assessments are potentially much more complex. As a result, there is a greater need for technical support but less opportunity for standardization. Some portions of the assessment problem are more readily adapted to standard analyses. For example, spatially explicit analyses of risks to humans or wildlife from contaminated soils have been implemented in systems such as DESYRE and SADA (Chapters 8, 11). However, portions of definitive assessments that weigh multiple lines of evidence or that include unusual exposure scenarios or receptors are less readily incorporated into a DSS.

2.4.2 *Weight of Evidence Approaches and Triad Schemes*

In the last decade, Weight of Evidence (WoE) approaches have become common in Environmental Risk Assessment, especially for Ecological Risk Assessment.

Definitions and interpretations of WoE vary broadly, and there are no standardized methods or regulatory guidance on how to conduct WoE studies.

As used in environmental studies, WoE can be defined as the determination of environmental risks by weighing multiple Lines of Evidence (LoE). This determination incorporates judgements concerning the quality, extent and congruence of the data contained in the different LoEs (Chapman et al. 2002). In this context, a LoE is information of a particular type (e.g., chemical analyses of water or laboratory acute lethality tests) that pertains to an important aspect of the environment (Smith et al. 2002).

WoE is useful because multiple LoEs are often available for environmental assessessment, and it is unreasonable to use only one and discard the others. In addition, different LoEs bear different realtionships to the assessment endpoint, and provide different types of evidence. Finally, the quality of data supporting the lines of evidence vary in ways that are not always apparent, so it is important to weigh all of the evidence so that errors or uncertainties are revealed by discrepancies among LoEs. This means that analysis of a single LoE is in most cases insufficient to achieve credible evaluation of environmental impairments and several lines are needed to adequately assess stressor exposure and effects. A single LoE can be useful as a screening tool but the possible conflicting results from different LoEs require a WoE assessment for final decision-making (Hall and Giddings 2000).

Each LoE can provide unique and useful information to the assessment process, and, in this sense, no LoE is redundant. However, a fundamental step in a Weight of Evidence consists in recognizing all the possible advantages and limitations presented by each LoE (Hall and Giddings 2000) in order to select an appropriate and complementary combination of LoEs. This evaluation strictly depends on the aims of the study and the analyzed situation; it takes place in the Problem Formulation phase of ERA, after developing the conceptual model in which relationships between targets and environmental stressors are made explicit.

The simplest form of assessment by WOE is: if most of the assessment results suggest impairment, then there is a greater likelihood that there truly is ecosystem impairment, but if most of the assessment results suggest no impairment, then impairment is unlikely (Burton et al. 2002a). The most common criticism of this simple approach is that not all LoEs are equally reliable or relevant, and high quality LoEs may be negated by more numerous low quality LoEs.

Perhaps the best-know approach for combining multiple LoEs is the Sediment Quality Triad (SQT) first articulated by Long and Chapman (1985). The Triad approach is based on a standard combination of three Lines of Evidence; in the SQT they are:

- sediment chemistry;
- benthic community structure; and
- sediment toxicity.

The assessment endpoint for the SQT is benthic community structure, and the observed or measured parameters included in each LoE are characteristics, related to the assessment endpoint, that change in response to a stressor to which they are exposed. These parameters are called measurement endpoints (US-EPA 1992) or measures of effects (US-EPA 1998).

These three LoEs have been selected because each of them can provide a distinct and complementary kind of information about the investigated environment, and, by means of their integration, a solid conclusion about the effective ecosystem impairment can be achieved (Chapman 1990).

The SQT is generally and widely used, but it is continuously improved (Chapman et al. 2002). It has been applied to numerous sediment quality assessments; however, it provides an overall interpretation scheme applicable also to other environmental media, including contaminated soil (Rutgers and Den Besten 2005). In fact, all WoE assessment methods based on the selection of three main LoEs (i.e. related to chemistry, ecology and toxicity) can be defined as Triad-like or Triad based approaches.

Triad-like approaches must address two main issues: (1) the proper selection of the measurement endpoints within each LoE in order to achieve an accurate assessment of impairment and (2) the method by which the different LoEs are combined and integrated into a WoE-based decision. Chapman (1990) provided a standard set of interpretations for each possible combination of positive or negative results for the three LoEs. For example, if the sediment is toxic and the benthic community is altered, but no chemical is measured at potentially toxic concentrations, then "unmeasured toxic chemicals are causing the degradation." Alternatively, Best Professional Judgement (BPJ) or Multi-Criteria Decision Analysis (MCDA) may be used when results are ambiguous or for some other reason triad results are not interpretable using Chapman's logic.

Triad based approaches and other WoE approaches, can be classified as qualitative or quantitative.

Qualitative approaches are defined as the "simple" combination of various LoEs results in a non-quantitative manner. Congruencies or disagreements among the different LoEs are evaluated, as in Chapman's standard interpretation, but no quantitative risk estimation is provided.

They includes all methods that use a matrix to integrate and summarize the information provided by different LoEs. A critical issue in this type of approach consists in the formulation of criteria for evaluating the relevance of impacts highlighted by each LoE, while the main advantages of matrix-based approach are that it has high degree of sensitivity, applicability and transparency (Burton et al. 2002b).

In general, qualitative approaches have the advantages of being easy to use and have a wide appropriateness but they may be characterized by low robustness (Burton et al. 2002b). However, if BPJ is used in place of clear inferential rules, qualitative infrerences are not transparent and thus cannot be easily understood by stakeholders.

Quantitative approaches combine different LoEs by developing and applying indices. A single index can be calculated for each LoE or an overall index can integrate information provided by all different LoEs.

Use of indices was criticized by Burton et al. (2002a), because it results in information compression. Indices have the great advantage to ease the communication to non-professional stakeholders, irrespective to the specific methods used. Moreover, the ranking of sites according to their values supports the identification of critical areas that need to be further investigated.

We have emphasized the triad approach to weighing multiple lines of evidence because of its popularity. However, other approaches have been developed that are more flexible. One is to numerical weighting and scoring, as in the Massachusetts method (Menzie et al. 1996). In such systems, the ourcomes are scored for each LoE (e.g., low, moderate and high responses receive scores of 1, 2, or 3), the scores are weighted based on the relevance and quality of the LoE, and the weighted scores are combined. Another approach developed for contaminated sites is abductive inference or "inference to the best conclusion" (Suter et al. 2000, Suter 2007). It combines all relevant information into a set of standard types: single chemical toxicity, ambient media toxicity, biological surveys, and biomarkers and pathologies. Results for each type of evidence are then evaluated against a set of issues and weighting considerations and a qualitative score, weight and explanation is assigned to each. Finally, a logical analysis is used to reach a conclusion concerning risks. It is similar to Chapman's original triad, but allows any number of types of evidence and it allows for ambiguous or uncertain information. Another application of "inference to the best conclusion" is found in the CADDIS system (Chapter 17).

2.4.3 Spatial Risk Assessment Approaches

Assessments of risks to human and ecological receptors increasingly use spatial information including the spatial distribution of stressors, receptors and effects (see also Chapter 4). Spatial analysis is crucial because regional-scale processes influence population processes and human impacts and because site boundaries and other jurisdictional units rarely, if ever, coincides with biologically or ecologically significant spatial units (Andersen et al. 2004). Moreover, the spatial distribution of multiple habitats, multiple sources, multiple stressors and multiple receptors as well as the characteristics of the landscape affect the risk estimate (Landis 2005). Similarly, the risk estimate is influenced by the spatial and temporal distribution of soil and hydrogeologic characteristics, the environmental settings, the current and anticipated use of land and the socio-economic situation (Bien et al. 2004).

According to the scale of the problem to be assessed, it is possible to identify two different approaches to deal with the spatial dimension of the risk assessment: the site-specific spatial risk assessment and the regional risk assessment.

Site-specific spatial risk assessments are performed at local scales using site-specific data to define the distribution of risks and thereby guide the distribution of remedial activities. Many scientists (e.g. Marinussen and Van der Zee 1996; Hope 2000; Korre et al. 2002; Linkov et al. 2002; Gaines et al. 2005; Makropoulos and Butler 2006) have described how exposure and risk are strongly influenced by the spatial distributions of both receptors and stressors. For instance, exposure to contaminants involves spatially complex situations due to the heterogeneity of contaminant distributions relative to habitat features and other environmental characteristics. Interest in site-specific spatial risk assessment is growing. It is defined as a methodology which combines quantitative risk assessment procedures and spatial distribution of stressors and receptors to produce an assessment of risks at the scale of the site to provide geographical risk maps. Such maps preserve the significant spatial dimension of the risk in order to facilitate understanding and the communication (Gay and Korre 2006). The objective of this approach is the spatial estimation of the risks posed by some stressors (mainly contaminated sites) to evaluate remedial alternatives and develop remedial plans.

Different site-specific risk assessment applications are available especially concerning the contaminated sites. Korre et al. (2002) coupled advanced geostatistics and exposure assessment to describe the spatial distribution of human health risk associated with ingestion of lead contaminated soil. Hope (2000, 2001, 2005) developed a random-walk model to estimate a receptor's exposure to one or more stressors taking into account the relative spatial position of both receptors and stressors and how that receptor is presumed to move within the studied area. Some site-specific risk assessment methodologies have also been implemented in dedicated software: HHRA-GIS (Morra et al. 2006), HIRET (Bien et al. 2004), NORISC (http://www.norisc.com/), SADA (http://www.tiem. utk.edu/~sada/help/) and DESYRE (Carlon et al. 2007). Finally, the integrated project NOMIRACLE (http://nomiracle.jrc.it), founded by the EU, addresses the issue of the pollutants transfer between different environmental compartments, and the impact of cumulative stressors, including chemical mixture to human health in a spatial way.

At regional scales, risk assessments deal with problems that affect large geographic areas and with the spatial relationship among multiple and widely distributed habitats, sources, stressors and endpoints (Hunsaker et al. 1990; Landis 2005). Regional risk assessment becomes important when policymakers are called to face problems caused by multiple sources of hazards, widely spread over a large area, which affect multiple endpoints of regional interest (Graham et al. 1991). Furthermore, in many cases, the limited economical resources prohibit remediation of all the identified risks to health and the environment in a region. As a result, methodologies are needed to rank risks within a region in order to select those to be investigated more thoroughly or to prioritize the remedial actions (Long and Fischhoff 2000).

For contaminated sites, assessments at regional or national levels are particularly useful to define priorities for action based on the risks and impacts of the contaminated sites. Accordingly, regional risk assessment can be applied to assess the extent of problems at regional scales in terms of sources of pollution and constraints on land use and redevelopment planning. Indeed, these methodologies have been widely used to classify and inventory of contaminated sites at regional and national levels in Europe. Moreover, regional risk assessment can identify significant causes of potential impacts on river basins and coastal areas to estimate the risk that water bodies will fail to achieve a good ecological status. This is done by modeling the relationship among sources, the distribution of stressors and locations of receptors in a watershed (see Chapter 18).

In comparison with traditional risk assessment concepts, one of the most important characteristic of the regional approach is the inclusion of the spatial characteristics of the regional landscape and any spatial characteristics associated with the exposure or the effects of the exposure in the risk assessment (Graham et al. 1991). Indeed, the spatiotemporal patterns of exposures depend on the spatial relationship between the hazard sources and the endpoints and these patterns influence the spatial distribution of the risks. Moreover, many cumulative effects, which are not evident at local scale, can be apparent at regional scale.

The main objectives of regional scale assessments are the evaluation of broader scale problems, their contribution and influence on local scale problems as well as the cumulative effects of local scale issues on regional endpoints in order to prioritize the risks present in the region of interest (Smith et al. 2000). Furthermore, the regional risk approach can support the formulation of risk reduction strategies in a spatially defined region, across a broad range of hazard sources (Gheorghe et al. 2000).

As discussed by Hunsaker and colleagues (Hunsaker et al. 1989, 1990; Suter 1990; Graham et al. 1991), in regional risk assessment, the concept of hazard is more nebulous and the interaction between the components of the problem formulation phase are often complex because source terms, endpoints and reference environments are all interdependent. Indeed, developing source terms can be difficult for regional hazards because they often involve multiple sources that vary in both space and time. Moreover in the analysis phase, regional assessment differs from local scale in two ways. First, the models used in the exposure and effect assessment must be regional: local models may have to be adapted to larger geographical regions or very different models developed. Second, the exposure or effects assessment must account for uncertainty that may arise because of spatial heterogeneity, a feature that may not be significant in local assessment.

Landis and colleagues (Landis and Wiegers 1997; Landis 2005) proposed ranking models to estimate the relative probability that some environmental negative effects, caused by anthropological activity, can occur. The criteria for setting ranks are developed case by case on the basis of the complexity of the system to be analyzed and the management needs. The objective of the relative regional risk assessment is to rank the different sources of stressors and the different endpoints (i.e. vulnerable habitats, species or resources) in order to define a relative priority for further characterization of those risk assessment components with a high relative risk rank. Moreover, the definition of relative scores allows the comparison of the risks posed by the different sources of hazards and the identification of the more vulnerable endpoints and more influential pathways. The relative risk estimates are obtained by integrating of the importance of the three components of the risk assessment (source rank, habitat rank and impact rank) and the interaction among them (spatial overlapping) (Hamamè 2002).

2.5 Challenges and New Direction of Environmental Risk Assessment

Based on our experience and reviews by Menzie (2002), Fairman et al. (1998) and van Leeuwen and Vermeire (2007), we consider the following to be the most important challenges for advancing the practice of environmental risk assessment: (a) integrating risk assessment and risk management; (b) dealing with spatial and temporal scales; (c) harmonization of methodologies and tools; (d) analysing uncertainties; and (e) communicating the risks and the decisions.

2.5.1 Integration of Risk Assessment and Risk Management

Although risk assessment exists to serve risk management, the proper relationship between them has been controversial. This controversy arises because risk assessment is a scientific enterprise which should provide unbiased estimates of risk, but risk management must include non-scientific considerations such as legal constraints, political policies, public preferences, and economic interests. Hence, if risk managers are involved in risk assessment, the scientific process of estimating risks will tend to become biased by other considerations. For this reason, the 1983 National Research Council red book called for a clear distinction between risk assessment and risk management (NRC 1983).

However, if risk assessment is completely separated from risk management, a new problem arises. Policy decisions related to the scope and goals of the assessment are made by technical experts (i.e., scientists or engineers) rather than the risk manager who has the appropriate authority. As a result, the risk assessments are unbiased by the political concerns of risk managers, but they become irrelevant, because they do not provide the information needed by the risk managers.

A solution to that quandary is to view risk assessment as a scientific practice embedded in a larger risk management context. For example, the US-EPA's (1998) guidelines for ecological risk assessment call for a planning stage prior to the problem formulation in which the risk manager and any involved stakeholders set goals and constraints. Then, after the risk assessment is completed, risk assessors should interact with the risk manager to assure appropriate communication of results. The guidance for ecological risk assessment for Superfund goes further by specifying six scientific/management decision points to ensure that the risk assessment produces useful and defensible results (US-EPA 1997).

Although such interactions with risk managers are not part of the human health risk assessment framework, the US-President/Congressional Commission on Risk Assessment and Management (CRARM 1997) provides the following principles for integrating human health risk assessment with risk management: (1) risk characterization should be a decision-driven activity, directed towards informing and solving problems; (2) coping with a risk situation requires a broad understanding of the relevant losses, harms or consequences to the interested and affected parties; (3) risk characterization is the outcomes of an analytical – deliberative process; (4) the analytical – deliberative process leading to a risk characterization should explicitly deal with problem formulation early on in the process; (5) the process should be mutual and recursive; (6) those responsible for the risk characterization should begin by developing a provisional diagnosis of the decision situation so they can better match the analytical – deliberative process to the needs of the decision, particularly in terms of the level and intensity of effort and representation of parties; (7) each organization responsible for making risk decision should work to build organizational capability to conform to the principles of sound risk characterization.

An efficient integration of environmental risk assessment and environmental risk management may be stimulated and supported by the development of risk based decision support systems (Bardos et al. 2001; Sullivan et al. 2000). That is, if risk managers are involved in the development of decision support systems, policy-based decisions can be made once for a nation or region and then implemented at all sites where the system is used. In that way, policy decisions are made by proper authorities rather than by technical experts. Then, when the DSS is implemented at a site, the influence of site-specific political and economic passions can be somewhat mitigated by invoking the standing policies. Examples of such decisions include, should risks to an intruder into a fenced site be included and should organism-level attributes such as mortality and fecundity be used as ecological assessment endpoints or should population-level endpoints be used instead?

In the absence of input by risk managers to the design of a DSS, groups of technical experts can develop their system based on precedents and interpretations of regulations and published policies. Examples of this approach include the Risk Based Corrective Action (RBCA) proposed by American Society for Testing and Materials (1998) and the Risk-Based Analysis, proposed by Frantzen (2002).

2.5.2 Dealing with Scales

According to Menzie (2002), the three major scales affecting environmental risk assessment results are: (1) spatial scales (for distribution of the stressors and receptors); (2) temporal scales (for release, transformation, sequestration rates as well as for biological processes including recovery); (3) effects scales (for type of effects and magnitude of the effects).

Issues of scale have particular importance at contaminated sites. Spatial scale is important because the areas within sites with particular types of contamination, and the areas that potentially provide habitat have spatial dimensions that limit the potential risks. For example, if the area contaminated is less than a hectare, that is less than the home range of most wildlife species. Temporal scale is important primarily because of the time required for contaminants to move off the site, for contaminants to degrade and for containment measures to fail. However, finer temporal scales are important to exposures if humans or wildlife such as migratory birds visit the site for only short time periods. Scales of effects determine whether the risks are significant. Adverse effects are more important if they are more frequent (e.g., more organisms die) and if they are more intense (e.g., if they reduce growth or fecundity by a greater proportion).

During the 1990s and early 2000s there has been increasing interest in developing tools that capture the spatial distribution of ecological entities, contaminants and the resulting risks (Menzie 2002; Clifford et al. 1995).

2.5.3 Harmonization of Methodologies and Tools

According to Fairman et al. (1998) and van Leeuwen and Vermeire (2007), the organizations currently carrying out risk assessment need to harmonise their programmes; to some extent, that is already occurring (McCutcheon 1996). Harmonisation of procedures does not require standardisation but is defined as "an understanding of the methods and practices used in various countries and organizations so as to develop confidence in, and acceptance of, assessments that use different approaches" (van Leeuwen et al. 1996).

Although many test protocols for chemicals have been harmonized, some areas need further work, particularly human health reproductive toxicity tests and test methods for the effects of chemicals influencing hormonal processes (van Leeuwen et al. 1996). The development and harmonization of toxicity testing methods for mixtures of chemicals has also been recommended (NRC 1996).

Finally, the development of a clear definition and classification of "assessment factors" (e.g., uncertainty factors) and guidance for their application in a defensible and harmonized manner, is advocated by many experts (Calabrese and Baldwin 1993; Suter et al. 2000). These factors are used to compensate for uncertain relationships between data and values to be estimated, requiring extrapolations within each field of risk assessment and across the areas of human health and ecological risk assessment. Uncertainty factors are applied mainly in the dose-response stage of a human health risk assessment or in the effects characterization of an ecological risk assessment, where they adjust for insufficient test data and to the extrapolate results across species and life stages. In addition, consensus should be developed on the use of more sophisticated methods for extrapolating between taxa, life stages and conditions such as allometric scaling, toxicokinetics and species sensitivity distributions (Suter 2007).

2.5.4 Analysing Uncertainties

Although the estimation of uncertainty received much attention during the 1990s, the actual experience applying quantitative uncertainty analyses in environmental risk assessments is still limited (Menzie 2002). While the state of practice is to list sources of uncertainty without ranking them or estimating their approximate magnitudes, the identified uncertainties should be quantified as far as is practical. For instance, uncertainties in the parameters due to measurement, extrapolation, and the fitting of any empirical models used in parameter derivation can be estimated by conventional statistics or expert judgment and propagated to estimate total uncertainty in the risk estimate, usually by Monte Carlo analysis. Although many techniques are already available, guidance is needed for their application.

The need to manage uncertainty and to ensure that decisions are environmentally protective led to the development of tiered approaches for evaluating environmental risk (US-EPA 1997; ASTM 1998). These approaches begin with simple conservative screening-level assessments and proceed to more sophisticated analyses as needed in order to make a decision. Within each tier, knowledge is increased and uncertainty reduced; sources and magnitudes of uncertainty are also better known and characterized to insure that they are properly managed. Quantitative uncertainty analysis is one of the tools that can be used at later tiers to provide this additional insight. Thus, the tiered assessment strategy is itself a tool for managing uncertainty. A tiered strategy also guides risk assessors and risk managers in allocating the resources needed for assessing risks.

2.5.5 Risk Communication

Environmental risk assessors face multiple communication challenges. The first is to understand what is expected of them. This requires that the assessors meet with the risk managers and appropriate stakeholders to ensure that their needs and concerns are properly translated into risk endpoints and scenarios.

Second, risk assessors must report their results to users such as economists performing cost-benefit analyses, to peer reviewers, and ultimately to the risk manager. This requires a report that presents the results in a useful way and that contains enough information concerning that data and analyses that they can understand the derivation of the results and judge the quality of the assessment. To that end, the Science Policy Council (2000) presented guidance on how to make a risk characterization clear, transparent, reasonable and consistent.

Finally, environmental risk assessors must communicate their results to the risk managers, stakeholders, and the public. Direct communication provides an opportunity to ensure that the results of the assessment are effective. However, risk communication can be difficult because audiences may have strong preconceptions about the risks posed by an environmental hazard and they typically have little knowledge of environmental science. Advice for communicating human health risks is relatively abundant (e.g., Lundgren and McMakin 1998), but advice for communicating ecological risks is rare (Suter 2007).

An overarching goal for engaging stakeholders in planning and problem formulation as well as in the rest of the risk assessment process is to achieve a common understanding on the objective and development of the assessment. Fisher (1998) identifies a "shared understanding" as the goal of the risk communication process. For example, the development or modification of understandable conceptual models provides useful means for developing a common understanding of the problem and supports the identification of the possible relations between stressors and ecological receptors.

Concerning "When should stakeholders be involved?", the following criteria should be considered: (1) the potential that they will be affected by the decision, (2) the potential that they have information important for the assessment and the decisions, (3) their level of interest, and (4) the magnitude of the potential problem. The answer to "How should stakeholders be involved?" will depend on the nature of the problem (Menzie 2002). In general, the risk managers determine when and how stakeholders should be engaged, and risk assessors play a supporting role.

2.6 Conclusions

Many authors (Menzie 2002; Suter 1997; Fairman et al. 1998; van Leeuwen and Vermeire 2007) envision increasing use of environmental risk assessment as a basic tool for environmental decision making. According to Menzie (2002), the evolution of risk assessment will especially involve the following topics: (a) integration of the assessment process into explicit decision frameworks through the development of risk based decision support systems, (b) refinement of tools to account for spatial, temporal and effects scales, (c) increased emphasis on population-level and ecosystem risks, (d) development of methodologies and tools to support probabilistic risk assessment, (e) better education and communication about environmental issues. We agree. In addition, we believe that decision support systems will be increasingly used to make environmental risk assessment more efficient, more legally and scientifically defensible, and more useful to risk managers.

Acknowledgment We thank Susan Cormier and Michael Griffith for their helpful reviews of this chapter.

References

American Society for Testing and Materials (ASTM) (1998) Emergency Standard Guide for Risk-Based Corrective Action Applied at Petroleum Release Sites. Designation PS 104; Philadelphia

Andersen MC, Thompson BC, Boykin KG (2004) Spatial risk assessment across large landscapes with varied land use: lessons from a conservation assessment of military lands. Risk Anal 24(5):1231–1242

Bardos RP, Mariotti C, Marot F, Sullivan T (2001) Framework for decision support used in contaminated land management in europe and north america. Land Contam Recl 9:149–163

Barnard RC (1994) Scientific method and risk assessment. Regul Toxicol Pharm 19:211–218

Bien JD, ter Meer J, Rulkens WH, Rijnaarts HHM (2004) A GIS-based approach for long-term prediction of human health risks at a contaminated sites. Environ Model Assess 9:221–226

Burton GA, Batley GE, Chapman PM, Forbes VE, Smith EP, Reynoldson T, Schlekat CE, Den Besten PJ, Bailer AJ, Green AS, Dweyer RL (2002a) A weight-of-evidence framework for assessment sediment (or other) contamination: improving certainty in the decision-making process. Hum Ecol Risk Assess 8(7):1675–1696

Burton GA, Chapman PM, Smith EP (2002b) Weight-of-evidence approaches for assessing ecosystem impairment. Hum Ecol Risk Assess 8(7):1657–1673

Calabrese EJ, Baldwin LA (1993) Performing Ecological Risk Assessments. Lewis Publishers, Chelsea, MI

CARACAS (1998) Risk Assessment for Contaminated Sites in Europe. Volume 1 – Scientific Basis. LQM Press, Nottingham

CARACAS (1999) Risk Assessment for Contaminated Sites in Europe. Volume 2 – Policy Frameworks. LQM Press, Nottingham

Carlon C, Critto A, Ramieri E, Marcomini A (2007) DESYRE decision support system for the rehabilitation of contaminated mega-sites. Integrated Environ Assess Manag 3:211–222

Chapman PM (1990) The sediment quality triad approach to determining pollution-induced degradation. Sci Total Environ 97/98:815–825

Chapman PM, McDonald BG, Lawrence GS (2002) Weight of evidence issues and framework for sediment quality (and other) assessment. Hum Ecol Risk Assess 8(7):1489–1515

Clarinet (2003) Sustainable management of contaminated land: an overview. A Report from the Contaminated Land Rehabilitation Network for Environmental Technologies. Umwelt-bundesamt GmbH, Vienna, p 115

Clifford PA, Barchers DE, Ludwig DF, Sielken RL, Klingensmith JS, Graham RV, Banton MI (1995) An approach to quantifying spatial components of exposure for ecological risk assessment. Environ Toxicol Chem 14:895–906

Covello VT, Merkhofer MW (1993) Risk Assessment Methods Approaches for Assessing Health and Environmental Risk. Plenum, New York

Crarm (1997) Framework for Environmental Health Risk Management. Commission on Risk Assessment and Risk Management, Washington, DC

European Commission (EC) (2001) White Paper on the Future of a Chemicals' Strategy. Commission of the European Communities, Brussels

European Commission (EC) (2003) Technical Guidance Document on Risk Assessment. Commission of the European Communities, Brussels

Fairman R, Mead CD, Williams WP (1998) Environmental Risk Assessment: Approaches, Experiences and Information Sources. European Environmental Agency, Copenhagen

Fisher A (1998) The challenges of communicating about health risks and ecological risks. Hum Ecol Risk Assess 4:623–626

Frantzen KA (2002) Risk-Based Analysis for Environmental Managers. Lewis Publisher, Boca Raton, Florida

Gaines KF, Porter DE, Punshon T, Brisbin Jr IL (2005) A spatially explicit model of the wild hog for ecological risk assessment activities at the department of energy's Savannah river site. Hum Ecol Risk Assess 11:567–589

Gay R, Korre A (2006) A spatially-evaluated methodology for assessing risk to a population form contaminated land. Environ Pollut 142:227–234

Gheorghe A, Mock R, Kröger W (2000) Risk Assessment of Regional Systems, in Reliability Engineering & System Safety 70. Elsevier Applied Science, Barking

Graham RL, Hunsaker CT, O'Neill RV, Jackson BL (1991) Ecological risk assessment at the regional scale. Ecol Applic 1:196–206

Hall LW, Giddings JM (2000) The need for multiple lines of evidence for predicting site specific ecological effects. Hum Ecol Risk Assess 7:459–466

Hamamè M (2002) Regional Risk Assesment in Northern Chile Report 2002: 1. Environmental System Analysis, Chalmers University of Thechnology, Gotegorg, Sweden

Hope BK (2000) Generating probabilistic spatially-explicit individual and population exposure estimates for ecological risk assessments. Risk Anal 20:573–589

Hope BK (2001) A case study comparing static and spatially explicit ecological exposure analysis methods. Risk Anal 21:1001–1010

Hope BK (2005) Performing spatially and temporally explicit ecological exposure assessments involving multiple stressors. Hum Ecol Risk Assess 11:539–565

Hunsaker CT, Graham RL, Suter GW, O'Neill RV, Jackson BL, Barnthouse LW (1989) Regional Ecological Risk Assessment: Theory and Demonstration. ORNL/TM–11128. Oak Ridge National Laboratory, Oak Ridge, Tennessee

Hunsaker CT, Graham RL, Suter II GW, O'Neill RV, Barnthouse LW, Gardner RH (1990) Assessing ecological risk on a regional scale. Environ Manage 14:325–332

Korre A, Durucan S, Koutroumani A (2002) Quantitative-spatial assessment of the risks associated with high Pb loads in soils around Lavrio, Greece. Appl Geochem 17:1029–1045

Landis WG (2005) Regional Scale Ecological Risk Assessment. Using the Relative Risk Model. CRC Press, Boca Raton

Landis WG, Wiegers JA (1997) Design considerations and a suggested approach for regional and comparative ecological risk assessment. Hum Ecol Risk Assess 3:287–297

Linkov I, Burmistrov D, Cura J, Bridges TS (2002) Risk-based management of contaminated sediments: consideration of spatial and temporal patterns in exposure modelling. Environ Sci Technol 36:238–246

Long ER, Chapman PM (1985) A sediment quality triad: measures of sediment contamination, toxicity and infaunal community composition in puget sound. Mar Pollut Bull 16:405–415

Long J, Fischhoff B (2000) Setting risk priorities: a formal model. Risk Anal 20:339–351

Lundgren R, McMakin A (1998) Risk Communication, A Handbook for Communicating Environmental, Safety and Health Risks. Battelle Press, Columbus, Ohio

Makropoulos CK, Butler D (2006) Spatial ordered weighted averaging: incorporating spatially variable attitude towards risk in spatial multi criteria decision marking. Environ Modell Softw 21:69–84

Marinussen MPJC, van der Zee SEATM (1996) Conceptual approach to estimating the effect of homerange size on the exposure of organisms to spatially variable soil contamination. Ecol Model 87:83–89

McCutcheon P (1996) Evaluation and control of chemical substances in the European Union. In: Richardson M (ed) Risk Reduction. Taylor and Francis, London, pp 311–324

Menzie CA (2002) The Evolution of ecological risk assessment during the 1990s: challenges and opportunities. In: Sunahara GI, Renoux AY, Thellen C, Gaudet C, Pilon A (eds) Environmental Analysis of Contaminated Sites. John Wiley & Sons, Chichester, UK, pp 281–299

Menzie C, Henning MH, Cura J, Finkelstein K, Gentile JH, Maughan J, Mitchell D, Petron S, Potocki B, Svirsky S, Tyler P (1996) Special report of the Massachusetts weight-of-evidence workgroup: a weight-of-evidence approach for evaluating ecological risks. Hum Ecol Risk Assess 2:277–304

Morra P, Bagli S, Spadoni G (2006) The analysis of human health risk with a detailed procedure operating in a GIS environment. Environ Int 32:444–454

National Research Council (NRC) (1996) Understanding Risk: Informing Decisions in a Democratic Society. National Research Council Committee on Risk Characterization, National Academy Press, Washington, DC

National Research Council of the National Academy of Science (NRC/NAS) (1983) Risk Assessment in the Federal Government: managing the process. National Academy Press, Washington, DC

Ross JF (1995) Where do real dangers lie? Smithsonian 26(8):43–53

Royal Society (1992) Risk Analysis, Perception and Management. The Royal Society, London

Rutgers M, Den Besten P (2005) Approach to legislation in a global context, B. The Netherlands perspective – soils and sediments. In: Thompson KC, Wadhia K, Loibner AP (eds) Environmental Toxicity Testing. Blackwell Publishing, CRC Press, Oxford, pp 269–289

Salgueiro A, Suter II GW, Antunes P, Santos R, Martinho S (2001) A framework to support insurers on the assessment of pollution risks. In: Linkov I, Palma Oliveira JM (eds) Assessment and Management of Environmental Risks. Kluewer Academic Pub, Amsterdam, pp 107–114

Science Policy Council (2000) Risk Characterization Handbook. EPA 100–B–00–002. US Environmental Protection Agency, Washington, DC

Smith EP, Lipkovich I, Ye K (2002) Weight of evidence: quantitative estimation of probability of impairment for individual and multiple Lines of Evidence. Hum Ecol Risk Assess 8(7):1585–1596

Smith ER, O'Neill RV, Wickham JD, Jones KB, Jackson L, Kilaru JV, Reuter R (2000) The US-EPA's Regional Vulnerability Assessment Program: A Research Strategy for 2001–2006. US Environmental Protection Agency, Office of Research and Development, Research Triangle Park, NC

Sullivan TM, Moskowitz PD, Gitten M (2000) Overview of environmental decision support software. In: Salem H, Olajos EJ (eds) Toxicology in Risk Assessment. Taylor and Francis, Ann Arbor, MI, pp 147–162

Suter GW II (1990) Endpoints for regional ecological assessments. Environ Manage 14:9–23

Suter GW II (1996) Toxicological benchmarks for screening contaminants of potential concern for effects on freshwater biota. Environ Toxicol Chem 15:1232–1241

Suter GW II (1997) Overview of the ecological risk assessment framework. In: Ingersoll CG, Dillon T, Biddinger GR (eds) Ecological Risk Assessment of Contaminated Sediments. SETAC Press, Pensacola, FL, pp 1–6

Suter GW II (2007) Ecological Risk Assessment, 2nd edition. CRC Press, Boca Raton. FL

Suter GW II, Efroymson RA, Sample BE, Jones DS (2000) Ecological Risk Assessment for Contaminated Sites. Lewis, Boca Raton, FL

Tarazona JV (2002) Trends in the risk assessment strategy of persistent organic pollutants. The European perspective. In: Proceeding of the 22nd International Symposium on Halogenated Environmental Organic Pollutants and POPs. CSIC, Barcelona, Vol 59 pp 107–110.

United States Environmental Protection Agency (US-EPA) (1986) Superfund Public Health Evaluation Manual. EPA/540/1–86/060. Office of Emergency and Remedial Response, Washington, DC

United States Environmental Protection Agency (US-EPA) (1989) Risk Assessment Guidance for Superfund (RAGS): Volume I – Human Health Evaluation Manual (HHEM) (Part A, baseline Risk assessment). EPA/540/1–89/002. Office of Emergency and Remedial Response, Washington, DC

United States Environmental Protection Agency (US-EPA) (1992) Framework for Ecological Risk Assessment. EPA/630/R–92/001. Risk Assessment Forum, Washington, DC

United States Environmental Protection Agency (US-EPA) (1996) Soil Screening Guidance: User's Guide. EPA 540/R–96/018. Office of Emergency and Remedial Response, Washington, DC

United States Environmental Protection Agency (US-EPA) (1997) Ecological Risk Management Guidance for Superfund: Process for Designing and Conducting Ecological Risk Assessments. EPA/540/R–97/006. Environmental Response Team, Edison, NJ

United States Environmental Protection Agency (US-EPA) (1998) Guidelines for Ecological Risk Assessment. EPA/630/R–95/002F. Risk Assessment Forum, Washington, DC

United States Environmental Protection Agency (US-EPA) (1999) Risk Assessment Guidance for Superfund: Volume 3 (Part A, Process for Conducting Probabilistic Risk Assessment). Draft, revision No. 5. Office of Solid Waste and Emergency Response, Washington, DC

United States Environmental Protection Agency (US-EPA) (2001) The Role of Screening-Level Risk Assessments and Refining Contaminants of Concern in Baseline Ecological Risk Assessments. EPA 540/F–01/014. Office of Solid Waste and Emergency Response, Washington, DC

United States Environmental Protection Agency (US-EPA) (2005) Guidance for Developing Ecological Soil Screening Levels, Revised. OSWER Directive 9285.7–55. Office of Solid Waste and Emergency Response, Washington, DC

van Leeuwen CJ, Vermeire TG (2007) Risk Assessment of Chemicals: An Introduction. Springer, Dordrecht

van Leeuwen CJ, Bro-Rasmussen F, Feijtel TCJ, Arndt R, Bussian BM, Calamari D, Glynn P, Grandy NJ, Hansen B, Van Hemmen JJ, Hurst P, King N, Koch R, Muller M, Solbé JF, Speijers GAB, Vermeire T (1996) Risk Assessment and Management of New and Existing Chemical Substances. Environ Toxicol Phar 2:243–299

Chapter 3
Decision Support Systems and Environment: Role of MCDA

Silvio Giove, Adriana Brancia, F. Kyle Satterstrom, and Igor Linkov

Abstract Decision Support Systems (DSS) are computer-based tools designed to support management decisions (Eom, 2001). Many environmental applications of DSS are reported in the current literature, including petroleum contamination detection (Geng et al., 2001), lake remediation (Gallego et al., 2004), soil decontamination (Zhiying et al., 2003), and many others. However, many of these DSS are in fact different models integrated to better visualize data or describe systems; they are not tailored to address specific decision problems or help decision makers in making inevitable trade-offs. Multicriteria Decision Analysis (MCDA), on the other hand, offers the ability to integrate policy preferences with the judgements of technical experts (Figueira et al., 2005; Linkov et al., 2007). MCDA methods enable simultaneous consideration of stakeholder interests and technical evaluations, utilizing rigorous scientific methods to process technical information. MCDA is especially important in situations of significant uncertainty and data scarcity, such as management and restoration of contaminated sites. This Chapter focuses on the conceptual background of MCDA, with particular attention paid to environmental DSS, and it discusses some of the most commonly used approaches, especially for multi-attribute decision problems (i.e. where both criteria and alternatives are finite in number).

Keywords DSS · Environmental risk assessment · Fuzzy logic · Management · MCDA

3.1 Introduction

In 1944, John von Neumann and Oskar Morgenstern published the *"Theory of Games and Economic Behaviour"*, a book that greatly influenced the development of modern decision theory. In 1970, Little discussed the use of scientific

S. Giove (✉)
Department of Applied Mathematics, University Ca' Foscari of Venice, Italy

A. Marcomini et al. (eds.), *Decision Support Systems for Risk-Based Management of Contaminated Sites*, DOI 10.1007/978-0-387-09722-0_3,
© Springer Science+Business Media, LLC 2009

models to help managers make decisions. In 1971, Gorry and Scott Morton more clearly defined a framework for DSS, an important step, since human decisions are generally expressed in a wide variety of terms and contexts. These procedures were developed, in particular, for *structured* decisions. An additional aspect concerned *strategic* decisions involving the long-range plan of organizations. Strategic decisions occur less often than operational decisions, but they affect the entire system, and they potentially involve large sums of money. For this reason, the risks involved when making them are higher. Decisions about managerial control also affect the structure of the support system. Based upon Simon's classification of problem structure and Anthony's classification of decision level, Gorry and Scott Morton developed a framework for computer-based decision support. Gorry and Scott Morton use the terms "*structured*", "*semi-structured*," and "*unstructured*," and they define a DSS as any computer system that deals with problems that have at least some unstructured components. They also point out that strategic planning problems have different information requirements than do information systems for structured problems. The information for strategic planning decisions is broad in scope, may come largely from external sources, is generally aggregated, often involves predictions of the future quite some time away, involves a lot of uncertainty, and is not as copious in quantity as information needed for operational control decisions. Keeney and Raiffa (1976) were first to integrate decision analysis theory with practical applications. Keen and Scott Morton (1978) considered DSS to be the mixture of individuals' intellectual resources with the potential of computers to improve decision quality. They point out that structured problems do not require much analysis by the decision maker, while unstructured problems are very difficult to manage with computers and models. Therefore, they conclude that DSS applications would be applied more often to semi-structured problem domains. Alter (1977) empirically investigated 56 applications of DSS. He describes seven types of DSS, which he organises into two groups:

- *Data-oriented*: use mainly a database that can analyse the data in one or more files using statistical procedures or accounting models
- *Model-oriented*: exploit management science models or expert systems that might suggest an "optimal" choice.

He found that, in practice, humans use computers in a variety of ways, not necessarily concerned with academic classification systems. He also noted that computer efficiency was distinctly less important than flexibility for DSS applications. Bonczek et al. (1980) studied the evolving role of models in DSS. They observed that users, models, and data compose a DSS. The critical interfaces between these three components were used as the basis for designing effective languages, both for computation and for data retrieval. These authors believed that DSS should be flexible and exploratory. Keen (1981) reviewed features and theoretical benefits of DSS, as well as a number of actual applications from the first decade of their use. Key points in the planning and evaluation of decision support systems are the appropriate style of developing these systems (reliance on prototypes as opposed to the system's development life cycle common to other MIS applications), the limitations of traditional means of economic evaluation (such as cost-benefit analysis), and the evolutionary nature of DSS development. Keen focused on what the manager would need to know to build a worthwhile decision support system, including how innovation could be encouraged while making sure money was well spent, and how the value of effectiveness, learning, and creativity could be assessed quantitatively. Keen also developed a list of the benefits of using DSS, such as an increased number of alternatives examined, the ability to carry out "*ad hoc*" analysis, better controls, better decisions, cost savings, and other factors. During the development of his studies, Keen also suggested the use of "*Value Analysis*" for assessing DSS proposals, which focuses on benefits and estimating the costs of obtaining those benefits. Rather than calculating a highly subjective benefit/cost ratio, the system can be evaluated in terms of improving the decision-making environment versus the monetary cost; this becomes the basis for the Decision Maker (DM) for making the decision.

Keen (1986) established also the principles of value analysis as:

- Separating benefit from cost
- Establishing what quantifiable benefits are worth (in monetary terms)
- Determining how much you would pay for that benefit
- Identifying qualitative benefits
- Rank ordering benefits (including both quantitative and qualitative benefits)
- Defining indicators with which qualitative benefits can be evaluated
- Identifying a rough estimation of needed benefits for project adoption, as well as likely benefits

Keen recommends the construction of a prototype so the benefits of the system can be evaluated by controlling what it can do in real terms. In this way, uncertainty and risk may also be better studied before developing the whole system. The final user can look at what the prototype does and, if necessary, suggest modifications before the development of the full version. According to Baker et al. (2001), a decision process should start by identifying the decision makers and stakeholders involved in the decision, reducing the possible disagreement about problem definition, requirements, goals, and criteria. Then, a general decision-making process can be divided into the following steps:

- Define the problem
- Determine requirements
- Establish objectives
- Identify alternatives
- Define criteria that should be:
  - able to discriminate among the alternatives and support the comparison of the performance of the alternatives
  - include all important aspects of the objectives
  - operational and concise

  o non-redundant with each criteria its own single concept
  o few in number
  o measurable, so that the alternative can be expressed on either a quantita-
    tive or a qualitative measurement scale

- Select a decision making tool
- Evaluate alternatives against criteria
- Validate solutions against the problem statement.

All these steps are part of both the **Value Tree Analysis** (top-down approach), which focuses on *"value thinking"*, as well as the bottom-up approach, which focuses on *"alternative thinking"*. These are integral parts in many DSS to define the structure of the decision.

The main protagonists in the decision analysis process are:

- A **decision-maker** is the person (or the organisation or any other decision-making entity) charged with finding solutions for the decision-making problem at hand.
- A **decision analyst** gives advices and clarification to the decision maker when he is unsure about which path to take. The main task of the decision analyst is to aid the decision maker in finding the best decision alternatives to bring to the development of the decision-making process.
- A **stakeholder** is normally a person (or group of persons) deeply concerned about the project, often about its economic aspects.

The relationships between the three main subjects can vary widely. The subjects may be completely independent, or the Decision Analyst may be the only person totally involved in the decision process, while the stakeholder and the decision maker have only partial influence. In another scenario, the decision analyst and the decision maker may be the same person and have more relevance than the stakeholder has in the decision process. It is also possible for a single person to cover all three roles. After identifying the main subject involved in the Decision Analysis Process (DAP), it is important to understand what it means to construct a DAP. The DAP is usually developed in four steps (Fig. 3.1) that provide a path for thinking about decisions, objectives, alternatives, creation of a hierarchy for the objectives and choice of the attributes in the *"collection of information"* phase which puts the decision problem in context, normally by asking a series of clarifying questions. When there is more than one objective, it is useful to establish an importance hierarchy and create the starting point for a second phase of analysis. In the problem-structuring phase, it is very important to assign attributes to the different alternatives as a means for discriminating between them. In the second phase, *"elicitation of decision maker's preference"*, the preferences of the decision maker are expressed by the weights that show the relative importance of the evaluation criteria from the point of view of the decision maker. This phase permits the elaboration of the hierarchy between the different alternatives and consequently the final choice of the best alternative. The final phase concerns the *"sensitivity analysis"* which

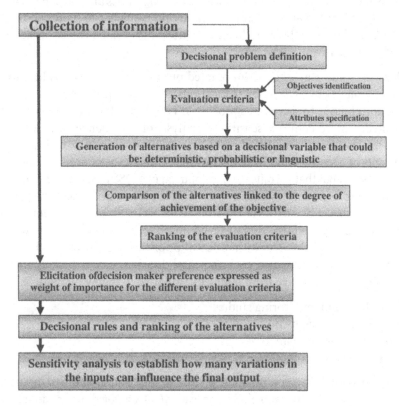

Fig. 3.1 Phases of the decision trees analysis process (DAP)

investigates the degree to which variations in the inputs influence the final result. For a wide class of multi-attribute decision models, Mareschal (1988) showed how to determine the stability intervals or regions for the weights of different criteria. These consist of the values that the weights of one or more criteria can take without altering the results given by the initial set of weights, all other weights being kept constant. Wolters and Mareschal (1995) proposed a linear programming model to find the minimum modification of the weights required to make a particular alternative rank first. Mészáros and Rapcsák (1996) presented a general and comprehensive methodology for a wide class of MAVT models where the aggregation is based on generalized means, including additive and multiplicative models as well. In this approach, the weights and the scores of the alternatives against the criteria can change simultaneously over given intervals.

The following questions were addressed:

- What are the intervals of the alternatives' final rankings with the restriction that the intervals of the weights and scores are given?
- What are the intervals of the weights and scores with the restriction that the final ranking of the alternatives does not change?

• Consider a subset of alternatives whose ranking values are allowed to change in a specific interval. In what intervals are the weights and scores allowed to vary, and how will these modifications affect the ranking values of the entire set of alternatives?

Mészáros and Rapcsák (1996) pointed out that these questions lead to the optimization of linear fractional functions over rectangles and proposed an efficient technique to solve these problems. Triantaphyllou and Sanchez (1997) presented a more complex sensitivity analysis of the change of alternatives' scores against the criteria. Ekárt and Németh (2005) recently extended some of the results of Mészáros and Rapcsák (1996) for more general decision functions.

One instrument that permits the development of a DSS is MCDA. It is capable of taking into consideration all of the variables present in the decision process. MCDA is especially useful for environmental management problems, which require balancing scientific findings with multifaceted, value-laden input from many different stakeholders with different priorities and objectives. Typically, the information presented to environmental decision makers falls into one of four categories that range from highly quantitative to highly qualitative:

• Modelling and monitoring studies
• Risk/impact assessments
• Cost or cost-benefit analysis
• Stakeholder preferences.

MCDA permits the combination of quantitative and qualitative inputs like risks, costs, benefits, and stakeholder views. MCDA algorithms are designed to synthesize a wide variety of information and raise awareness of the tradeoffs that must be made between competing projects objectives. They also provide a systematic approach for integrating risk levels, uncertainty, and technical valuations. However, few MCDA approaches are specifically designed to incorporate multiple stakeholder perspectives. In situations with multiple stakeholders, the decision process may be more intensive, often incorporating aspects of group decision making. One of the advantages of an MCDA approach in group decisions is the capacity for calling attention to similarities or potential areas of conflict between stakeholders, resulting in a more complete understanding of the values held by others (Linkov et al., 2006). In a group situation, the different roles of the decision-maker(s) and the expert(s) should be emphasized and described. The expert's judgments should have a scientific and technical basis, while the decision-maker's judgments are usually based on more subjective political and managerial considerations. The process-based schedule is another important item for the efficiency and generalization of the decision strategy. A versatile framework, the *People-Process-Tool,* has been proposed for this task (Linkov, 2004).

This paper is organized as follows: Section 3.2 gives a general presentation of Multicriteria Decision problems; Section 3.3 considers in particular detail Multi-Attribute problems and methods for solving them.

3.2 Multicriteria Decision Analysis (MCDA)

MCDA includes a large class of methods for the evaluation and ranking or selection of different alternatives that considers all the aspects of a decision problem involving many actors. A structural platform common to almost all the decision problems includes the following items:

- An objective or target function (to be optimized)
- A set A of alternatives, in the finite case: $A = \{\alpha_j : j = 1, 2, \ldots m\}$
- The decision maker, a conceptual figure, a single person or a group of persons or an entity
- A countable family of criteria, $K = \{k_i : i = 1, 2, \ldots n\}$ with their attributes; the criteria can be organized into a hierarchical structure, a decision tree where the root is the objective function whose leaves are the first-level criteria, each of them split again into second-level criteria (sub-criteria), and so on to the last level, whose terminal leaves are the indicators (or the last level sub-criteria) formed by the available information (data or judgements)
- The decision maker's preferences for the different evaluation of the criteria
- An algorithmic tool designed to optimize the objective function, considering all the above information.

The information concerned with decision maker preferences can be expressed with different methods, among them lexicographic order, minimum needs, aims levels, average systems, trade-off weights, ordered weighted averaging (OWA), or more complex aggregation function. Moreover, MCDA problems can be categorized in the following ways:

- *Multi-Attribute Decision Analysis* (MADA) that focuses on a *finite* number of pre-existing alternative choices, versus *Multi-Objective Decision Analysis* (MODA) that considers an infinite number of alternatives
- *Group decision maker* versus *individual decision maker* problems
- *Single step* versus *Multi-step* evaluation procedures

A typical MCDA classification is described below (Fig. 3.2) (Vincke, 1992):

- **MAUT/MAVT** (Multi-Attribute Utility/Value Theory). The criterion values, normalized into a common numerical scale by means of a suitable transformation function (or Utility/Value Function), are aggregated using an aggregation operator, a function which satisfies a set of rationality axioms. Using a bottom-up approach, this operation is repeated for all the nodes in the decision tree (if the problem is hierarchically structured) and for all the alternatives. At the tree root (the objective) a single numerical value is then computed, which is the score of the proposed alternatives. The alternatives can be rated and ranked, since MAUT/MAVT produces a total ordering, and the best one can be selected.
- **Outranking.** This group of methods constructs an "outranking relationship", stating that an alternative *may* have a *degree of dominance* over another one.

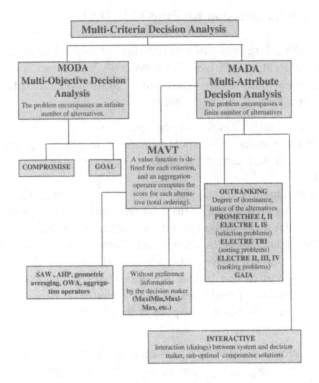

Fig. 3.2 A classification of MCDA problems and methods

Relationships are neither complete nor transitive. In this way, only a *partial* ordering is produced, implicitly admitting that comparable alternatives may exist.

- **Interactive methods.** These methods consist of iterated steps. At first, a rough solution is proposed to the decision maker, which can be accepted or rejected. In the latter case, after the acquisition of new data or information (for instance, extra information concerning a decision maker's preferences) the system computes and proposes a new solution to the decision maker. These steps, elicitation of preferences and re-computation, are repeated, creating successive compromise solutions, until the satisfaction of the decision maker is reached.

The main difference between **MADA** and **MODA** is that: in the first case we can speak of a decision in a *"discrete environment"*, in which the decisions are selected from a finite number of possible alternatives; in the second, we are in a *"continuous environment"*, in which a linear function is created and optimized for reaching the proposed objective. MADA considers the *"attributes"* that are measurable values, expressed as a nominal scale, ordinal scale, or comparison scale. MODA, however, considers *"objectives"* that represent the improving level of the attributes, in this case maximizing or minimizing the functions that are concerned with the attributes (minimizing costs or maximizing earnings, for example).

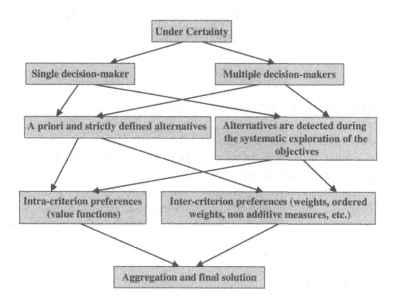

Fig. 3.3 A classification of MCDA problems under certainty

Another classification is decision under *certainty* or *uncertainty*. In the first case, the decision maker has detailed knowledge of the environment in which the decision will be made as well as exhaustive information of the decision process. For classification with "certainty", the steps of the decision process are sketched in Fig. 3.3. The management of a decision under "*uncertainty*" may be due to a variety of causes. Uncertainty can be caused by the lack of knowledge of all or part of the parameters that influence action; uncertainty can also be caused by internal indecision (about the alternatives, criteria importance, or other factor). For situations with uncertainty due to internal indecision, it is possible to construct scenarios for possible values (*fuzzy* decision-making) or to treat probabilistically stochastic events (*probabilistic* decision-making). Another distinction is *compensatory* and *non-compensatory* weighting methods: in the former case, an interaction among attributes is possible (a "good" value is substituted for a "no good" value). Complete compensation is not usually a requested property for environmental applications. For instance, in computing a sustainable development index, a critical environmental impact due to chemical toxicity cannot be offset by high economic development. This is the reason why the linear score, obtained by a convex combination of the normalized criterion value by means of a set of weights (acting as a compensative aggregation operator), is not advisable for such applications, even if widely used.

Three general models can be defined: *scoring* models (a global score is defined), *compromising* models (evaluation of the proximity in comparison with the ideal result) and *concordance* models (evaluation based on the concordance with decision maker judgements). These models include the TRADE-OFF method

(Keeney and Raiffa, 1976), the SWING method (Von Winterfeld and Edwards, 1986), the RESISTENCE TO CHANGE (Rogers and Bruen, 1998), and MACH-BETH (Bana, Costa and Vansnick, 1994).

3.3 Multi-Attribute Decision Analysis

Multi-Attribute Decision Analysis (MADA) is used when the number of alternatives is finite, while Multi-Objective Decision Analysis (MODA) is used in other cases, for instance in optimizing a portfolio or maximizing a utility function. As mentioned above, a possible classification of MADA methods includes MAUT/MAVT, outranking and interactive methods. In what follows, we briefly describe some of the most common approaches for each of the three families.

3.3.1 MAUT/MAVT

Multiple attribute utility/value theory (MAVT) constructs a utility/*value* function for each criterion, usually a monotonic function whose co-domain is included in the closed interval [0, 1]. Given that the assignment of such functions is subjective, even if guided by a suitable software interface, and depends on the user's preference structure or perception about the criterion impact, the normalization problem is solved without resorting to a data-driven formula such as subdivision by maximum. Any data-driven normalization algorithm is quite sensitive to outliers and can induce distortion in the final scoring if the sampled data are dense around an average value. Therefore, rescaling all the available criterion data into a common closed numerical scale easily and more correctly solves the normalization problem. The value functions (sometimes also called "utility" functions) convert the attribute values into a common scale, and then these numerical values are aggregated into the final score. At the end of the process, a *complete order* of all the alternatives is obtained.[1] The MAUT/MAVT approach hypothesizes the *rationality* and the *consistency* properties for the decision maker (Bridges et al., 2004) and implies the existence of the value function for each criterion or sub-criterion. Many methods exist to define the value functions (Keeney, 1976), but their description is beyond the aim of this chapter.

The next step consists of the aggregation of this normalized data into a single numerical output, the score of the alternative, or of an intermediate level node of the decision tree, if a hierarchical structure is defined. To this purpose, an Aggregation Operator needs to be defined, (Klement et al., 2000), that is a multi-dimensional function that satisfies a set of rationality axioms. The most popular is the (simple) Weighted Averaging (WA) approach that is a

---

[1] On the other side, outranking methods usually originate a *partial pre-order*.

compensative method and requires independence of the criteria. Many other methods have been proposed in the literature to solve MADA problems. They include:

- *Geometric Averaging* (GA) (which has been usefully applied in strongly conservative cases, since it gives a null global score if only one criterion is null, thereby impeding compensation);
- *Ordered Weighted Averaging* (OWA) operators (Yager, 1988);
- *Compensation operator* introduced by Zimmermann (Von Altrock, 1995);
- Methods based on *non-additive measures* such as the Choquet integral.

A different approach is based on the scoring system obtained by a Fuzzy Inference System (Von Altrock, 1995), wherein the implicit knowledge about the system is implemented with a suitable set of inference rules based on linguistic attributes of natural language, making it a suitable linguistic interface. Also, we include in the family of MAVT, the Analytic Hierarchy Process (AHP) and the Ideal Point methods, which are usually considered alone given their peculiar characteristics, but do in fact compute a complete ordering of the alternative by means of a suitable score. In the next sub-paragraphs, we furnish a short description of these methods, in particular of WA, OWA, the Choquet integral, the AHP methodology, and the Ideal Point Method.

3.3.1.1 Simple Additive Weighting

Because of its easiness to understand and compute, the most popular Aggregation Operator function is the (simple) Weighted Averaging approach, the computation of a weighted average of each criterion's score for each alternative. This very popular additive model computes the aggregated valued as (Eq. 3.1):

$$V(a) = \sum_i w_i * v_i(a) \tag{3.1}$$

where V(a) is the total value associated with each alternative a and w_i is the weight linked with the ith criterion, usually selected by the decision maker, under the constraints: $w_i \geq 0, \sum w_i = 1$.

It is a compensative approach, and no interaction among the criteria can be modelled, since the Independent Preference axiom is assumed (Keeney and Raiffa, 1993). Simple Additive Weighting is based on the weighted average and it is the most classic of the MADA methods; it is also known as a weighted linear combination or scoring method (Malczewski, 1997; Janssen, 1992; Eastman, 1993). The steps for using this method are:

- The decision maker must give a weight to establish the importance of each attribute
- A score is built for each alternative
- The alternative with the highest score is chosen.

It is a simple and intuitive approach, but it is based on a strong independence hypothesis about the criteria (no interactions are admitted); thus, its applicability should be limited solely to the cases where independency is satisfied (i.e. compensation). Nevertheless, given its easy comprehension, it is often used for real world applications. Another problem regards the interpretation of the weights (*relative importance* or *trade-off*), and their assignment. For this purpose, the AHP methodology may be applied, though many other tools have also been proposed (Lootsma, 1999).

Many applications exist of the WA approach for environmental problems, even if checking the Preferential Independence axiom is rarely proposed. One notable example includes the Battelle-Columbus project, elaborated in the USA for planning and management of water basins (Battelle-Columbus Laboratories, 1972). It can be used to asses the environmental impact of different projects, concerned with water basins, and to plan projects for the medium and longer terms with the minimal environmental impact. The method is based on a list of *"impact indicators"*, with 78 parameters or environmental factors that represent a significant environmental impact aspect (Ecology, Contamination, Aesthetic and Human-interest). The rough data are transformed into the equivalent corresponding *"environmental quality index"* using a suitable *value function*, and the WA method is applied to calculate the environmental quality index, following the philosophy of the MAVT.

3.3.1.2 Ordered Weighted Average (OWA)

The OWA method has been developed in the context of fuzzy set theory (Yager, 1988). It includes, as particular cases, weighted averaging, and, as extreme situations, the Max and Min operators. If the weights are obtained by a non-monotonic quantifier (Yager, 1993), the OWA operator implements linguistic statements as "at least", "at most", "at least the half" and so on.[2] As soon as the OWA weights are defined in such a way to compute a value close to the Min operator, we say that the decision maker has a pessimistic tendency, and the weights define a ANDness scenario. In the opposite case, the decision maker is characterized by an optimistic behaviour. The degree of pessimism (optimism) is measured by the following index (Eq. 3.2):

$$\text{ORness}(w) = 1n - 1 \sum_r (n - r) w_r \qquad (3.2)$$

In OWA operators the coefficients "w" express the relevance given to the best or worst attribute, they are useful to give more or less importance to the best criteria. OWA operators are particularly interesting because they permit the definition of the *"decisional behaviour"* of the decision maker through the distribution of the coefficient "w". Indeed, if the coefficients are concentrated

---

[2] Indeed, the OWA operator is a particular case of the Choquet integral, see later.

in the first position, the best scores will be highlighted in comparison to the worst. As an extreme case, MAX assumes the best criteria as representative of the total judgement assigned to the evaluated alternative; it is a very optimistic method and shows that the decision maker is comfortable with risk. On the other hand, when the coefficients are concentrated on the last positions, the opposite occurs. In the extreme case of the MIN, only the worst criterion is considered: this is the case of the conservative approach, in which the decision maker is pessimistic and risk-averse.

3.3.1.3 Non-Additive Measures and the Choquet Integral

More recently, the introduction of methods based on non-additive measures helped to solve many theoretical and cumbersome aggregation problems. These methods are recognized as the most mathematically well-founded MAVT approach. A non-additive measure assigns a positive weight to every possible subset of the criteria, instead of to a single criterion only, as WA does. In so doing, the global importance of two, three or more criteria can be greater, equal, or less than the sum of the importance of each single criterion. A simple algorithm (Choquet integral or similar ones such as the *multi-linear* algorithm) computes the score of the alternatives, simply averaging the values of all the possible subsets of criteria. With respect to WA or to OWA, this method requires parameters to be assigned for all the possible interactions of criteria. If no interaction exists among the criteria, the method degenerates to WA. Formally speaking, given $N = \{1,2,3,\ldots,n\}$ the set of the criteria (for each node in the hierarchy), a *non-additive measure* is a set function: $m : S \subseteq N \rightarrow [0,1]$, so that, $\forall S, T \subseteq N$ the following condition holds (Marichal, 1998) (Eq. 3.3):

$$m(\emptyset) = 0, \forall S, T \subseteq N : S \subseteq T \Rightarrow m(S) \leq m(T), m(N) = 1 \qquad (3.3)$$

The second condition is a monotonic constraint, a quite intuitive rational condition. A non-additive measure is named:

- additive if: $m(S \cup T) = m(S) + m(T)$
- sub-additive if: $m(S \cup T) < m(S) + m(T)$
- super-additive if: $m(S \cup T) > m(S) + m(T)$

where $S \cap T = \emptyset$.

Let now $(x_1, \ldots, x_n)$ be the normalized values of the criteria (normalized by means of suitable value functions). If $(x_1, \ldots, x_n)$ is an index permutation so that: $x_{(1)} \leq \ldots \leq x_{(n)}$, $A_{(i)} = \{i, \ldots, n\}$, $A_{(n+1)} = \emptyset$, the Choquet integral of the vector is defined as (Eq. 3.4):

$$C_M(x_1, \ldots, x_n) = \sum_{i=1}^{n} \left(x_{(i)} - x_{(i-1)} \right) \cdot m(A_{(i)}) \qquad (3.4)$$

It coincides with the WA operator if the measure is additive (Marichal and Roubens 2000). Moreover, every OWA operator is a Choquet integral if every subset of the same cardinality has the same measure, i.e. (Eq. 3.5):

$$m(A) = \sum_{j=0}^{i-1} w_{n-j}, \forall A : |A| = i \qquad (3.5)$$

Many other properties can be defined for it, like the Möbius transform, the reduced order models, and the *andness* and *orness* indices, having the same meaning as the ones defined for the OWA aggregation operators (Grabisch, 1997, Grabisch et al., 2001, 2003).

3.3.1.4 Analytic Hierarchy Process (AHP)

Even if usually not strictly considered a MAVT approach, given its particular characteristics (the hierarchical decomposition and the pairwise comparison of alternatives with respect to each criterion for intangible evaluations), the computational algorithm of the AHP methodology is in fact an iteration of linear combination of the criterion values. We thus prefer to include this approach in the MAVT family, even if not "*strictu sensu*". AHP is a very popular method, originally developed by Saaty (Saaty, 2000) and widely reviewed and applied in environmental applications (Ramanathan, 2001). It makes use of pairwise comparison of alternatives, allowing the production of numerical values even from intangible criteria. For this reason, AHP is recognized as one of the most robust approaches for structuring complex problem and obtaining a significant score for the alternatives.

The **AHP** basic model is structured on the following four steps:

1. **Structuring the problem**: generating a hierarchical decision tree
2. **Comparison of judgments**: to compute the relative importance of the variables belonging to the same level, and relative to each of the associated variables. For each pair of attributes, the Expert answers question like "How much more important" in which one attribute is presented with respect to another one, using a Likert scale, or the natural 1,2,…,9 point scale. This scale has a semantic interpretation: 1 = equally important, 3 = weakly preferred, 5 = preferred, 7 = strongly preferred, 9 = totally preferred, with even numbers (2, 4, 6, 8) used in the case of uncertainty between two adjacent linguistic terms. Other scales have been proposed, but the Likert scale is generally accepted (Harker and Vargas, 1987; Lootsma, 1999). Moreover, for a comparison matrix with N alternatives, only N(N–1)/2 comparisons are required, given the symmetry of the comparison matrix (the elements in the diagonal are all equal to 1).
3. **Consistency Analysis** in which a set of the Expert's pairwise comparisons is *consistent* if the transitivity property is satisfied (Saaty, 2000). Nevertheless,

a limited amount of inconsistency has to be accepted, given the uncertainty characteristics of human thinking. The *ordinal* consistency index is obtained using the principal eigenvalue of the comparison matrix, and the judgements are considered acceptable if the CI is inferior than the average of the consistency indices of many randomly generated reciprocal matrices. It is preferable to avoid an intransitivite preference cycle, because it is possible for a reciprocal matrix to exhibit a good consistency index even with irrational preference cycles. Thus suitable algorithms have been developed to check for *ordinal* consistency (Kwiesielewicz and Van Uden, 2004). We note that other consistency measures can be defined, such as the *max-min transitivity*, $a_{ij} \geq \max_h (\min(a_{ih}, a_{hj}))$, for every a_{ij}, a_{ih}, a_{hj} of the comparison matrix.

4. **Analysis of priorities** leads, through suitable aggregation tools, to a final ranking of the alternatives. The original aggregation tool, based on the *principal eigenvector* method, is still the most commonly used in real applications, but others are available (Lootsma, 1999), including OWA-based AHP (Yager, 1999). AHP has been intensively applied even for Group Decision problems, (see Van Den Honert and Lootsma, 1996 and the reference therein). Schmoldt (Schmoldt et al., 2001) reports a set of environmental case studies solved through AHP. Despite its great popularity, though, AHP has received criticism. For instance, the phenomenon of *rank reversal* is troubling, as is the exponential growth of computation and comparisons with respect to the number of the alternatives. Rank reversal is observed when, upon adding or deleting an alternative, at least two of the other alternatives invert their position in the ranking (Belton and Gear, 1983). Rank reversal depends on the absence of trade-offs among the criteria (Wedley et al., 2001). This undesired phenomenon can be solved using the super-matrix approach and the AHP network. To reduce the great amount of data required, a limited number of elements may be filled into the comparison matrix, and algorithms may be implemented to compute the missing ones (Fedrizzi and Giove, 2007), a solution still not widely applied in practice. Alternatively, a node may be decomposed into more than one level if it has too many criteria (although some information may be lost). Finally, we remark that, for the WA approach, many other methods exist to compute weights, including SMART, SWING, the Direct Rating approach, and the PCT approach (Rogers, 1998).

3.3.1.5 The Ideal Point Methods

Ideal Point methods, strictly speaking, are also not classified as MAVT. However, because they generally furnish a total order and a score of the alternatives, we prefer to include them in the same family. The most commonly used method is TOPSIS (Yoon and Hwang, 1995) in which each criterion is represented along an axis in n-dimensional Euclidean space. In this space, each alternative is assigned a position based on its performance related to each criterion. A hypothetical optimum solution is positioned in the same way, and the distance from the optimal is then calculated for all of the alternatives. This produces a ranking of alternatives, with the best being the alternative closest to the optimal solution.

3.3.2 Outranking Methods

Outranking methods are also known as concordance methods Their main characteristic is the pairwise comparison of the alternatives of each criterion to create partial binary relations that explain which alternative is preferred but not by how much. Weights are given to the preferences on a normalized scale, from 0 to 1, where zero is no preference and one is strict preference. These weights primarily indicate the degree of dominance of one alternative over another and the partial preference ranking of alternatives, but they do not give a cardinal measurement of preference relationships. These methods are largely used in environmental impact assessment (EIA). The most-applied methods are ELECTRE I (Roy, 1968), with its variations ELECTRE II (Roy and Bertier, 1973) and ELECTRTE II (Roy, 1978), and PROMETHEE I (Brans and Vincke, 1985). In brief, concordance methods suffer from a few theoretical issues, as well as from difficulty in assigning the necessary parameters for understanding the meaning of the model as soon as the number of variables reaches real values. Thus, in our opinion, they can be applied only in particular cases and with close attention; therefore, we have not chosen to discuss them in depth.

3.3.3 Interactive Methods

In interactive methods, the decision maker gives information about his preferences while exploring different solutions. Normally the information corresponds to local trade-offs in comparison with the present solution and this information is used to determine a new solution.

The advantages of these methods are that they:

- do not require any "*a priori*" information
- allow for decision maker learning
- include local preferences
- may lead to the best acceptance of the final solution
- allow a less restrictive hypothesis.

Despite these advantges, these methods are rarely applied to environmental applications.

3.3.3.1 Final Remarks and Conclusion

Multi-criteria tools are a well-recognized and efficient way to describe and formalize a complex decision problem, and to suggest the best decision. One advantage is that the reasons for the preferences for one alternative over another can be made clear and revealed to stakeholders and the public. Moreover, multicriteria tools are particularly suitable for resolving problems that are difficult to formalize in purely economic terms or are characterized by intangible criteria. This is a typical situation for environmental decision problems, where a multitude of often non-economic factors increases the complexity of the decision process, including the necessity of balancing between sustainable development and environment protection.

MCDA problems are usually partitioned into two main classes: Multi-Attribute (MADA) and Multi-objective (MODA) decision analysis. MADA problems consider a finite number of criteria and alternatives, and it is the most often used in environmental applications. Conversely, MODA problems consider non-linear, high-dimensional, optimization problems, difficult to be analytically solved, and vector optimization, like Goal Programming (Charnes and Cooper, 1961). In this paper we presented methods and tools of MADA problems. For what concerns MODA problems, we limit to remark that a specialized mathematical literature exists, together with many commercial available solving packages.

Environmental complex problems are sometimes addressed by typical economic methods such *Cost-Benefit Analysis,* but this approach requires that costs and benefits to be expressed in monetary terms, and sometimes this is impossible, mainly for intangible criteria. Other econometric approaches, based on revealed preferences, such as travelling costs, conjoint analysis, and so on, suffer from other disadvantages, the most important of them being the high costs involved in obtaining data. Multi-criteria methods, even if strongly dependent on the decision makers' decision preference structure, seem to be the most appropriate tools for solving complex environmental problems. Nonetheless, further methodological research is advisable, since some problems still need to be solved. For instance, how does one justify the choice of a particular approach? Which properties are required for the application of one particular method or another? In fact, many reports using multi-criteria analysis do not justify the adoption of a particular tool or method that was used in their application. To obtain a satisfactory answer, the underlying assumptions of any method should clearly stated. In this vein, it is our conviction that methods based on *measurement* theory (Grabisch, 1997; Marischal, 2000) will become more prevalent in the future providing a more satisfactory basis of and performance by Decision Support Systems for environmental applications. Other aspects, which will be important in the future, include consensus measuring and managing between all the actors involved in the decision process. Even though we do not address proposals and results in the Group Decision literature (see for instance Kacprzyk et al., 1992 and the reference therein), these will surely be of crucial importance in the future, and they will also be further advanced by the improvement of communication and hardware tools.

Other promising approaches make use of Expert System for complex decision problems. We have not addressed this very important class of quantitative tools based on Artificial Intelligence methods, which have been applied recently to MCDA problems. Another class of innovative tools for quantitative analysis depends on a *Data Mining* approach. Roughly speaking, these methods try to extract general rules from sampled data, avoiding pre-determined quantitative models of the observed phenomenon. Examples of this class include neural nets, advanced cluster methods, nearest neighbour, kernel method, fuzzy logic, genetic algorithms, rough sets theory, and other approaches.[3] Nevertheless, Data Mining usually requires a great amount of data for the *learning* phase (knowledge acquisition by input-output data) and for its validation (*test* phase). Furthermore, it is inspired by a different philosophy than MCDA, which recognizes the need to efficiently acquire limited data to determine the preference structure of experts or decision makers.

Nonetheless, we feel that the DSS developers will gain much more from the integration of MCDA, Data Mining, Group Decision Theory, and participatory model using a common platform. The design of high performance integration and management of such tools will be the next challenge for complex environmental decision problem solving that when available will enable stakeholders, technical experts, politician decision makers, and citizens to engage in mutual and synergic cooperation thus benefiting everyone, but especially emerging countries and economies and future generations.

References

Alter SL (Fall 1977). *A taxonomy of decision support systems*. Sloan Management Review, vol 19 no 1, pp 39–56

Antoine J (1998). *Information technology and decision-support systems in AGL*. Land and Water Development Division, FAO, Rome.

Baker D, Hunter R, Johnson G, Krupa J, Murphy J, Sorenson K (2001). *Guidebook to decision-making methods*. US Department of Energy, Washington DC, WSRC-IM-2002-00002

Bana e Costa C, Vansnick J (1994). *MACBETH – an interactive path toward the constructions of cardinal value functions*. International Transactions in Operational Research, vol 1, pp 489–500

Bonczek RH, Holsapple CW, Whinston AB (1980). *The evolving roles of models in decision support systems*. Decision Sciences, vol 11 no 2, pp 337–356

Brans J, Vincke P (1985). *A preference ranking organization method: the PROMETHEE method for multiple criteria decision making*. Management Science, vol 31, pp 647–656

Bridges T, Apul D, Cura J, Kiker G, Linkov I (2004). *Towards using comparative risk assessment to manage contaminated sites*. In "Strategic management of Marine Ecosystems", Edited by Levner E, Linkov I, Proth JM. Kluwer, Amsterdam

Charnes A, Cooper WW (1961). *Management models and industrial applications of linear programming*. Wiley, New York.

Chankong V, Haimes Y (1983). *Multiobjective decision making: theory and methodology*. North Holland, Amsterdam.

---

[3] Among all the other ones, we limit to quote the ClusDM approach (Clustering for Decision Making) (Valls, 2003).

Ekárt A, Németh SZ (2005). *Stability analysis of tree structured decision functions*. European Journal of Operational Research, vol 160 no 3, pp 676–695

Eom SB (2001). *Decision support systems*. In "International Encyclopaedia of Business and Management", 2nd Edition, Edited by Malcolm Warner. International Thomson Business Publishing Co, London, England

Fedrizzi M, Giove S (2007). *Incomplete pairwise comparison and consistency optimization*. European Journal of Operational Research, vol 183, pp 303–313.

Figueira J, Ehrgott M, Greco S, (2005). *Multiple criteria decision analysis: state of the art surveys*. Springer, Berlin.

Gallego E, Jiménez A, Mateos A, Ríos-Insua S, Sazykina T (2004). *Application of multi-attribute analysis (MAA) to search for optimum remedial strategies for contaminated lakes with the MOIRA system*. Paper Presented at the 11th Annual Meeting of the International Radiation Protection Association 23–28 May, Madrid.

Grabisch M (1997). *K-order additive discrete fuzzy measures and their representation*. Fuzzy Sets and Systems, vol 92, pp 167–189

Grabisch M, Labreuche C, Vansnick JC (2001). *Construction of a decision model in the presence of interacting criteria*. Proceedings of AGOP 2001, Asturias, Spain, pp 28–33

Grabisch M, Lebreusche C, Vasnick JC (2003). *On the extension of pseudo-boolean functions for the aggregation of interacting criteria*. European Journal of Operational Research, vol 148 n° 1, pp 28–47

Halen H, Maes E, Moutier M, (2004). *Gestion durable des terrains affectés par les anciennes activités industrielles en Wallonie: les enjeux et le défis posés par l'évaluation des risques dans le cadre des nouveaux développements réglementaires sur la pollution locale des sols*. Biotechnologie Agronomie Société et Environnement. 8 n°2, pp 101–109

Harker PT, Vargas LG (1987). The *theory of ratio scale estimation: Saaty's analytic hierarchy process*. Management Science, vol 33, pp 1383–1403.

Hwang C, Yoon K (1981). Multiple attribute decision making: methods and applications. Springer, Berlin.

Janssen R, Van Herwijnen M (1992). *Multiobjective decision support for environmental management*. Kluwer Academic, Boston, MA.

Jeffreys I (2002). *A Multi-Objective Decision-Support System (MODSS) with Stakeholders and Experts in the Hodgson Creek Catchment*. A report for the RIRDC/Land & Water Australia/FWPRDC: Joint Venture Agroforestry Program, July 2002, Edited by Harrison, S and Herbohn, J.

Kacprzyk J, Fedrizzi M, Nurmi H (1992). *Group decision making and consensus under fuzzy preferences and fuzzy majority*. Fuzzy Sets and Systems, vol 49, pp 21–31

Keen Peter GW, Scott Morton MS (1978). *Decision support systems: an organizational perspective*. Addison-Wesley, Reading, MA, USA.

Keen Peter GW (1981). *Value analysis: Justifying decision support systems*. MIS Quarterly, vol 5 no 1, pp 1–16

Keeney R, Raiffa H (1976). *Decision with multiple objectives. Preferences and value trade-off*. Wiley, New York, pp 589.

Keeney RL, Raiffa H (1976). *Decisions with multiple objectives: preferences and value trade-off*. Wiley, New York

Keeney RL, Raiffa H (1993). *Decisions with multiple objectives: preferences and value trade-offs*. Cambridge University Press, Cambridge.

Klement EP, Mesiar R, Pap E (2000). *Triangular norms*. Kluwer Academic Publishers, Netherlands.

Kwiesielewicz M, Van Uden E (2004). *Inconsistent and contradictory judgements in pairwise comparison method in the AHP*. Computer & Operations Research, vol 31, pp 713–719

Linkov I, Bridges TS, Jamil S, Kiker GA, Seager TP, Varghese A (2004). *Multi-criteria decision analysis: Framework for applications in remedial planning for contaminated sites*. Kluwer, Amsterdam, pp 15–54

Linkov I, Figueira JR, Levchenko A, Tervonen T, Tkachuk A, Satterstrom FK, Seager TP (2007). *A multi-criteria decision analysis approach for prioritization of performance metrics: U.S. government performance and response act and oil spill response*. In "Managing Critical Infrastructure Risks", pp 261–298, Springer, Netherlands.

Linkov I, Belluck DA, Bridges T, Gardner KH, Kiker G, Meyer A, Rogers SH Satterstrom FK, Seager TP (2006). *Multicriteria decision analysis: a comprehensive decision approach for management of contaminated sediments*. Risk Analysis, vol 26 n°1, pp 61–78

Lootsma FA (1999). *Multi-criteria decision analysis via ratio and difference judgement*. Kluwer, Dordrecht

Marichal JL (1998). *Dependence between criteria and multiple criteria decision aid*. Proceedings of 2nd International Workshop on Preference and Decisions TRENTO'98, Trento

Malczewski J, Bojórquez-Tapia L, Moreno-Sánchez R, Ongay-Delhumeau E (1997). *Multicriteria group decision-making model for environmental conflict analysis in the Cape Region, Mexico*. Journal of Environmental Planning and Management, vol 40, pp 349–374

Mareschal B (1988). *Weight stability intervals in multicriteria decision aid*. European Journal of Operational Research, vol 33 no 1, pp 54–64.

Marichal JL, Roubens M (2000). *Determination of weight of interacting criteria from a reference set*. European Journal of Operational Research, vol 124, pp 641–650

Mészáros CS, Rapcsák T (1996). *On sensitivity analysis for a class of decision systems*. Decision Support Systems, vol 16 no 3, pp 231–240

Ramanathan R (2001). *A note on the use of the analytic hierarchy process for environmental impact assessment*. Journal of Environmental Management, vol 63, pp 27–35

Rogers M, Bruen M (1998). *Choosing realistic values of indifference, preference and veto thresholds for use with environmental criteria within ELECTRE*. European Journal of Operational Research, vol 107, pp 542–551

Rogers M, Bruen M (1998). *A new system for weighting environmental criteria for use within ELECTRE III*. European Journal of Operational Research, vol 107, pp 552–563

Roy B (1968). *Classement et choix en présence de points de vue multiple (la méthode ELEC-TRE)*. Revue Française d'Informatique et de Recherche Opérationnelle, vol 8, pp 57–75

Roy B, Bertier P (1973). *La méthode ELECTRE II: une application au médiaplanning*. OR'72, Edited by Ross M. pp 291–302. North Holland, Amsterdam

Roy B (1978). *ELECTRE III: un algorithme de classement fondé sur une représentation floue des préférences en présence de critères multiples*. Cahiers Centre d'Etudes de Recherche Opérationnelle, vol 20, pp 3–24

Saaty TL (2000). *Fundamentals of decision making and priority theory with the analytic hierachy process*. RWS Pubblications, Pittsburg

Schmoldt DL, Kangas J, Mendoza GA, Pesonen M (2001). *The analytic hierarchy process in natural resource and environmental decision making*. Kluwer Academic Publishers

Tryantaphyllou E, Sanchez A (1997). *A sensitivity analysis approach for some deterministic multicriteria decision making methods*. Decision Science, vol 28 no 1, pp 151–194

Valls A (2003). *ClusDM: a MCDM Method for Heterogeneous Data Sets*. Bellaterra, IIIA-CSIC monographies. ISBN:84-00-08154-4.

Van Den Honert RC, Lootsma FA (1996). *Group preference aggregation in the multiplicative AHP. The model of the group decision process and Pareto optimality*. European Journal of Operational Research, vol 96, pp 363–370

Vincke P (1992). *Multi-criteria decision aid*. John Wiley and Sons, Chichester.

Von Altrock C (1995). *Fuzzy logic and neuro-fuzzy applications explained*, Prentice Hall PTR, Upper Saddle River, NJ

Von Winterfeldt D, Edwards W (1986). *Decision analysis and behavioural research*. Cambridge University Press, Cambridge

Wedley WC, Choo EU, Schoner B (2001). *Magnitude adjustment for AHP benefit/cost ratio*. European Journal of Operational Research, vol 133, pp 342–351

Wolters WTM, Mareschal B (1995). *Novel types of sensitivity analysis for additive MCDM methods*. European Journal of Operational Research, vol 81 no 2, pp 281–290

Yager RR (1988). *On ordered weighted averaging aggregation operators in multicriteria decision-making*. IEEE Transactions on Systems, Man and Cybernetics, vol 18 n° 1, pp 183, 190

Yager RR (1993). *Families of OWA operators*. Fuzzy sets and Systems, vol 59 no 1, pp 25–148

Yager RR (1999). *An extension of the analytical hierarchy process using OWA operators*. Journal of Intelligent and Fuzzy Systems, vol 7, pp 401–417

Yoon K, Hwang C (1995). *Multi-attribute decision-making: an introduction*. Sage Publications, London

Chapter 4
Spatial Analytical Techniques for Risk Based Decision Support Systems

Mark S. Johnson, Marek Korcz, Katherine von Stackelberg and Bruce K. Hope

Abstract Interactions of biological entities within the environment occur on spatial and temporal scales. Likewise, the spatial and temporal distributions of contamination within the environment affect the degree to which plants, animals, and humans are exposed and how they respond. These interactions can be complex, however, through the recent advances in geographical information systems (GIS) and other models that integrate spatial considerations, estimates of risk can be more accurately described. Moreover, presentation of contaminated sites on spatial scales allow for a clearer understanding of the problem. This chapter will explore the recent advances in these models that integrate spatial attributes of contamination; from the latest in GIS techniques describing the nature and extent of contamination, to improved population-based exposure models to predict risk, to models that then integrate risk from contamination exposure relative to other stressors in a multi-stressor analysis. Examples of these advances will be described, including considerations from a decision-making perspective.

4.1 Introduction

Spatial analysis is an important component of risk assessment and risk management for contaminated sites, because the problems addressed are inherently spatial. Chemicals in the environment are rarely distributed in uniform concentrations. Fate and transport of chemicals occur relative to time and space. Additionally, interactions of receptors such as humans, wildlife, and fish species within the environment occur in biased, heterogeneous ways, often directed by demographic or habitat preferences. Together, these attributes suggest that predictive risk assessment depends upon the spatial examination of contamination

M.S. Johnson (✉)
U.S. Army Center for Health Promotion and Preventive Medicine, Health Effects Research Program, 5158 Blackhawk Road, Aberdeen Proving Ground, MD 21010-5403, USA
e-mail: mark.s.johnson@us.army.mil

A. Marcomini et al. (eds.), *Decision Support Systems for Risk-Based Management of Contaminated Sites*, DOI 10.1007/978-0-387-09722-0_4,
© Springer Science+Business Media LLC 2009

distributions and the relationships of receptors to them. Effective risk management practices depend on how these distributions affect risk and how resources could be best used to reduce the probability of unacceptable outcomes. Risk assessors and managers must consider spatial aspects of the following:

- Sources of contamination including spills, buried wastes, and leaking tanks are unevenly distributed on sites.
- Transport and transformation processes redistribute the waste constituents unevenly within and outside the site.
- Organisms and their activities are unevenly distributed, so the distribution of exposure is not the same as the distribution of contaminants.
- Effects of exposure to contaminants depend on the movement of organisms among locations with different types and levels of contamination and on the interaction of affected organisms with other members of their population or community which are unevenly distributed across the landscape.
- Remedial actions must be distributed in space in a way that meets goals for risk reduction and facilitation of future uses.

Because assessment and management processes must deal with these spatial problems, it is important for decision support systems to include spatial analysis capabilities. Tools for spatial analysis have different uses in different steps in the process. Examples include:

1. Data management and mapping tools can be used to organize records of waste disposal, prior analyses of contaminated media, and other information to clarify the nature and extent of the problem during the scoping phase.
2. Spatial analysis can be used to design additional sampling schemes that optimize the information obtained.
3. Spatial modeling can be used to simulate movement of contaminants to identify secondarily contaminated areas, to estimate contaminant concentrations in unsampled areas, and to predict contaminant concentrations in the future under different management scenarios.
4. Spatial modeling can also be used to estimate exposures of organisms based on their distributions relative to the distributions of contaminants.
5. Exposure models can be used for risk characterization by adding toxicological exposure-response models.
6. Risks and benefits of alternative actions can be estimated by using spatial modeling to estimate the risks associated with different extents and locations of remedial actions.
7. Maps can be used to display results that inform decision makers, stakeholders, and the public.
8. Virtual clean up exercises can be conducted and used to maximize resources while minimizing the amount of environmental disturbance.

Current risk assessment practices often rely on a mixture of conservative assumptions and variability estimation in reducing the complex nature of spatial heterogeneity. Although these techniques can simplify the process, important information is lost and risk calculations are rarely predictive. Spatial techniques can greatly assist in making more accurate risk predictions as well as providing tools for management in a virtual context.

This chapter describes the capabilities of Geographic Information Systems (GISs) and other specific exposure models that use spatial techniques, to meet these assessment and management needs. It then discusses in some depth the principal use of spatial analysis in ecological risk assessments of contaminated sites, namely, the estimation of exposures of fish and wildlife.

4.2 Geographical Information Systems

In the process of contaminated land risk analysis, geographic information systems (GIS) play an important role that supports the following groups of tasks:

- management of spatial data and information,
- spatial data preprocessing and modeling,
- spatial data and information visualization.

The data used in risk analysis are of different spatial, temporal, and thematic resolutions. The amount of data and information that have to be analyzed during the process may be enormous. Data are of different forms and types: e.g. features, tables, rasters, relations, reports. These data can be managed by separate databases systems (DBMS) or by DBMS which operate as integral components within a GIS. Recent GIS products offer the possibility of sharing data with regional repositories. This can greatly decrease cost and the time involved in data gathering. The recent thematic GIS tools integrate a higher degree of environmental spatial data processing than the earlier, more typical GIS tools.[1] The ability to network the spatial information systems dedicated to different problems change the role of GISs used at local or regional scales. At the stage of systematical planning, it is possible to establish the links between regional and other local GISs to virtually collect all data necessary to establish a conceptual site model (CSM) for a megasite[2] or site. At this stage, the role of a local GIS dedicated to contaminated land risk analysis is searching for and analyzing appropriate spatial data and information. A typical example of a GIS application that implements such an approach is BASINS (Better Assessment Science Integrating Point & Nonpoint Sources).[3] The list of data necessary to construct a CSM is limited to:

---

[1] For example compare the ARC/INFO 7.4 with ARC GIS 9.2

[2] Is a large area (indicative size: 5–500 km^2) with multiple contaminant sources related to (former) industrial activities, with a considerable impact on the environment, through groundwater, surface water and/or air migration – http://www.euwelcome.nl/kims/glossary/index.php?l=J.

[3] http://www.epa.gov/waterscience/basins/

- management of spatial data and information
- boundaries of administrative units
- boundaries of protected natural and cultural objects (existing or designed)
- boundaries of natural objects such as watersheds, groundwater bodies
- registers of monitoring stations
- register of potential sources of contaminants
- land use/land cover maps as surrogates of receptors' habitats boundaries.

At this stage of conceptual site model development, GIS tools are used to establish the boundaries of site or megasite and the extent of the analysis. These two activities differ because one is a simple boundary of all potential sources and the second represents a boundary of potential impact on the environment. This second boundary is a logical sum of boundaries of all natural objects that are in spatial contact with potential sources. The analysis of sources, type of contaminants potentially released from these sources, as well as the rough assessment of potential receptors allow the analyst to distinguish the most important (structural) objects within analyzed extent and to focus further analysis on these objects and their spatial relations. In contaminated land management, the core object is usually a single parcel which may contain one or several sources. In the case of megasites, the set of parcels is usually embedded in a region of potential impact.

The next step of CSM development is related to establishing the potential routes of the migration of contaminants from potential sources to the potential receptors. Here, the stage of CSM development can be treated as the stage of problem redefinition in space. From this moment, the risk analysis is focused on improvement of basic data quality and on development of the information necessary to reduce the uncertainty of further decision making. This requires that each series of relationships be analyzed that were preliminarily established to provide necessary data about contaminant behavior from release point to the potential receptor. The potential fate and transport routes have to be described in a quantitative way. This requires more detailed knowledge about all environmental media that are often converted into migration models.

As previously mentioned, recent sophisticated GIS applications can be linked with specialized models that enable the analysis of contaminants movement in space and time. These GIS applications contain many tools that are specialized models or simple interfaces to specialized models. The most important are hydrologic models, hydrogeological models, as well as models of air contamination (Maidment and Djokic 2000, Chiang and Kinzelbach 2001). The types of data that need to be collected depend on model implementation. The analysis integrated within GIS information layers allows the assessor to evaluate the completeness of data while taking into account the applied model.

The quantity of necessary data and other qualitative information increases with each step of CSM development. The types of data that increase in complexity include the following (Korcz et al. 2003):

- management of spatial data and information
- land surface geomorphology (e.g. Digital Elevation Model (DEM) and its derivatives)
- soil type or characteristics (the main environmental interface for most site contaminants)
- land use
- land cover (e.g. vegetation maps, human and animal population maps)
- geology (e.g. lithological map, boreholes logs, the elevation models of top and bottom layers of distinct lithological units)
- hydrogeology (e.g. water tables, elevation models of aquifers' boundaries derived from lithological data, springs, wells, directions of water movement)
- meteorology and climate (e.g. precipitation, temperatures, wind fields, and location of meteorological posts).

Each of these groups is composed from several layers of information that can be linked spatially or can be organized hierarchically. Part of this information can be integrated into a GIS framework as primary data or as secondary data, generated as a result of simulations conducted within GIS models or external models.

The data in GIS are managed as information layers organized thematically. The primary set includes layers each of a single set of geographic features in 2D space. A third dimension (e.g. depth) can be extracted from a database and presented as a separate layer. Time is represented by a set of information layers developed for separate time slices or intervals of time based on time series data, usually gathered from a data base.

Based on a set of information layers, a hypothesis concerning the extent of contamination on the land surface can be generated. This is the last step of preliminary CSM development.

The next step is verification of the extent of contamination by direct field measurements. Generation of an optimal measurement network is supported by many GIS tools. The measurement points are located near the existing or former installations that are regarded as potential sources of contaminant emissions as well as on the suspected directions of contaminant movements. Generally, a regular or quasi regular measuring network is preferred and sampling is conducted in a way that minimizes the uncertainty of omitting a part of a contaminated area. The second important criterion is the minimization of estimation error. The total quantity of sampling points depends on the size of the investigated site area and its geometry, suspected variability of contaminants, minimum volume (or area) of contaminated space that can be treated as negligible, as well as the budget for the assessment.

The individual results of a site characterization representing 4D space can be treated in different ways, deterministic or probabilistic. A single value can represent a given location and given time. This value in a deterministic approach is treated as certain. In the probabilistic approach, this same value is uncertain due to the error of determining locations and due to the measurement error of a given parameter. The measured value describes the part of time-space interaction with an assumed probability. The measured value is one of the possible realizations of random variables at a given point in 4D space. As a consequence, we can determine uncertainty of a given feature in the assessment starting from an assumed distribution of this random variable. The more recent GIS applications are capable of collecting data from databases of all forms to describe the 4D space. Thanks to such recent GIS developments, import/export capabilities are enhanced so it is now possible to combine analysis in a classical GIS environment sense with specialized models that handle multidimensional and multivariable data. An example is netCDF (Network Common Data Form).[4] Using net CDF, four spatiotemporal dimensions can be coded along with values of selected environmental parameters describing real geographic features or phenomena.

The scale of decision making is usually smaller than the scale used in describing the spatial dimensions. This dichotomy creates a new role for GISs, i.e. proper aggregation of observation data in space and time to characterize the given objects in a scale appropriate for decision making.

Many tools in contemporary GIS are applicable to the analysis of data and information. Typical GISs allow transformation of geographic object attribute data with application of scalar, classification, overlay, connectivity, and neighborhood operations. More sophisticated operations include statistical and mathematical modeling. All of these operations are intensively used during CSM development. The verification of the model results concerning the extent and characterization of contamination of the site can be conducted for whole areas of concern or for one measurement point depending on applied decisions rules.

The verification for a single prediction is a simple task that is solved by the comparison of measured values (with or without confidence limits depending on applied approach) with the model result. Real, multidimensional objects require a more sophisticated approach where the contamination at measurement points should be evaluated as well as between them. The assessment can be conducted for whole objects after aggregation of point data with application of classical statistical tools or by converting the point data into a continuous field and applying further statistical treatment. This last task is easily conducted with application of traditional GIS tools and using different interpolators (usually 2D) – beginning with the use of typical geometrical tools such as Inverse Distance Weighted (IDW) or spline and finishing on statistical interpolators, such as trend or kriging.

The aggregation procedure produces one estimate for each object. The interpolation allows for a discreet delineation of the contaminated and uncontaminated

---

[4] netCDF is a machine-independent, self-describing, binary data format standard for exchanging scientific data developed within the Unidata program at the University Corporation for Atmospheric Research (UCAR). The format is an open standard. http://www.unidata.ucar.edu/software/netcdf/

zones within each object. The selection of the interpolators strongly depends on the quantity of samples, as a consequence of the measurement strategy. The verification of hypothesis about object contamination usually requires a small quantity of samples. It is assumed that a hypothesis concerning the contamination is confirmed if the concentration of a chemical in at least one sample is higher than the threshold value. In such a case, the object is qualified as contaminated. An object represents an area of concern to be evaluated by the risk assessment.

The use of various interpolation methods should be phased on the CSM process in that they are often dependant on the number of discrete samples. During the detailed scale of the site specific risk assessment, GIS tools are focused on delineation of contamination zones within analyzed objects (Wcislo et al. 2005). For small objects or when the quantity of samples is small, geometrical interpolators are applied. One of the more robust methods resistant on interpolation errors is the use of Thiessen polygons, which provide spatial weights for each measurement location. Geostatistical interpolator methods such as kriging are useful for objects with dense sampling networks (i.e. more than 80 samples for one soil layer at a given time). Only advanced GIS tools are useful for geostatistical analysis.

Maps of contamination can be further classified using risk based standards. The sum of maps that show different contaminant levels can be used to delineate zones of contaminated and uncontaminated land for a given scenario or used to produce maps that illustrate zones of high or negligible risk, that each vary relative to time.

Finally, it is most important for decision makers to have an appreciation for the level of uncertainty associated with each variable. Uncertainty here can be described as a sum of measurement variability (sampling and analytical) plus the real variability associated with true geochemical differences. This latter attribute can be decomposed into space and time dependent variability (autocorrelation) and error that are frequently treated as a measure of pure uncertainty. Measurement errors can be minimized by improving the measurement techniques. Uncertainty can be minimized through densification of the observation network as well as by the use of other parameters (auxiliary variables). The recognition of total uncertainty structure is one of most critical tasks conditioning the costs of uncertainty reduction. The measurement uncertainty as an evaluation method for precision requirements in applied geochemistry was proposed by Ramsey (1992, 1998), Ramsey and Argyraki (1997). This approach was implemented in contaminated site characterization process by Demetriades (1999, 2006) as a base for the probabilistic classification of contaminated land as well as for the probabilistic delineation of risk zones.

Typical GISs contain many tools for visualization of primary data as well as for visualization of final risk assessment results. The simplest is a map of measurement locations. The map of classified risk zones along with diagrams and tables that contain risk characteristics of a site of concern is a typical product of GIS analysis. The results of 3D modeling can be presented as several static 2D images or as a movie that enhances the reception and perception of results by user. All results can be transparently superimposed on topographical maps or air/satellite images.

4.3 State of the Art – Spatially Explicit Exposure/Risk Models

The probability of a receptor contacting a stressor is governed by the distribution and movement of both the stressor and the receptor; that is, both may be dynamic. There are several tools for describing the distribution of stressors in the environment. There are far fewer tools for doing the same for receptors, either human or ecological; and even fewer for doing this for several, different types of stressors that occur together. Typically, the level of a stressor is statically estimated at locations on a landscape and a receptor is assumed to be static in those locations for some period of time. Neither move.

Levels of chemicals and other stressors at static locations can be described using the 95% upper confidence level (UCL) of the mean value or through a variety of more sophisticated geospatial (e.g. kriging, polygons) and statistical techniques (Burmaster and Thompson 1997, Gilbert 1987). Stressor movement can be described with a variety of transport and fate models, but this can make for very complex models. Although there may be concerns over details, these approaches generally have regulatory acceptance. By making a simplifying assumption of equal and random receptor access, the UCL method allows samples at discrete points to be coalesced into a mean concentration (or its upper bound confidence limit) to represent the concentration over the area encompassing the sampling locations (USEPA 2002). This may be somewhat applicable to human receptors (given the wide latitude they have in habitat choices), but is an unrealistic assumption for ecological receptors due to the greater constraints typically placed on them by habitat requirements. The UCL method may also use samples collected within discrete polygons (e.g. areas of low, medium, and high concentration). Even subdividing areas at three relative levels of contamination can provide more accurate estimates of exposure.

Quantification of habitat/receptor interactions began relatively recently. Clifford et al. (1995) moved a fixed forage area over a static landscape of stressor concentrations. Freshman and Menzie (1996) were the first to apply an individual based model (IBM) to move receptors across a static landscape of contaminant levels. In these models, stressors are static but receptors moved as individuals. In IBMs, organisms are represented individually, making it possible to explore how aggregate system properties (e.g. population-level effects) can emerge from interactions among these fundamental units, the landscape, and stressors.

This was followed by additional, more sophisticated IBM models for terrestrial mammals (Hope 2000, 2001a, 2001b, 2004) and fish (Linkov et al. 2002, Dortch and Gerald 2004). Hope (2005) used an IBM to expose moving receptors to multiple chemical and biological stressors, some of which varied in time and space.

Greater predictability of exposure can be gained when habitat preferences are considered with geospatial delineation of contamination (Johnson et al. 2007). Habitat-area weighted averaging is a similarly simple approach. It requires understanding an ecological receptor's habitat requirements and mapping them relative to sampling locations, but it uses a more ecologically representative approach. However, it requires a willingness to invest some time and effort in understanding life cycles and niche-specific differences among species.

The Spatially Explicit Exposure Model (SEEM) is a model developed to improve on methods to characterize exposure to vagile terrestrial vertebrate species (birds, mammals, reptiles) in a heterogeneous environment (Wickwire et al. 2004, http://chppm-www.apgea.army.mil/tox/HERP.aspx). An important feature of the model is that it allows for exposure/toxicant interaction within ecological habitat preferences in space and time. It incorporates a user-guided interface and a simple map/polygon drawing tool that allows for spatial delineation of contamination as well as habitat preferences. Each receptor/ contaminant risk profile is dealt with separately. Receptor exposure inputs are recorded and printed, and contaminant profiles are added in a spatial context through a user-defined polygon tool. Each contaminant polygon assumes a value (often the 95% UCL) representing the mean value of the media contaminant concentration. Each habitat polygon receives a user-defined habitat suitability weighting (unitless value between 0 and 1), which requires on-site truthing, consistent with the Habitat Suitability Index approach (USF&WS 1981, Kapustka et al. 2004). Often habitat suitability values can be obtained through information provided in aerial photography. Daily exposure estimates are determined for a user defined period of time for a number of individuals up to 1000. A schematic of input variables is presented in Fig. 4.1.

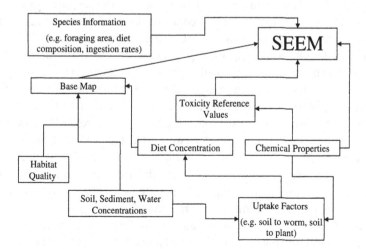

Fig. 4.1 Input model schematic describing user-defined criteria for the Spatially Explicit Exposure Model (SEEM)

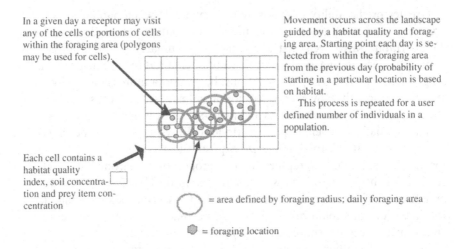

In a given day a receptor may visit any of the cells or portions of cells within the foraging area (polygons may be used for cells).

Movement occurs across the landscape guided by a habitat quality and foraging area. Starting point each day is selected from within the foraging area from the previous day (probability of starting in a particular location is based on habitat.

This process is repeated for a user defined number of individuals in a population.

Each cell contains a habitat quality index, soil concentration and prey item concentration

⬭ = area defined by foraging radius; daily foraging area

⬡ = foraging location

Fig. 4.2 Simplified conceptualization for the free ranging foraging exposure model option for the Spatially Explicit Exposure Model (SEEM)

SEEM contains two alternative exposure models. The static home range model places individuals randomly in the defined landscape and allows for up to 15 foraging events for each individual within each individual's home range. The free ranging model allows individuals to move across the landscape, preferentially in areas with a greater habitat suitability weighting (Fig. 4.2).

Mean daily contaminant exposure is calculated for each individual, compared with a static toxicity benchmark, and displayed as a population-based metric in both a graphical (Figs 4.3 and 4.4) and tablature context (data not shown).

As few as three levels of resolution can provide a significantly greater level of predictability in risk estimates when compared with values developed using a single value to describe media-specific concentrations of chemicals in the environment. In one evaluation, results of SEEM were compared with those of using conventional deterministic (HQ approach) models for several songbird species at two small arms ranges where lead exposure was a concern (Johnson et al. 2007). These results were then compared with site-specific estimates of lead exposure using blood lead data collected from the sites. Three spatial levels of lead soil concentrations were used (95% UCL) as were three habitat suitability weights. Identical exposure parameters were used for both models. Risk estimates from SEEM were within a 3-fold difference of blood lead levels where sublethal effects may occur, whereas deterministic hazard quotients ranged from 10 to 100 times higher, suggesting that the crude introduction of three levels of delineation greatly enhanced prediction capabilities.

FISHRAND is an example of an aquatic analog to the terrestrial wildlife exposure models. Drawing from user-defined input parameters, FISHRAND is a tool for estimating body burdens of organic chemicals in fish under current and future exposure scenarios incorporating temporal and spatial variability and

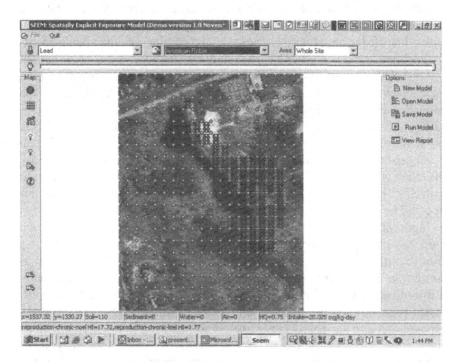

Fig. 4.3 User interface of SEEM. Imported map is aerial photography. Cross hatching indicates areas of highest modeled risk estimates from soil lead exposure to the American robin (*Turdus migratorius*)

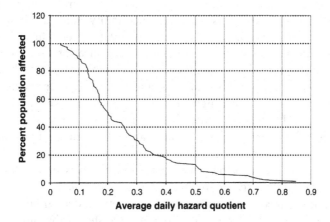

Fig. 4.4 Graphical output of SEEM for a population of individuals. Curve is created from mean individual risk estimates over a user-defined period of exposure as a ratio of exposure and toxicity using space and habitat as variables. For example, the model projects that 90% of individuals in the population have an HQ < 1.0 and > 14% would be expected to have a HQ greater than 0.5

uncertainty. Based on the Gobas (1993, 1995) modeling approach, FISHRAND is a mechanistic, time-varying exposure model that employs mass balance principles, knowledge about chemical uptake, and elimination and species-specific foraging preferences to generate probability distributions of fish tissue concentrations. The estimated body burden concentrations are generated using parameters such as biota-sediment accumulation factors for infaunal organisms, lipid content, chemical concentrations in associated physical media, total organic carbon in sediment, chemical assimilation efficiency, residence time, and octanol-water partition coefficients. FISHRAND also incorporates aquatic habitat heterogeneity, fish migration and the spatial distribution of chemicals to improve the reliability of exposure estimates. These body burdens can then be compared with effect-based body burden data to estimate risk to fish populations over time or could be used as input into human health exposure models.

Input variables can be described by distributions or point estimates, and users can specify whether parameters should be considered as "variable" (e.g. contributing directly to the population distribution of concentrations) or "uncertain" (e.g. contributing to the uncertainty bounds around the population distribution). There is "true" uncertainty (e.g. lack of knowledge) in the estimated concentrations of sediment and water to which aquatic organisms are exposed and also variability in parameters contributing to contaminant bioaccumulation. Uncertainty and variability should be viewed separately in risk assessment because they have different implications to regulators and decision makers (Thompson and Graham 1996, Morgan and Henrion 1990). Variability is a population measure that provides a context for a deterministic point estimate (e.g. average or reasonable maximum exposure). Variability typically cannot be reduced, only better characterized and understood. In contrast, uncertainty represents unknown but often measurable quantities. Typically, uncertainty can be reduced by obtaining additional measurements of the uncertain quantity. Quantitatively separating uncertainty and variability allows an analyst to determine the fractile of the population for which a specified risk occurs and the uncertainty bounds or confidence interval around that predicted risk. If uncertainty is large relative to variability (i.e. it is the primary contributor to the range of risk estimates) and if the differences in cost among management alternatives are high, additional collection and evaluation of information can be recommended before making management decisions regarding risks from exposures to contaminants. Also, including variability in risk estimates allows decision makers to quantitatively evaluate the likelihood of risks both above and below selected reference values or conditions (for example, average risks as compared to 95th percentile risks).

Characterizing uncertainty and variability in any model parameter requires informed and experienced judgment. Studies have shown that in some cases, based on management goals and data availability, it is appropriate to "parse" input variables as predominantly uncertain or variable (for example see Stackelberg von et al. 2002a, Kelly and Campbell 2000, Cullen 1995). Figure 4.5 provides a schematic of the nested modeling framework.

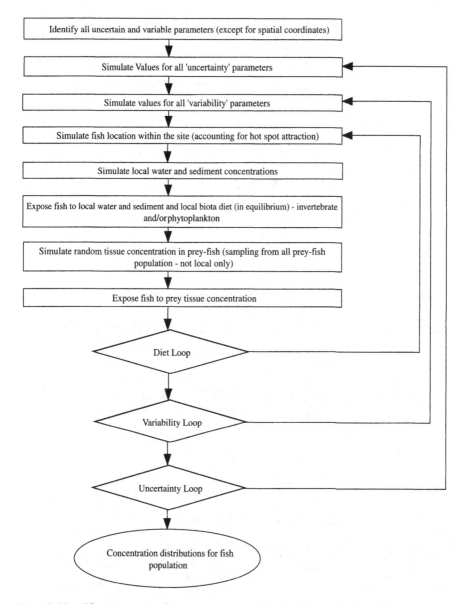

Fig. 4.5 Simplified schematic of input parameters and calculations within FISHRAND

The spatial submodel of FISHRAND is described in detail in Linkov et al. (2002) and Stackelberg von et al. (2002b). It uses variables that describe fish foraging behaviors to calculate the probability that a fish will be exposed to a chemical concentration in water or sediment. The spatial submodel uses time-varying sediment and water chemical concentrations, size of the site and hotspots (using GIS-based inputs), attraction factors, migration habits of the fish, fish foraging area sizes and habitat sizes to calculate the probability that a fish will be exposed to chemicals in the site.

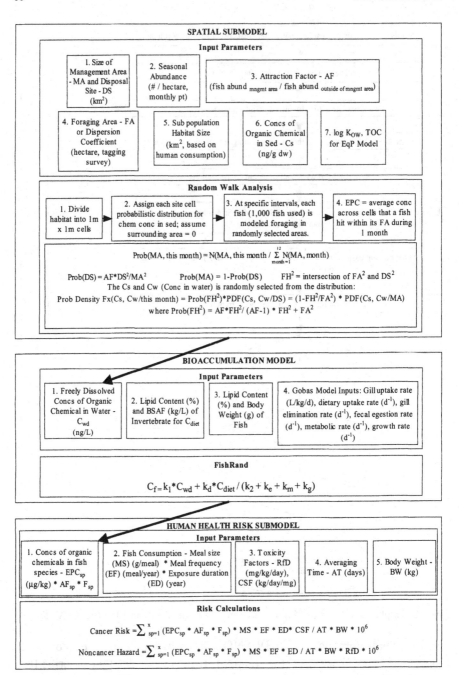

Fig. 4.6 Modeling and mathematical framework for FISHRAND

The management area or site is divided into background areas and up to 10 hotspots. The model requires that all hotspots be located within the management area and that they do not overlap. Each area is defined by minimum and maximum X and Y coordinates and the user provides the water and sediment chemical concentrations, organic carbon content of sediments, and water temperature for each area. These inputs can be point estimates or, preferably, distributions. Different areas may have the same sediment and water concentration, organic carbon content, and temperature or one or all of these values may differ among areas. When a fish is not located within the site, the water and sediment exposures as well as exposure from diet, are assumed to be zero.

Essentially, FISHRAND starts with a large number of fish (e.g. the number of simulations in the variability loop of the model) and scatters them randomly over the modeling grid. The modeling grid is defined by a GIS-based map of the management area with spatially-defined exposure concentrations in sediment and water. These can be defined in as much spatial and temporal detail as is available, including hot spots, background concentrations, and changes in concentrations over different time periods. These fish then move and forage according to their user-specified feeding preferences and foraging areas over the time interval specified in the model (typically 1 week, although it could be as little as 1 day or as much as a season). As the fish engage in these individual behaviors, they are exposed to sediment, water, and benthic invertebrate concentrations relative to the underlying modeling grid. Figure 4.6 presents a schematic of the modeling equations and the mathematical connections that link model components. In addition to capturing the impact of migratory behaviors on exposure, the spatial submodel also incorporates the impact of heterogeneous chemical distribution across the site.

Model output is presented in a number of different ways, including tabular and graphical. The basic form of the model results are individual percentiles of variability (concentrations across the population) and associated uncertainty for each population percentile for each time period and species. The user can select individual percentiles for plotting, or can average the data in different ways (e.g. seasonal or annual average with associated uncertainty). Figure 4.7 presents an example of one method for visualizing output from the FISHRAND model.

Fig. 4.7 Example graphical FISHRAND output showing mean PCB tissue concentration in Atlantic croaker. *Solid line* represents mean value; *dotted line* is upper 95% confidence interval

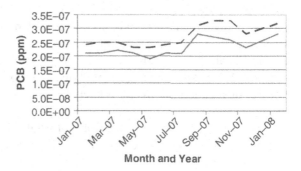

4.4 Conclusions

Because contaminants, receptors and the processes that connect them are unevenly distributed in space, spatial analysis is an important aspect of risk analysis for contaminated sites. Spatial analytical tools should be incorporated in Decision Support Systems (DSSs) for a variety of uses including organizing data, planning sample collection and analysis, modeling exposures and risks, and presenting results to decision makers and stakeholders. Geographic information systems (GISs) and the like, provide powerful tools for all of these purposes.

Quantitative analysis of exposure in a dynamic landscape is relatively complex. It is essential that those using such models understand how they work and are able to confidently describe what they mean to concerned stakeholders and risk managers. Additionally, like all models, a greater reliance on verifying such risk estimates will help ensure a greater acceptance. Significant challenges include providing the life-history, demographic data needed as input parameters for many receptors. A keen knowledge of factors most important in regulation for each population is also needed. Moreover, an integration of chemical-related risks with those from other stressors (e.g. climatic changes, habitat alterations, interspecific competition) is needed to help ascertain the relative impact of decision choices, such as remediation, with those caused by the impact of the proposed remedy.

While the availability of such models intended for ecological assessments suggests there is a greater knowledge of the factors that influence exposure and effects for environmental receptors, there is no evidence that we are aware of that this is true. Conversely, the availability of exposure factors for humans (USEPA 1997) and the relative abundance of non-clinical rodent studies used to develop toxicity benchmarks far outnumber studies and datasets for wildlife. The discrepancy in the lack of similar models intended to estimate risk to humans may be due to the political aspect of communicating the details and interpretation of risk estimates.

The scientific and regulatory communities now generally recognize that the relative spatial positions of receptors and stressors can strongly influence estimates of exposure and hence of risk at contaminated land sites. Yet it is still rare to find a risk assessment that explicitly considers such spatial relationships. The typical practice in USEPA human health and ecological risk assessments conducted at Superfund (CERCLA) sites has been to assume that: (a) an exposed individual moves randomly across an exposure area, thus allowing the area-averaged media concentration to represent the true average concentration contacted over time, and (b) that equal time is spent in different parts of the site (USEPA 2002). Applying these simplifying assumptions eliminates all information about spatial patterns and is likely to overstate exposure in human health risk assessments.

Currently, the modeling of exposures of wildlife and fish is the most relevant and most technically advanced use of spatial analysis in DSSs for risk assessment of contaminated sites. Exposures have typically been determined without spatial tools, particularly in screening assessments and other early tier risk assessments (USEPA 2002). Simple nonspatial analyses are appropriate in screening risk assessments. However, in many cases, the magnitude of risk and the necessary extent of remediation cannot be reasonably determined without analysis of the distributions of contamination and receptors and of transport and exposure processes (Burmaster and Thompson 1997). As a result there is an increasing interest in spatial analysis of exposures and risks and more tools available for this purpose, particularly with respect to ecological receptors (DeMott et al. 2005, USEPA 2004, Woodbury 2003, Zakikhani et al. 2006). We believe that judicious application of spatially-explicit methods to human and ecological risk assessments would allow decision makers the opportunity to identify contaminated land management options that are reasonable and cost-effective, as well as protective.

References

Burmaster DE, Thompson KM (1997) Estimating exposure point concentrations for surface soils for use in deterministic and probabilistic risk assessments. Hum Ecol Risk Assess 3:363–384

Chiang WH, Kinzelbach W (2001) 3D-Groundwater Modeling with PMWIN. Springer-Verlag, New York, p 346

Clifford PA, Barchers DE, Ludwig DF, Sielken RL, Klingensmith JS, Graham RV, Banton MI. 1995. An approach to quantifying spatial components for ecological risk assessment. Environ Toxicol Chem 15:895–906

Cullen AC (1995) The sensitivity of probabilistic risk assessment results to alternative model structures: A case study of municipal waste incineration. J Air Waste Manage Assoc 45:538–546

Demetriades A (1999) Geochemical Atlas of the Lavrion Urban Area for Environmental Protection and Planning: Explanatory Text. Institute of Geology and Mineral Exploration, Athens, Greece, Open File Report, Vol. 1

Demetriades A (2006) National Inventory of Potential Sources of Soil Contamination in Cyprus. Part 4 – Quality Assurance and Quality Control, Estimation of Measurement Uncertainty and Compilation of Probability Risk Assessment Maps. Institute of Geology and Mineral Exploration, Athens, Greece

DeMott RP, Balaraman A, Sorensen MT (2005) The future direction of ecological risk assessment in the United States: Reflecting on the U.S. Environmental Protection Agency's "Examination of Risk Assessment Practices and Principles". Integr Environ Assess Manag 1:77–82

Dortch MS, Gerald JA (2004) Recent advances in the army risk assessment modeling system. In: Whelan G, (ed.) Brownfields, Multimedia Modeling and Assessment, WIT Press, Southampton, UK

Freshman JS, Menzie CA (1996) Two wildlife exposure models to assess impacts at the individual and population levels and the efficacy of remedial actions. Hum Ecol Risk Assess 2:481–498

Gilbert RO (1987) Statistical Methods for Environmental Pollution Monitoring. Van Nostrand Reinhold, New York

Gobas FAPC (1993) A model for predicting the bioaccumulation of hydrophobic organic chemicals in aquatic food-webs: application to Lake Ontario. Ecol Model 69: 1–17.

Gobas FAPC, Z'Graggen MN, Zhang X (1995) Time response of the Lake Ontario ecosystem to virtual elimination of PCBs. Environ Sci Technol 29(8): 2038–2046

Hope BK (2000) Generating probabilistic spatially-explicit individual and population exposure estimates for ecological risk assessments. Risk Anal 20:573–588

Hope BK (2001a) A case study comparing static and spatially explicit ecological exposure analysis methods. Risk Anal 21:1001–1010

Hope BK (2001b) A spatially and bioenergetically explicit terrestrial ecological exposure model. Toxicol Ind Health 17:322–332

Hope BK (2004) Approaches to spatially-explicit, multi-stressor ecological exposure estimation. In: Kapustka LA, Gilbraith H, Luxon M, Biddinger GR (eds) Landscape Ecology and Wildlife Habitat Evaluation: Critical Information for Ecological Risk Assessment, Land-Use Management Activities, and Biodiversity Enhancement Practices. ASTM STP 1458 (American Society for Testing and Materials International), West Conshohocken, PA, pp 311–323

Hope BK (2005) Performing spatially and temporarily explicit ecological exposure assessments involving multiple stressors. Hum Ecol Risk Assess 11:539–565

Johnson MS, Wickwire WT, Quinn MJ Jr, Ziolkowski DJ Jr, Burmistrov D, Menzie CA, Geraghty C, Minnich M, Parsons PJ (2007) Are songbirds at risk from lead at small arms ranges? An application of the spatially explicit exposure model (SEEM). Environ Toxicol Chem 26:2215–2225

Kapustka LA, Galbraith H, Luxon M, Yocum J, Adams WJ. (2004) Application of suitability index values to modify exposure estimates in characterizing ecological risk. In: Kapustka LA, Galbraith H, Luxon M, Biddinger GR (eds), Landscape Ecology and Wildlife Habitat Evaluation: Critical Information for Ecological Risk Assessment, Land-Use Management Activities and Biodiversity Enhancement Practices. ASTM STP 1458. American Standards for Testing and Materials International, West Conshohocken, PA, pp 169–201

Kelly EJ, Campbell K (2000) Separating variability and uncertainty in environmental risk assessment – making choices. Human Ecol Risk Assess 6:1–13

Korcz M, Bronder J, Sowikowski D, Dugosz J (2003) The role of GIS in post-industrial site management on example of Tarnowskie Góry megasite. In: Procc ConSoil 2003, 8th International FZK/TNO Conference on Contaminated Soil Gent, Belgium, pp 3113–3119, http://www.euwelcome.nl/kims/about/index.php?index = 10

Linkov I, Burmistrov D, Cura J, Bridges TS (2002) Risk-based management of contaminated sediments: consideration of spatial and temporal patterns in exposure modeling. Environ Sci Technol 36:238–246

Maidment DR, Djokic D (2000) Hydrologic and Hydraulic Modeling Support with Geographic Information Systems, ESRI Press, Redlands, CA

Morgan MG, Henrion M (1990) Uncertainty: A Guide to Dealing with Quantitative Risk and Policy Analysis. Cambridge University Press, New York

Ramsey MH (1992) Sampling and analytical quality control (SAX) for improved estimation in the measurement of Pb in the environment using robust analysis of variance. Applied Geochem, Suppl Issue No. 2, pp 149–153

Ramsey MH (1998) Sampling as a source of measurement uncertainty: techniques for quantification and comparison with analytical sources. J Anal At Spectrom 13:97–104

Ramsey MH, Argyraki A (1997) Estimation of measurement uncertainty from field sampling: implications for the classification of contaminated land. Sci Total Environ 198:243–257

Stackelberg von K, Vorhees D, Linkov I, Burmistrov D, Bridges T. (2002a) Importance of uncertainty and variability to predicted risks from trophic transfer of contaminants in dredged sediments. Risk Anal 22:499–512

Stackelberg von K, Burmistrov D, Linkov I, Cura J, Bridges TS (2002b) The use of spatial modeling in an aquatic food web to estimate exposure and risk. Sci Tot Environ 288(1–2):97–110

Thompson KM, Graham JD (1996) Going beyond the single number: Using probabilistic risk assessment to improve risk management. Hum Ecol Risk Assess 2:1008–1034

United States Environmental Protection Agency (USEPA) (1997) Exposure Factors Handbook, Vol. 1–3, EPA/600/P-95/002Fa-c, Office of Research and Development, Washington DC

United States Environmental Protection Agency (USEPA) (2002) Calculating Upper Confidence Limits for Exposure Point Concentrations at Hazardous Waste Sites. OSWER Directive 9285.6-10. Office of Solid Waste and Emergency Response, Washington DC

United States Environmental Protection Agency (USEPA) (2004) An Examination of EPA Risk Assessment Principles and Practices. EPA/100/B-04/001. Office of the Science Advisor, U.S. Environmental Protection Agency, Washington DC

United States Fish and Wildlife Service (USF&WS) (1981) Standards for the Development of Habitat Suitability Index Models for Use in the Habitat Evaluation Procedures. ESM 103 U.S. Department of the Interior, Fish and Wildlife Service. Division of Ecological Services, Washington DC

Wcislo E, Dlugosz J, Korcz M (2005) A human health risk assessment software for facilitating management of urban contaminated sites: A case study: The Massa Site, Tuscany, Italy. Hum Ecol Risk Assess 11(5):1005–1024

Wickwire WT, Menzie CA, Burmistrov D, Hope BK (2004) Incorporating spatial data into ecological risk assessments: the spatially explicit exposure module (SEEM) for ARAMS. In: Kapustka LA, Galbraith H, Luxon M, Biddinger GR (eds) Landscape Ecology and Wildlife Habitat Evaluation: Critical Information for Ecological Risk Assessment, Land-Use Management Activities and Biodiversity Enhancement Practices. ASTM STP 1458. American Society for Testing and Materials International, West Conshohocken, PA, pp 297–310

Woodbury P (2003) Do's and don'ts of spatially explicit ecological risk assessments. Environ Toxicol Chem 22:977–982

Zakikhani M, Brandon DL, Dortch MS, Gerald JA (2006) Demonstration applications of ARAMS for aquatic and terrestrial ecological risk assessment. ERDC/EL TR-06-1. Environmental Laboratory, US Army Engineer Research and Development Center; Vicksburg, Mississippi

Chapter 5
Indicators and Endpoints for Risk-Based Decision Processes with Decision Support Systems

Paola Agostini, Glenn W. Suter II, Stefania Gottardo, and Elisa Giubilato

Abstract The decision process for contaminated sites is composed of two important phases, assessment and management. In the first phase, relevant information is collected and processed by experts on all the involved aspects of the decision problem; in the second phase, the same information is evaluated, weighted and communicated by decision-makers and stakeholders. The complexity of both phases may be reduced by adopting suitable indicators and endpoints, and including them in Decision Support Systems (DSSs).

In the assessment phase, indicators can support the definition and description of the information to be analysed. The input information may range from environmental to socio-economic, and may have different levels of integration and detail. Indicators can be used to reduce abundant and diverse information into a manageable and organized scheme. In the management phase, indicators can be used by decision-makers and stakeholders to characterize predicted scenarios and to evaluate management alternatives (e.g. for the selection of remediation technologies). Moreover, indicators can be valid instruments for communication with the public. They serve to compact and interpret assessment results and provide them in easily understood forms. In all these functions, the utility of indicators is enhanced by inclusion in DSSs.

This chapter is an overview of the theory and practise of indicators and endpoints for the assessment and management of contaminated sites. It illustrates the state-of-the-art of indicators development and proposes definitions and methodological approaches. It then presents the indicators used in the assessment phase, as analytical instruments that define relevant aspects, such as environmental assessment endpoints, environmental quality (for both water and soil) and human health. Finally, indicators for the management phase, where the elaboration and evaluation of management alternatives is the central objective, are discussed.

P. Agostini (✉)
Consorzio Venezia Ricerche, via della Libertà 12, 30175 Marghera, Venice, Italy

A. Marcomini et al. (eds.), *Decision Support Systems for Risk-Based Management of Contaminated Sites*, DOI 10.1007/978-0-387-09722-0_5,
© Springer Science+Business Media, LLC 2009

5.1 Introduction

Decision making for the management of environmental resources is complex and requires information from many disciplines. Economics, social studies, environmental sciences, engineering, statistics, and information technology may all provide tools for the definition of alternative management options and for the selection of the best solution. However, the major roles in the decision making process are played by policy makers and stakeholders, who need technical information from all relevant disciplines but are not experts in any or all of those disciplines. Indicators are positioned at the science-policy interface (Turnhout et al., 2007). On one hand, the scientific community uses indicators to study and define the state, the functioning and the evolution of natural or human systems, looking for pieces of information that can summarize the complexity of the natural systems and their interactions with the human activities.

On the other hand, policy makers, concerned with practical considerations, claim that, by reducing complexity, these instruments help them to compare alternative decisions and communicate the bases for decisions to a wider audience (Turnhout et al., 2007).

The combination of these two different approaches and needs and the difficulty of combining types of information without losing critical information content results in difficulties of defining the good indicators. As a result, the quality and utility of particular indicators are unclear and therefore many alternatives are currently available.

In general terms, an indicator is a value that represents a phenomenon being studied and that may aggregate different types of data (EEA, 1999; OECD, 2002). Some simple examples of indicators are those that describe the releases of substances (e.g. CO_2 emissions or phosphorous and sulphur releases in water systems), the use of resources (e.g. the amount of land used for roads) or the quantity and the quality of physical and chemical phenomena (such as temperature or CO_2 concentrations) (EEA, 1999). Indicators may also provide information about more complex phenomena, such as sustainability; this category includes the UN Development Programme Human Development Index, the World Economic Forum Environmental Sustainability Index, or the Ecological Footprint (OECD, 2002).

However, an indicator is representative of a phenomenon in the sense that often the phenomenon cannot be or has not been directly measured. Environmental indicators are derived by identifying an environmental property of concern such as water quality or biotic integrity, some environmental parameters (also called metrics) that are related to the property, and some relationship between them. The relationship may be a simple identity (e.g. oligochaete abundance is an indicator of low dissolved oxygen and vitellogenin in male fish is an indicator of estrogenic compounds in water).

Alternatively, an indicator may be an aggregate of multiple parameters such as the Index of Biotic Integrity (IBI) which is an arithmetic combination of 13 normalized fish community metrics (Karr et al., 1986).

Parameters that represent valued properties of the environment and are estimated in risk assessments are termed assessment endpoints. They are commonly used to drive environmental management decisions, but, like other parameters, they may be aggregated to generate an indicator.

Indicators usually have three relevant functions: they may reduce the number of parameters that normally would be required to represent a situation, they may simplify the process of results communication to the users, and they may quantify abstract concepts such as ecosystem health or biotic integrity that are not measurable.

In general, communication is the main function of indicators: they have to provide clear information to all beneficiaries for whom they are developed. Environmental indicators may also be used to raise public awareness on environmental issues, enhancing the support for current environmental policy (EEA, 2005).

If communication and support in decision making is the most important role of indicators and endpoints, their construction and use should be planned very carefully in order to meet the problem solving objectives.

Different international authorities (OECD, 1993; EEA, 1999) have identified criteria for the selection of indicators. We believe the most important criteria are:

- **Policy relevance**. The indicator should address priority issues and provide a representative picture of conditions, pressures on the environment or society's responses.
- **Analytical soundness**. The indicator should be theoretically well founded and based on international standards, so that its validity may be supported on an international level.
- **Measurability**. The data required to support the indicator should be reliable, readily available and adequately documented.
- **Monitor progress toward the quantified targets**. The indicator should provide a clear representation of environmental conditions, showing trends over time; in addition, it should have a threshold or reference value against which it can be compared, so that users can assess the significance of the value associated with it.
- **Time coverage**. The indicator should be updated at regular intervals in accordance with reliable procedures.
- **Be understandable and simple**. The indicator should be simple and easy to interpret.
- **Be timely**. The indicator should be produced in reasonable and useful time for decision-making.
- **Be well documented and of known quality**. The indicator should be produced transparently, by a clear procedure and with defined quality.

In the case of ecological indicators, which specifically address problems of environmental management and could be useful information within processes of contamination management, some additional criteria proposed by Dale and Beyeler (2001) are:

- Be sensitive to stresses on the system, which means that the indicator should be responsive to anthropogenic stressors and therefore suggest reduced system integrity.
- Respond to stress in a predictable manner, so that every indicator response is unambiguous and predictable.
- Have a known response to disturbances, anthropogenic stresses and changes over time, which means that the indicator should have a well-documented reaction to both natural disturbances and anthropogenic stresses in the system.
- Have a low variability in response, since indicators that have a small range in response to particular stresses allow changes in the response value to be better distinguished from background variability.

Finally, another aspect to consider is the integration, both in terms of frameworks connecting different indicators in an assessment and/or management scheme, and in terms of aggregation of different indicators.

OECD (1993) proposed a linking framework for different indicators called the "Pressure-State-Response" (PSR) model, composed of the human activities that cause pressures on the environment, the quality and the quantity of natural resources (state) and the society responses to the produced changes, through specific economic and environmental policies.

EEA (1999) has enhanced this model by adding two other components, in the "Driving forces-Pressures-State-Impact-Responses" (DPSIR) framework, reported in Fig. 5.1. According to this framework, social and economic developments (the driving forces or drivers) exert pressures on the environment and consequently change the state of the environment. This leads to different impacts on human health, ecosystems and materials that may require a societal response, directed to the management of the initial driving forces, or to the state or the impacts. Each of the five components of the DPSIR framework can be analysed through the use of suitable indicators, in such a way that the complexity of the environmental dynamics, without loosing its own flexibility, is well described. Therefore, the framework can be very useful for structuring reports on environmental issues.

Fig. 5.1 The Driving forces-Pressures-State-Impact-Responses (DPSIR) framework shows how the environment is modified and managed (EEA, 1999). Indicators may be developed for each node in the framework

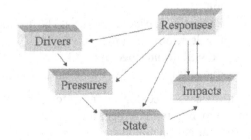

The DPSIR framework is currently embedded in many Decision Support Systems that address contamination assessment and management. In the following chapters of this book, many examples will be provided of tools that have been constructed around this framework, which allows clear and structured analysis of the decision process, thus fitting perfectly with the functionalities and goals of a decision support system. Examples of DSSs that organize indicators in terms of the DPSIR framework include the MULINO DSS, described in Chapter 15 and the MODELKEY DSS, described in Chapter 16.

The second issue of integration is the combination of different indicators (OECD, 2002; Niemeijer, 2002). For this purpose, mathematical or statistical tools are used for transformation, weighting and aggregation.

The purpose of transformation is to reduce all the indicators with different units of measurement into a common scale to make them comparable. Usually, normalization (measured value divided by some benchmark value of the same variable) or standardization are used. For example, metrics may be scaled 0–1 by converting them to proportions of their maximum values and concentrations of chemicals may be expressed as proportions of a regulatory standard or a common toxic response (i.e. toxic units).

When aggregating, the weighting should reflect the relative importance of each component indicator to the overall final indicator. Aggregation methods include: linear aggregation, where the results are proportional to weights of their indicators; geometric aggregation that emphasizes the indicators with the highest weight, and multi-criteria aggregation, which is described in detail in Chapter 3 of this book.

The aggregation of information and data is a critical aspect of the decision making process, due to its function of reducing the complexity of the examined problem into manageable elements of discussion for decision makers and stakeholders. Therefore, aggregation is a component of many Decision Support Systems for risk-based management of contaminated sites, as reported in many of the discussions about DSSs in this book.

It is also important to be able to disaggregate the indicators into their component parameters. Individual response parameters are needed for determining the causes of impairments (see also Chapter 17), to develop models of the functional relationships between pollutants and effects, and to explain exactly what effects are occurring when decision makers or stakeholders want more details. Hence, even after indicators have been derived, the original parameters should be retained in a readily available form.

The following sections of this chapter are organized in terms of the environmental assessment and management phases of the decision process for contaminated sites. Section 5.2 focuses on indicators developed for the support in the assessment phase. In Section 5.3, the requirements of the management phase are briefly described and examples of indicators that support this part of the process are introduced.

5.2 Indicators for Environmental Analysis

This section focuses on the assessment phase of the decision making process for contaminated sites. This critical phase is aimed at generating all the information that is required to correctly and efficiently manage the contamination. Therefore, the major role in this phase is played by experts in scientific disciplines who use their expertise and experience to estimate the risks posed by the considered site.

As explained in Chapter 2, the risk-based assessment process is composed of three main steps: problem formulation, analysis and risk characterization. In the first step, it is essential to define the assessment endpoints, which are the values to be protected, and upon which the subsequent decisions are taken. In the second step, in order to assess the environmental quality and the effects on human health, other indicators can be developed.

In compliance with this scheme, the next subsections deal with environmental assessment endpoints, environmental quality (for both water and soil) and human health aspects respectively.

5.2.1 Environmental Assessment Endpoints

Assessment endpoints are explicit expressions of the environmental values that are to be protected (US EPA, 1998; Suter, 1989). They consist of an entity and a valued attribute of that entity. Examples include the primary production of a meadow, the species richness of a soil invertebrate community, the frequency of deformities in a fish population, or, for humans, the probability of cancer or the frequency of asthma attacks.

Assessment endpoints differ from indicators in that they are not indicative of something, they are significant things themselves. For example, the U.S. Clean Water Act requires that the biological integrity of the Nation's waters be protected. The term biological integrity is ambiguous. One response to that ambiguity is to develop an indicator such as the Index of Biotic Integrity (Karr et al., 1986). Alternatively, one can decide what set of entities and attributes constitute biological integrity for the site or region of concern and use them as assessment endpoints. Because the endpoints are components of the real world with real units, they have many advantages (Suter, 2001). They can be measured, tested, and modeled so their responses can be predicted; they can be explained to decision makers and the public; and they can be balanced against other considerations in a transparent manner.

The selection of endpoints for an assessment requires balancing and, as far as possible, reconciling the following criteria (EPA, 1998; Suter, 1989):

Policy Goals and Societal Values — Because the risks to the assessment endpoint are the basis for decision-making, the choice of endpoint should reflect the policy goals and societal values that the risk manager intends to protect.

Ecological Relevance — Entities and attributes that are significant determinants of the attributes of the system of which they are a part are more worthy of consideration than those that could be added or removed without significant system-level consequences. Examples include the presence of ecological engineer species such as beavers or species that provide habitat structure such as eel grass.

Susceptibility — Entities that are potentially highly exposed and responsive to the exposure should be preferred.

Operationally Definable — An operational definition is one that clearly specifies what must be measured or modeled in the assessment. Without clear operational definitions of the assessment endpoints, the results of the assessment would be too vague to be useful.

Appropriate Scale — Ecological assessment endpoints should have a scale appropriate to the site or action being assessed. Populations with large ranges relative to the site have low exposures. In addition, the contamination or responses of organisms that are wide-ranging relative to the scale of an assessment may be due to sources outside the scope of the assessment.

Practicality — Some potential assessment endpoints are impractical because good techniques are not available for use by the risk assessor. For example, there are few toxicity data available to assess effects of contaminants on lizards, no standard toxicity tests for any reptile are available, and lizards may be difficult to quantitatively survey. Therefore, lizards may have a lower priority than other, better known taxa. Practicality should be considered only after the other criteria are evaluated. If, for example, lizards are included because of evidence of particular sensitivity or policy goals (e.g. presence of an endangered lizard species), then some means should be found to deal with the practical difficulties.

If a decision support system actually supports the estimation of risks to endpoints (such as ARAMS or SADA in Chapter 11) rather than some component of risk assessment such as exposure analysis (such as BASINS, Chapter 18), they must define assessment endpoints and provide the data and models needed to estimate their exposures and responses. The selection of assessment endpoints for a DSS must take into consideration the six criteria described above. In addition, the endpoints must be broadly applicable. That is, they must be specified in such a way that they can be applied to various sites with differing contamination, ecological conditions and societal or policy concerns. One way to accomplish that is to incorporate the endpoints that are most commonly used. Generic endpoints have been identified for the U.S. (USEPA, 2003), for some states (ADEC, 2000), and for units within a very large site (Myers, 1999, Suter et al., 1994). An alternative is to include endpoints for which required data are available. For example, SADA (Chapter 11) includes six endpoint species for terrestrial wildlife for which exposure and effects data are available in the Wildlife Exposure Factors Handbook (USEPA, 1993) and the ecological soil screening levels (OSWER, 2005).

The problem with incorporating commonly used or well studied generic endpoints is that, in many cases, an uncommon or poorly known assessment endpoint is important at a particular site. For example, populations of salamanders or unionid mussels are not commonly used as endpoint entities, but they can be very important in some cases. To accommodate these less common endpoints, a DSS should allow the user to bypass the standard endpoints and provide exposure and effects information ad hoc.

5.2.2 Environmental Quality Indicators

Current environmental policies (e.g. the European Water Framework Directive) are centred on the assessment and sustainable management of the quality of environmental resources, including air, soil, water and biota. This general consideration assumes a higher importance in relation to the assessment and management of contaminated sites, since the major objectives of the decision process in these applications are the restoration of environmental quality and the reduction of risks. In order to reach the quality objectives of the remediation processes, the analytical functions of tools such as DSSs should be integrated with specific instruments that allow the assessment of the current environmental quality. To this end, indicators for soil and water quality are widely used and included in DSSs devoted to contaminated sites.

With regard to environmental resources quality, terrestrial and aquatic ecosystems under stress undergo changes in both structure and function (Fränzle, 2003). Structure is determined by the abundance, composition and spatial distribution of biological communities, while function refers to processes such as nutrient cycling, primary production, and organic matter degradation. Indicators for environmental analysis have been developed and used in order to detect and describe such changes as well as to diagnose causes of impairment, thus giving a picture of status and impacts in water and soil compartments. Appropriate management objectives may be developed from this information.

Aquatic environments are generally more open and dynamic in time and space than soils, and for this reason they are more able to restore their equilibrium after disturbance events. Moreover, specifically for rivers, cause-effect relationships have a spatial dimension due to the unidirectional flow of water and matter from the catchment to the mouth, making any evaluation of water quality more complex (Lorenz, 2003).

In the past, water quality evaluation was mainly based on single physicochemical indicators such as temperature, pH, dissolved oxygen, nutrients as well as chemical concentrations of the most important toxicants. Recently, both in Europe (Water Framework Directive; EC, 2000) and the USA (USEPA, 1998; Barbour et al., 1999; USEPA, 2000) a strong emphasis is given to bioassessment, as the various biological communities reflect the overall ecological functioning of aquatic environments, integrating over time the effects of multiple stressors (e.g. toxicants, eutrophication, physical alterations).

Generally, the use of single metrics such as abundance or species richness of communities provides information on general degradation of aquatic environments while the application of indicators including species sensitivity scores allows assessors to detect impacts due to specific stressors. In particular, multimetric indices combining different indicators (or metrics) into a single index allowing an overall quality evaluation are widely used in USA (Barbour et al., 1999) and have been recently adopted in Europe (Buffagni et al., 2006). The most critical issue in bioassessment is the need of establishing reference conditions consisting of pristine or minimally impaired sites characterised by high quality biological communities.

Multiple biological communities living in aquatic environments should be monitored since each one is sensitive to specific pressures with different reaction times. Specifically, there is a long and well-established tradition of considering macroinvertebrates assemblages as indicators for organic pollution and toxic pressure at local scales, because they are sedentary and encompass a wide range of trophic levels and tolerance levels (Barbour et al., 1999). Two typologies of indicators are applied: saprobic indices evaluating "self-purification" capacity of rivers based on macroinvertebrates species composition, such as the Saprobic Index used in Germany (SI; DEV, 1992), and biotic indices based on species sensitivity, e.g. the Biological Monitoring Working Party (BMWP; Armitage et al., 1983). Periphyton is a good indicator of short-term impacts and as primary producer is directly affected by physico-chemical factors as well as by herbicides (Barbour et al., 1999).

Specific indices based on diatoms communities are commonly applied for rivers quality evaluation in Europe, e.g. Trophic Diatoms Index (TDI; Hofmann, 1994; Kelly, 1998). Moreover, other plant communities can be used as indicators for eutrophication processes: macrophytes in rivers and lakes, e.g. Mean Trophic Index (MTI; Holmes et al., 1999; Kelly and Whitton, 1998), and phytoplankton in lakes and coastal waters, e.g. the Brettum multimetric index (Brettum, 1989). Often the phytoplankton biomass measured in terms of chlorophyll a is considered together with physico-chemical attributes of water in order to evaluate trophic conditions of lakes, e.g. the Trophic Status Index (TSI; Carlson, 1977), and coastal waters, e.g. the TRIX index (Vollenweider et al., 1998). Finally, as fish communities represent the last trophic level of the aquatic food web, they are good indicators of long-term effects and broad habitat conditions, e.g. the multimetric Index of Biological Integrity (IBI; Karr, 1981).

In this context, abiotic attributes (i.e. physico-chemical and hydromorphological characteristics) should be considered as "preconditions" (Lorenz, 2003) making aquatic environments able to support life providing acceptable thermal and oxygenation conditions as well as habitat quality. In particular, it has been recognized that the riparian zone plays an important role in enhancing self purification processes and water quality in rivers, because it has a direct influence on aquatic communities by providing additional habitats and it functions as a buffer zone retaining amounts of nutrients and pollutants draining from the surrounding area (Ghetti, 1999). For this reason, hydromorphological indices, that take into account both riverbed and riparian zone conditions, should be used when performing an overall water quality evaluation, e.g. the River

Habitat Survey applied in UK (RHS; Raven et al., 1998) and the Australian River Assessment System (AUSRAS; Parsons et al., 2002).

Once impairment is detected by means of biological indicators, chemical and toxicological information on water and sediments are useful in order to diagnose the causes. Generally, chemical concentrations of individual hazardous substances are monitored and evaluated by comparison with Environmental Quality Standards (EQS) expressing a level of acceptable potential risk. Toxic pressure of mixtures can be estimated by means of Species Sensitive Distributions (SSD; Posthuma et al., 2002) or measured by performing toxicological tests (i.e. bioassays) on water and sediment samples. However, exceedence of standards or benchmarks does not demonstrate that the contaminants in question are the cause of a particular observed biological impairment. A causal analysis system such as CADDIS (Chapter 17) should be used.

In another example, MODELKEY DSS (Chapter 16) considers any available biological, physico-chemical, chemical, toxicological and hydromorphological indicator for the river basin of interest and integrates such information by means of a Weight of Evidence approach (WOE; Burton et al., 2002) and by applying Multi Criteria Decision Analysis methods (MCDA; Kiker et al., 2005) in order to evaluate rivers quality and to identify the most relevant causes of ecological impairment (i.e. eutrophication, organic pollution, toxic pressure, acidification, and altered hydromorphology).

It is relatively difficult to select appropriate indicators of soil quality, because changes in soil status are not readily observable and the consequences of any decline in soil quality may not be experienced immediately. In fact, many changes in soil may occur over long periods of time, and decline may be obvious only when sufficient cumulative impacts have occurred (Nortcliff, 2002).

While fitness for drinking and for breathing may be good indicators of water and air quality, for soil there is no equivalent criterion.

Breure and colleagues (2005) reviewed indicators of soil quality. Some are concerned with microbial communities, such as the quotient of microbial carbon in the biomass to organic carbon content as an indicator of C-dynamics in soil (Kaiser et al., 1992); the metabolic quotient as an indicator of energetic efficiency (Insam and Haselwandter, 1989); and the respiratory activation quotient as an indicator of the presence of contaminants (ISO, International Organization for Standardization, 2001). In recent years, more efforts have been devoted to the study of structural aspects of the microbial communities. There are also schemes concerned with invertebrate classification, such as SOILPACS, a proposal based on site properties and qualitative occurrence of selected invertebrate groups, used for the assessment of heavy metal contaminated sites in Wales, UK (Weeks et al., 1998). Other indicators are concerned with ecotopes defined as soil fauna communities (Sinnige et al., 1992) or with decomposer communities (Graefe and Schmelz, 1999).

In many cases, a description of the functioning of the soil ecosystem and the possibility to differentiate reference conditions between different categories of soil type and use is the most critical issue. It is addressed for instance by the

German soil biological site classification concept (BBSK; Römbke et al., 1997). The Biological Indicator for Soil Quality (BISQ) approach developed in the Netherlands, integrates data on physical, chemical and biological characteristics, types of land use, soil food web structure, and relationships between food web structure and soil micro- and meso-fauna (Schouten et al., 2001).

In general, any ecological classification and assessment system for soil quality should take into consideration not only the soil and site properties analysed by physical and chemical attributes or indicators (such as soil texture, porosity, depth, pH, salinity, organic matter content) but also and mostly biological attributes, including abundance of organisms, respiration rate, structure of the community and so on. These biological attributes are obviously the most controversial to be considered, and they must be differentiated in consideration of the other soil characteristics. In fact, it is difficult to select the combination of those parameters, and as a consequence those indicators, that are relevant for specific regions. The characterized relevant combinations represent the reference sites, which in many studies are then used for the comparative evaluation of the quality of a soil (Breure et al., 2005). Moreover, as for any other environmental indicator, soil quality indicators should follow the same criteria for selection mentioned above.

An example of a DSS concerned with site-specific ecological risk assessment for contaminated land is provided in Chapter 10, which describes the ERA-MANIA DSS. In this system, quantitative and qualitative evaluations of soil quality is performed in the Integrated Ecological Risk Indexes module. A comprehensive evaluation of the impairment of the terrestrial ecosystem is performed by taking into consideration indicators equally of conditions (richness, abundance, toxicity) of taxonomic groups and of specific functionalities that the same taxonomic groups support in the terrestrial environment.

5.2.3 Human Health Indicators

The assessment and management of contaminated sites can be driven by the need for evaluating and reducing the possible impacts of polluted environment on human health. In this context, characterized by the complex relationships linking environment and health, indicators can be useful tools for analysing, managing and communicating information.

An environmental health indicator may be defined as "an expression of the link between the environment and human health, targeted at an issue of specific policy or management concern, and presented in a form which facilitates interpretation for effective decision-making" (Corvalàn et al., 1997). In this sense, environmental health indicators are more than either environmental indicators or health indicators because they are chosen and used for the known or supposed causal relationship (the link) existing between environmental risk factors and health outcomes.

A specific framework for human health purposes is the DPSEEA (Driving Force-Pressure-State-Exposure-Effect-Action) proposed by Corvalàn and colleagues (1996) and endorsed by WHO. This framework, like the DPSIR

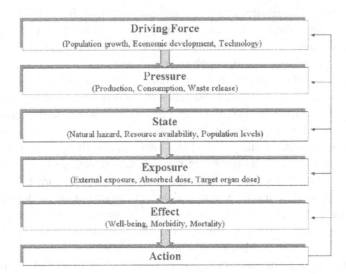

Fig. 5.2 The Driving Force-Pressure-State-Exposure-Effect-Action (DPSEEA) framework shows how humans come to be affected by the environment and how management actions may address any step in the process (WHO, 2002). Indicators may be developed for any component

framework described in Section 5.1, links different components in order to represent how various driving forces generate pressures that affect the environment and finally human health (Fig. 5.2).

The most significant and widely used indicators related to environmental health are indicators of exposure and effects.

Exposure indicators should be effective in describing the actual exposure of human beings to environmental hazards, taking into account all possible exposure pathways which can determine the contact between humans and environmental stressors. The dispersal of the population at risk over time and the time elapsing between the exposure and the first appearance of adverse health effects (especially in the case of carcinogenic effects) can make the reconstruction of effective exposure complex and problematic (WHO, 2002).

Exposure can be characterized and measured in distinct ways: indirectly as the concentration of a stressor in the environment, as an estimate of the amount that actually enters the human body, or as the amount that actually reaches a target organ where a health effect may occur (WHO, 2002). Considerable difficulties often exist in quantifying exposure conditions directly; therefore, it could be necessary to rely on proxies of exposure (e.g. environmental concentration of pollution or pressure indicators such as emission rates).

Some examples of exposure indicators are the proportion of a population living in proximity of sources of air pollution (indirect indicator), the proportion of a population whose drinking-water supplies do not meet health standards or the proportion of children with blood lead levels >10 µg/dl.

Indicators of human effects should refer only to effects which can be in some way connected with exposure to environmental factors. Depending on the type of hazard, level of exposure and other factors, a wide range of health effects may be measured, from sub-clinical effects to illness, morbidity, or mortality. Mortality and morbidity associated with chronic respiratory disease or environmentally-related cancers are examples of effect indicators.

Mortality data are easily accessible through national and regional registries, but data on incidence (number of new cases) or prevalence (number of existing cases) of disease at national or local level can provide better information about population health status and are most useful for properly analysing health impacts of environmental hazards.

These specific health effects are assessment endpoints in health risk assessments. In addition, multiple health effects may be combined and given common units by generating a public health indicator. Examples include Quality-Adjusted Life Years (QALYs) and Disability-Adjusted Life Years (DALYs) (Haddix et al., 1996).

When assessing the impacts of environmental contamination on human health, special attention should be paid to vulnerable groups (WHO, 2002), i.e. to all those sub-populations that for specific reasons are more exposed to hazards or are more susceptible to adverse health outcomes (children, elderly people, pregnant women). For example, children are particularly vulnerable, because they take up a greater amount of contaminants in relation to their body weight than adults, and they have an immature and thus more susceptible physiology. Appropriate indicators should be developed for vulnerable subgroups, with the objective of addressing targets-specific issues and therefore supporting effective protective interventions (Briggs, 2003; Pond et al., 2007).

5.3 Indicators for Overall Management and Decision-Making

The next phase of the decision process, the management phase, requires integration of the information from the environmental assessment with other considerations. In the management phase, the experts from the natural sciences leave the floor to the decision makers and stakeholders, who have to implement political and societal responses to address the problem of contaminated sites. A new set of experts may be engaged such as economists and decision analysts. In this phase, indicators assume the role of information and decision support, including economic, technological, social and planning considerations. Therefore, the same DSSs usually enlarge their assessment support capabilities by processing relevant information in order to produce indicators helpful in answering stakeholders and decision-makers relevant questions and in supporting their final decisions.

The following chapters of this book present examples of complex decision-making processes for contaminated sites management. They demonstrate the necessity of providing the involved authorities, stakeholders and general public with instruments to find solutions to contamination, which are cost-effective, risk-reductive, socially acceptable, sustainable in the long term, and in compliance with laws and policies.

For contaminated sites redevelopment, Vik and colleagues (2001) propose to take into account some central issues: drivers and goals for the remediation, sustainable development, risk management, cost-effectiveness, technical suitability and feasibility, and stakeholders' views. These categories may be further explained as: redevelopment objectives in consideration of local socio-economic plans; risk-based approach for the environmental restoration; capability of technologies in dealing with a specific problem and their feasibility in the specific conditions; valuation of costs versus benefits; consideration of multiple stakeholders' preference profiles.

Taking into account these needs, suitable indicators and indices should be developed, to describe risk, costs, benefits, technological measures, spatial features and so on. Until now, few approaches and frameworks for the remediation of contaminated sites including relevant indicators have been developed. Many management frameworks are centred on the cost-effectiveness of risk reduction through a cost-benefit analysis procedure, and therefore their analysis is limited to the evaluation of risk reduction, cost minimisation and regulatory compliance (Khadam and Kaluarachchi, 2003; Aven and Kørte, 2003). Other researchers have considered additional aspects such as life cycle costs or cultural and historical resources (Bonano et al., 2000). The REC system (Risk reduction, Environmental merit and Costs), presented in the review of Chapter 7, includes integrated indices for the analysis and evaluation of possible clean-up strategies at contaminated sites that evaluate the three major aspects (Nijboer, 1998).

A good example of indicators used in the overall process of contaminated sites management is offered by the DESYRE decision support system (Chapter 8). DESYRE indicators include a socio-economic indicator that help to identify the best option for reuse of the site; an environmental impact indicator that helps to estimate the wider effects on the environment of the possible solutions; technological indicators that provide information on remediation efficiency of possible technical solutions; and risk indicators, that provide indications of reduction in levels and areas of risk.

At a higher level of management and policy, indicators developed by the European Union, and specifically the European Environment Agency, are suitable elements to evaluate the trends and effectiveness of policies toward contaminated site redevelopment (EEA, 2007). Within this category of indicators, a set of indicators is used to evaluate the progress in the management of contaminated sites, by answering specific key policy questions, such as which sectors contribute most to soil contamination, how much is being spent on cleaning-up soil contamination, how much progress is being achieved in the management and control of local soil contamination or how the problem of contaminated sites is being addressed. These indicators may include analyses of industrial and commercial activities causing soil contamination by European countries, the annual national expenditures for management of contaminated sites per unit of GDP in each country, the remediation technologies applied in the surveyed European countries as percentages of number of sites per type of treatment, or finally the status of investigation and clean-up of contaminated sites in Europe (EEA, 2007). The purpose of these indicators is to provide information about the effectiveness of implemented policies and help in their correction, if appropriate. The indicators provide key condensed messages, easily understandable but appropriate for the scope of the survey.

5.4 Conclusions

The chapter provided a wide overview of the theory and application of indicators and endpoints in the assessment and management of contaminated sites.

More efforts are still needed in order to improve the inclusion of these instruments in Decision Support Systems, with specific regard to their selection, aggregation and validation.

One controversy concerns the adoption of a single indicator versus the use of a set of indicators. In fact, policy makers, the main beneficiaries of indicators, often prefer to have very compacted and easily interpreted information for decision-making, but the aggregation of different information through transformation and weighting to obtain indicators is a delicate and uncertain process. Hence, a set of different indicators that are not aggregated may be a preferable option. Moreover, particularly in DSSs that support analyses at larger scales, aggregation is often dependent on spatial issues (e.g. indicators developed for regional or national studies), which may pose significant problems of correct integration.

A final consideration concerns the validation of indicators, which is particularly important when these instruments are embedded in Decision Support Systems. As observed by Bockstaller and Girardin (2003), validation can be performed for the indicator design, for the output of the indicator and for the use by decision-makers. The last validation may be performed by the end-users or by organizations that audit the outcomes of management actions. This ultimate validation is essential to assure the usefulness of an indicator as a benchmark for decision making and not only as a scientific product.

References

ADEC (Alaska Department of Environmental Conservation) (2000) User's Guide for Selection and Application of Default Assessment Endpoints and Indicator Species in Alaska Ecoregions. Available at http://www.state.ak.us/local/akpages/EN.CONSERV/dspar/csites/ind docs.htm.

Armitage, P.D., Moss, D., Wright, J.F., Furse, M.Y. (1983) The performance of a new biological water quality score system based on macroinvertebrates over a wide range of unpolluted running waters. Water Research 17:333–347.

Aven, T., Kørte, J. (2003) On the use of risk and decision analysis to support decision-making. Reliability Engineering and System Safety 79, 289–299.

Barbour, M.T., Gerritsen, J., Snyder, B.D., Stribling, J.B. (1999) Rapid Bioassessment Protocols for Use in Streams and Wadeble Rivers: Peryphyton, Benthic Macroinvertebrates, and Fish. Second Edition. US Environmental Protection Agency, Washington, DC.

Bockstaller C., Girardin, P. (2003) How to validate environmental indicators. Agricultural Systems 76:639–653.

Bonano, E.J., Apostolakis, G.E., Salter, P.F., Ghassemi, A., Jennings, S. (2000) Application of risk assessment and decision analysis to the evaluation, ranking and selection of environmental remediation alternatives. Journal of Hazardous Materials 71:35–57.

Brettum, P., (1989) Planteplankton som indicator pá vannkvalitet I norke innsjøer. Planteplankton (Algae as indicators of water quality in Norwegian lakes). Niva Rapport 0-86116:111p.

Breure, A.M., Mulder, C., Boembke, J., Ruf, A. (2005) Ecological classification and assessment concepts in soil protection. Ecotoxicology and Environmental Safety 62:211–229.

Briggs, D. (2003). Making a Difference: Indicators to Improve Children's Environmental Health. World Health Organization, Geneva.

Buffagni, A., Erba, S., Cazzola, M., Murray Bligh, J., Soszka, H., Genoni, P. (2006) The STAR common metrisc approach to the WFD intercalibration process: full application for small, lowland rivers in three European countries. Hydrobiologia (special issue) 566(1):379–399.

Burton, G.A., Chapman, P.M., Smith, E.P. (2002) Weight-of-Evidence approaches for assessing ecosystem impairment. Human and Ecological Risk Assessment 8:1657–1673.

Carlson, R.E. (1977) A trophic state index for lakes. Limnology and Oceanography 22:361–369.

Corvalàn, C., Briggs, D., Kjellstrom, T. (1996) Development of environmental health indicators. In: Briggs, D., Corvalàn, C., Nurminem, M. eds, Linkage Methods for Environmental and Health Analysis. General Guidelines. World Health Organization, Geneva, pp. 19–53.

Corvalàn, C., Kjellström, J., Briggs, D. (1997) Health and environmental indicators in relation to sustainable development. In: Moldan, B., Billharz, S. eds,. Sustainable Indicators. Report on the Project on Indicators of Sustainable Development, Scientific Committee on Problems of the Environment (SCOPE), Wiley.

Dale, V.H., Beyeler, S.C. (2001) Challenges in the development and use of ecological indicators. Ecological Indicators 1:3–10.

DEV (1992) Biologische-oekologische Gewaessergueteuntersuchung: Bestimmung des Saprobienindex (M2). In Deutsche Einheisverfahren zur Wasser-, Abwasser- und Schlammuntersuchung. Deutsche Institut fur Normung, VCH Verlagsgesellschaft mbH, Weinheim, pp. 1–13.

EEA (European Environment Agency) (1999) Environmnetal Indicators: Typology and Overview. Technical Report n. 25. Copenaghen, Denmark.

EEA (European Environmental Agency) (2005) EEA Core Set of Indicators. Technical Report n.1/2005. ISSN 1725-2237. European Environmental Agency, Office for Official Publications of the European Communities, Luxembourg.

EEA (European Environment Agency) (2007) Progress in Management of Contaminated Sites (CSI 015) – Assessment Published in August 2007. Copenaghen, Denmark.

European Commission (2000) Directive 2000/60/CE of the European Parliament and of the Council of 23 October 2000 establishing a framework for community action in the field of water policy. Official Journal of the European Communities L327, 22/12/2000.

Fränzle, O., (2003) Bioindicators and environmental stress assessment. In: Markert, B.A., Zechmeister, A.M. eds, Bioindicators & Biomanitors. Principles, Concepts and Applications, Elsevier, Amsterdam (The Netherland), pp. 41–84.

Ghetti, P.F. (1999) Le reti ecologiche: struttura e funzioni. In Provincia di Milano, Atti del seminario Reti Ecologiche in Aree Urbanizzate, Milan, Franco Angeli Editore, pp. 19–21, in italian.

Graefe, U., Schmelz, R. (1999) Tabellarische Zusammenstellung der ökologischen Ansprüche und Lebensformtypen terrestrischer Enchytraeenarten. Newsletter on Enchytraeidae 6:59–68.

Haddix AC, Teutsch SM, Shaffer PA, Dunet DO (1996) Prevention Effectiveness: A Guide to Decision Analysis and Economic Evaluation, Oxford University Press, UK.

Hofmann, G. (1994) Aufwuchs-Diatomeen in Seen und ihre Eignung als Indikatoren der Trophie. Bibliotheca Diatomologica 30, Cramer, Berlin, 241 pp.

Holmes, N.T.H., Newman, J.R., Chadd, S., Rouen, K.J., Saint, L., Dawson, F.H. (1999). Mean Trophic Rank: A User's Manual. Environment Agency R&D Technical Report E38.

Insam, H., Haselwandter, K. (1989) Metabolic quotient of the soil microflora in relation to plant succession. Oecologia 79:174–178.

ISO (International Organisation for Standardisation) (2001) Soil Quality: Determination of Abundance and Activity of Soil Microflora Using Respiration Curves. ISO/DIS 17155.

Kaiser, E.-A., Müller, T., Joergensen, R.G., Insam, H., Heinemeyer, O. (1992) Evaluation of methods to estimate the soil microbial biomass and the relationship with soil texture and organic matter. Soil Biology and Biochemistry 24:675–683.

Karr, J.R. (1981) Assessment of biotic integrity using fish communities. Fisheries 6:21–27.

Karr JR, Fausch KD, Angermeier PL, Yant PR, Schlosser IJ (1986) Assessing biological integrity in running waters; a method and its rationale. Illinois Natural History Survey Special Pub. 5., Champaigne, IL.

Kelly, M.G. (1998) Use of the trophic diatom index to monitor eutrophication in rivers. Water Research 32:236–242.

Kelly, M.G., Whitton, B.A. (1998). Biological monitoring of eutrophication in rivers. Hydrobiologia 384:55–67.

Khadam, I., Kaluarachchi, J.J. (2003) Applicability of risk-based management and the need for risk-based economic decision analysis at hazardous waste contaminated sites. Environment International 29:503–519.

Kiker, G.A., Bridges, T.S., Varghese, A., Seager, T.P., Linkov, I. (2005) Application of multicriteria decision analysis in environmental decision making. Integrated Environmental Assessment and Management 2:95–108.

Lorenz, V. (2003). Bioindicators for ecosystem management, with special reference to freshwater ecosystems. In: Markert, B.A., Zechmeister, A.M. eds, Bioindicators & Biomanitors. Principles, Concepts and Applications, Elsevier, Amsterdam (The Netherland), pp. 123–152.

Myers, O.B. (1999) On aggregating species for risk assessment. Human and Ecological Risk Assessment 5(3):559–574

Niemeijer, D. (2002) Developing indicators for environmental policy: data-driven and theory-driven approaches examine by example. Environmental Science and Policy 5:91–103.

Nijboer, M.N. (1998) REC: a decision support system for comparing soil remediation options based on risk reduction, environmental merit and costs. Contaminated Soil. Thomas Telford, London, 1173–1174.

Nortcliff, S. (2002) Standardisation of soil quality attributes. Agriculture, Ecosystems and Environment 88:161–168.

OECD (1993) Core set of indicators for environmental performance review. Environment Monographs n. 83, OCDE/GD(93)179, Organization for Economic Co-operation and Development, Paris, France.

OECD (2002) Environmental indices. ENV/EPOC/SE(2001)2/FINAL, Organization for Economic Co-operation and Development, Paris, France.

OSWER (2005) Guidance for developing ecological soil screening levels, revised. OSWER Directive 9285.7–55. U.S. Environmental Protection Agency, Washington, DC.

Parsons, M., Thoms, M., Norris, R. (2002) Australian River Assessment System: AusRivAS Physical Assessment Protocol. Monitoring River Heath Initiative Technical Report no 22, Commonwealth of Australia and University of Canberra, Canberra.

Pond, K., Kim, R., Carroquino, M., Pirard, P., Gore, F., Cucu, A., Nemer, L., MacKay, M., Smedje, G., Georgellis, A., Dalbokova, D., Krzyzanowsi, M. (2007) Workgroup report: developing environmental health indicators for European children: World Health Organization working group. Environmental Health Perspectives 115(9):1376–1382.

Posthuma, L., Suter, G.W. II, Traas, T.P. (2002) Species Sensitivity Distributions in Ecotoxicology, Lewis Publisher, Boca Raton, FL, USA.

Raven, P.J., Holmes, N.T.H., Dawson, F.H., Fox, P.J.A., Eeverard, M., Fozzard, I.R., Rouen, K.J. (1998). River Habitat Quality – The Physical Character of Rivers and Streams in the UK and Isle of Man. River Habitat Survey Report No. 2. May 1998. Bristol (Environment Agency).

Römbke, J., Beck, L., Förster, B., Fründ, H.C., Horak, F., Ruf, A., Rosciczewski, K., Scheurig, M., Woas, S. (1997) Boden als Lebensraum für Bodenorganismen und die bodenbiologische Standortklassifikation: Eine Literaturstudie. Texte und Berichte zum Bodenschutz 4/97. Landesanstalt Umweltschutz Baden- Württemberg (Karlsruhe).

Schouten, A.J., Bloem, J., Breure, A.M., Didden, W.A.M., Van Esbroek, M., De Ruiter, P.C., Rutgers, M., Siepel, H., Velvis, H. (2001). Pilotproject Bodembiologische Indicator voor Life Support Functies van de bodem. RIVM Report 607604001.

Sinnige, N., Tamis, W., Klijn, F. (1992) Indeling van Bodemfauna in ecologische Soortgroepen. Centrum voor Milieukunde, Rijksuni- versiteit Leiden Report No. 80.

Suter, G.W., II (1989) Ecological endpoints. In: Warren-Hicks, W., Parkhurst, B.R., Baker, S.S. Jr. eds, Ecological Assessment of Harardous Waste Sites: A Field and Laboratory Reference Document. EPA 600/3-89/013. Corvallis Environmental Research Laboratory, Corvallis, OR, pp. 2-1–2-28.

Suter, G.W., II (2001) Applicability of indicator monitoring to ecological risk assessment. Ecological Indicators 1:101–112.

Suter, G.W., II, Sample, B.E., Jones, D.S., Ashwood, T.L. (1994) Approach and strategy for performing ecological risk assessments for the Department of Energy's Oak Ridge Reservation. ES/ER/TM-33/R1. Environmental Restoration Division, Oak Ridge National Laboratory, Oak Ridge, TN.

Turnhout, E., Hisschemoeller, M., Eijsacker, H. (2007) Ecological indicators: between the two fires of science and policy. Ecological indicators 7:215–228.

USEPA (US Environmental Protection Agency) (1993) Wildlife exposure factors handbook. EPA/600/R-93/187. U.S. Environmental Protection Agency, Office of Health and Environmental Assessment, Washington, DC.

USEPA (U.S. Environmental Protection Agency) (1998) Guidelines for ecological risk assessment. EPA/630/R-95/002F. Risk Assessment Forum, Washington, DC.

USEPA (U.S. Environmental Protection Agency) (2000). National Water Quality Inventory. EPA/305b/2000 Washington DC.

USEPA (U.S. Environmental Protection Agency) (2003) Generic Ecological Assessment Endpoints (GEAEs) for Ecological Risk Assessment. EPA/630/P-02/004B. U.S. Environmental Protection Agency, Risk Assessment Forum, Washington, DC.

Vik, E.A., Bardos, P., Brogan, J., Edwards, D., Gondi, F., Henrysson, T., Jensen, B.K., Jorge, C., Mariotti, C., Nathanail, P., Papassiopi, N. (2001) Towards a framework for selecting remediation technologies for contaminated sites. Land Contamination and Reclamation 9(1):119–127.

Vollenweider, R.A., Giovanardi, F., Montanari, G., Rinaldi, A. (1998). Characterization of the trophic conditions of marine coastal waters with special reference the NW Adriatic Sea: Proposal for a trophic scale, turbidity and generalized water quality index. Environmetrics 9:329–357.

Weeks, J.M., Hopkin, S.P., Wright, J.F., Black, H., Eversham, B.C., Roy, D., Svendsen, C. (1998) A Demonstration of the Feasibility of SOILPACS. HMIP/CPR2/41/1/247.

WHO (2002) Health in Sustainable Development Planning: The Role of Indicators. World Health Organization, Geneva.

Chapter 6
Contaminated Land: A Multi-Dimensional Problem

Claudio Carlon, Bruce Hope, and Francesca Quercia

Abstract The Chapter addresses the problem of contaminated land by analyzing its main dimensions in both European Union and USA context. After the introductory definitions of contaminated land, Brownfield and remediation, five dimensions of the contaminated land problem are identified: liability, risk, technological, socio-economic, stakeholder. The liability dimension is discussed underlying the difficulties raised by its determination. The risk dimension is centered on the human health and ecological risk assessment. The technological dimension requires to take into consideration different aspects, such as the variety of technologies, the technical practicability, the cost/benefit ratio. Examples of technological solutions in Europe and US are presented. The socio-economic dimension concerns the holistic approach to the management of contaminated land and brownfields. The stakeholder dimension specifies the need to involve the different stakeholders and include their perspectives and needs in the management process. The Chapter ends with the discussion on need and role of Decision Support Systems in proposing valuable solutions for the management of this multi-dimensional problem.

6.1 Introduction

6.1.1 What is Contaminated Land?

6.1.1.1 Contaminated Land in the European Union

In the European Union (EU), there is no commonly accepted definition for what constitutes contaminated land, although the vast majority of EU Member States define it in relation to the potential risk posed by land contaminants to

C. Carlon (✉)
European Commission Joint Research Centre, Institute for Environment
and Sustainability, Rural Water and Ecosystem Resources Unit, Via E. Fermi 1,
I-21020 Ispra (VA), Italy
e-mail: claudio.CARLON@echa.europa.eu

A. Marcomini et al. (eds.), *Decision Support Systems for Risk-Based Management*
of Contaminated Sites, DOI 10.1007/978-0-387-09722-0_6,
© Springer Science+Business Media, LLC 2009

human health and the environment (Carlon 2007). Because policy and legislation are the prerogatives of national governments, current responses to contaminated land across the EU are dependent on the policy concerns, the political system, and the physical features of each state. Remediation standards and guidelines on risk assessment tend to be set at the national level and most national governments also maintain contaminated land registers and provide some financing for remediation. In federal nations, regions are often instigators of contaminated land policy, so that regional authorities are the main policy implementers, with responsibilities for all aspects of contaminated land management. The day-to-day management and monitoring of contaminated land usually occurs at the local authority level. Local authorities are frequently the first link in the identification of contaminated sites and often have responsibilities to take emergency action. Belgium and the UK are anomalous in their contaminated land administration: all Belgian policy and legislation come from the regions, while in the UK, most implementation comes from local authorities (Christie and Teeuw 2000). The number of potentially contaminated sites in Europe is estimated at approximately 3.5 million with 0.5 million sites being significantly contaminated and needing remediation (European Commission 2006). This rough estimate is affected by the slightly different definitions for contaminated sites used in EU member states, but correctly reflects the magnitude of the problem. Although the largest and most affected areas are concentrated around heavily industrialised regions, contaminated sites exist everywhere throughout the continent. The per site cost of remediation has been estimated between € 19,500 and 73,500, with a cost for remediation of all sites of approximately € 28 billion (European Commission 2006). Even though many countries apply the "polluter pays" principle, public funding supplies a considerable amount of total remediation costs. Furthermore, although considerable financial resources have already been spent on remediation activities, the European Environmental Agency estimates that this represents only about 8% of total estimated remediation costs. Since the early 1980s, technical procedures for registering and ranking the hundreds or thousands of contaminated sites have been in place at regional or national levels. So far, these registries of contaminated sites are still managed at regional or national levels and not at the level of the EU. More recently, the European Environmental Agency has proposed a system, named PRAMS, to rank areas in Europe at risk from soil contamination (EEA 2006).

6.1.1.2 Contaminated Land in the United States

The U.S. Environmental Protection Agency (USEPA) has defined "contaminated land" as ground that has been polluted with hazardous materials and requires cleanup or remediation, and which may contain both polluted objects (e.g., buildings, machinery) and land (e.g., soil, sediments, and plants). At the federal level, USEPA manages uncontrolled contaminated land under the statuatory authority granted by the Comprehensive Environmental Response, Compensation and Liability Act (CERCLA; 42 U.S.C. § 9601–9675) of 1980 (commonly referred to as "Superfund"). "Superfund" is the common name for the 1980 CERCLA act, and its associated amendment the Superfund Amendment and Reauthorization Act (SARA) of 1986. CERCLA was enacted by the U.S. Congress to address the problem of cleaning up abandoned toxic waste dump sites. CERCLA provides broad federal authority to cleanup releases or threatened releases of hazardous substances that may endanger public health or the environment. Under CERCLA, liability for people or companies associated with a contaminated property is "strict, joint and several, and retroactive", meaning that USEPA can sue one of several contributors to pollution, can sue for pollution that occurred before environmental regulations were in place, and does not need to prove negligence in order to hold a party responsible. The fund was created through taxes on petroleum and other chemical industries to finance litigation and cleanup of severely contaminated sites; however, its taxing authority expired several years ago, and Superfund now continues only because of special funding appropriated annually by the U.S. Congress.

Uncontrolled hazardous waste sites that are polluted above a minimally acceptable hazard ranking score are placed on the National Priorities List (NPL). USEPA uses a Hazard Ranking System (HRS) to determine whether a contaminated site should be placed on the NPL. The HRS is a numerically based screening system that uses information from initial, limited investigations to assess the relative potential of sites to pose a threat to human health or the environment. Inclusion of a site on the NPL does not in itself require potentially liable parties to initiate action to cleanup the site, nor does it assign liability to any person. The NPL serves primarily informational purposes, identifying for the States and the public those sites or other releases that appear to warrant remedial actions and helping the Agency prioritize sites for cleanup. As of July 2007, there were a total of 61 proposed and 1243 final sites on the USEPA NPL. Because of differing definitions, the number of contaminated sites being managed by the States is difficult to determine, but may be in the thousands.

6.1.2 What is a Brownfield?

In the U.S., the term "Brownfield" refers to "real property, the expansion, redevelopment, or reuse of which may be complicated by the presence or potential presence of a hazardous substance, pollutant, or contaminant. Cleaning up and reinvesting in these properties takes development pressures off of undeveloped, open land, and both improves and protects the environment." These sites are not subject to the liability framework set up by CERCLA. The purpose for this designation is to reap the economic and environmental benefits of returning contaminated land to productive uses by limiting or eliminating the threat of liability under CERCLA. In Europe the term Brownfield has no institutional recognition, and simply refers to previously developed land burdened with real or perceived contamination. This notwithstanding, the need to develop specific approaches for the re-use and re-development of brownfields has been widely recognized and analysed within the context of EU stakeholder networks (e.g. the CABERNET network). In contrast to brownfields, the term greenfields refers to areas that have not been intensively developed and where a certain extent of environmental quality has been preserved. It follows that one benefit of re-using brownfields is that consumption of greenfields can be reduced. There is no single "European" way of dealing with the issue of brownfields. The complexity of the problem, the differences in industrial development, and varying political and legal frameworks have hindered a uniform approach. This incoherence, however, has led to a range of methods and expertise that can be turned into a resource if brought together, evaluated, and fed into the development of an overall strategy (Ostertag 2003).

6.1.3 What is Remediation?

Remediation is generally defined as corrective action to clean up an environmentally contaminated site to eliminate contamination or reduce it to an acceptable level. It includes a variety of actions taken to lower exposure to hazardous substances to below some agreed upon acceptable level, typically through removal or various in situ or ex situ treatment technologies. In situ technologies avoid expensive excavation of soils but may not be practical or effective in all situations. In these circumstances, ex situ technologies have to be used on-site or at an off-site specialised installation. In some instances, it may be possible to eliminate exposure altogether. The establishment by decision makers of necessary trade-offs between the degree of remediation desired and that which is technically attainable is guided by knowledge of both the level of the threat posed by the contamination (often from a risk assessment), results of cost-benefit and technical analyses, and other factors unique to the decision context. Remediation can encompass two basic types of response actions: short-term "removal" actions or long-term "remedial" actions. Short-term "removal" actions are taken to address releases or threatened releases requiring an immediate or otherwise prompt response. Such actions can be classified as: (1) emergency; (2) time-critical; or (3) non-time critical. Removal responses are generally used to address localized risks such as abandoned drums containing hazardous substances, contaminated surface soils posing acute risks to human health or the environment, etc. Most long-term "remedial" actions are intended to permanently and significantly reduce unacceptable risk associated with releases, or threats of releases, of hazardous substances but lack the time-criticality indicative of a removal action. Remedial actions may also include such measures as preventing the migration of pollutants and neutralization of toxic substances.

6.2 Contaminated Land: A Multi-Dimensional Problem

Despite the seeming lack of a commonly accepted definition of contaminated land, as well as differing approaches by various jurisdictions to the issues of assessing impacts and devising remedial responses, there are some common themes related to the resolution of a contaminated land problem. These may be succinctly stated as the need to: (1) identify the presence, as well as the nature and extent, of specific hazardous substances (typically chemical substances), (2) determine the level of contamination and whether this level is posing or could pose a problem (as defined by the governing jurisdiction), (3) if so, choose an implementable (i.e., technically, economically, and politically workable) response (i.e., remedial) action capable of rectifying as much of the problem as is practical, and (4) recover the cost for both investigating and responding to the contamination, preferably from the party or parties responsible for causing it. Thus the successful resolution of a contaminated land problem (i.e., cost-effective minimization of threats posed to health or the environment, as well as a potential return to productive use) requires that managers and decision makers simultaneously address the following key dimensions of the problem:

- *Liability*: identifying those responsible for the problem and those who will bear the cost of its resolution. The determination of liability can be difficult when those responsible for the pollution are not the same as those who own the land that was polluted;
- *Risk*: assessing the human health and ecological risks posed by site-related contaminants;
- *Technological*: determining the availability, performance and feasibility of remediation techniques;
- *Socio-economic*: evaluating impacts and benefits to the land owner (e.g., estimated remediation costs versus land value), and to society at various levels (e.g., impacts on local and regional redevelopment projects, lower consumption of greenfields at national level), in consideration of a sustainable reuse of the site;
- *Stakeholder*: promoting and facilitating communication and consensus building among the potentially conflicting interests of various stakeholders.

The multiple dimensions involved in managing a contaminated land problem are described in further detail in the following sections. In the final section, we consider the need for, and expected functionalities of, decision support systems (DSSs) to support analysis and decision making with respect to such a multi-dimensional problem.

6.2.1 The Liability Dimension

As illustrated in Fig. 6.1, every site will have a "decision context" with both technical and legal/political dimensions, often moving in parallel, defined by jurisdiction-specific requirements. It will generally be necessary to know what

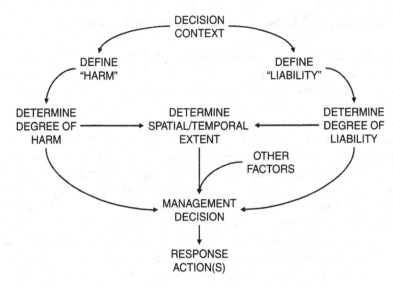

Fig. 6.1 Liability in the context of the problem of contaminated land

constitutes both harm (as impact or risk) and liability, as well as the means (i.e., technical, legal, political) by which such determinations will be made. The spatial and temporal extent of both harm due to the contamination and liability for that harm are also key considerations. For example, in the USEPA Superfund program, responsibility for contamination often extends to the "locality of the facility", defined as any point where a receptor (human or environmental) might contact or be reasonably likely to contact a site-related hazardous substance. Information on harm and liability, as well as on other factors (e.g., social, cultural, economic, etc.), is used by the decision making authority (typically a regulatory body) to decide whether a problem exists and whether a response is necessary and, if so, the characteristics of that response (including who will pay for that response).

6.2.2 The Risk Dimension

Contamination typically involves soil, groundwater and surface water. The main sources of soil contamination are disposal of municipal and industrial wastes, accidents during industrial activities, non-operating or abandoned industrial plants and mining operations, or former military sites. Industrial activities that typically make a greater contribution to soil contamination include the metal working, chemical, oil and wood industries. The most common contaminants found on contaminated land are:

– Some heavy metals and metalloids, like Lead, Manganese, Arsenic, Copper, Nickel, Cadmium, Chromium, Zinc, Barium, Mercury, Antimony, Beryllium;

- Aromatic hydrocarbons, like Benzene, Toluene;
- Polycyclic aromatic hydrocarbons (PAHs), like Naphthalene and Benzo(a)pyrene;
- Chlorinated hydrocarbons, like Vinyl Chloride, Dichloroethane, Trichloroethane, Trichloroethylene, Tetrachloroethylene, Carbon Tetrachloride;
- Polychlorinated Biphenyls (PCBs);
- Dioxins.

There are several technical methods available for assessing the nature and extent of the problem posed to either human or environmental receptors by hazardous chemical substances on contaminated land. Risk assessments, of varying degrees of complexity, have become a commonly encountered tool for making such determinations. In the U.S., natural resource damage assessments may also be used, during or after a risk assessment, to determine whether contaminated land has impacted environmental resources.

In general, risk assessments evaluate the probability that adverse effects may occur to a receptor (human or ecological) as a result of exposure to chemical (e.g., release of hazardous substances) or physical (e.g., site cleanup activities) stressors. Risk is assumed to emerge from the coexistence of a contaminant source (which creates a concentration above a certain level), a human or ecological receptor (but the latter may exclude agricultural crops and livestock), and a potential exposure pathway between the contaminant and the receptor. The elimination or the absence of any one of these three components is enough to declare that there is no risk. Risk to human health from contaminated soils relates to the likelihood and magnitude of adverse effects on human health due to direct or indirect exposure to toxic substances in soil. Human health effects that are commonly of concern include those related to chronic (long term) exposure, including mutagenic and carcinogenic effects, according to the toxicological properties of contaminants. Potential pathways by which humans may be exposed to soil contaminants include: direct (dermal) contact, inhalation, incidental ingestion of particles and consumption of contaminated groundwater. The most commonly encountered exposure pathways are: ingestion of contaminated soil, soil vapour and dust accumulation in enclosed spaces (through wall and pavement breaks), ingestion of contaminated groundwater, and bioaccumulation in home-grown vegetables, depending on the physico-chemical properties of contaminants. Assessment of ecological risk due to soil contaminants can consider adverse effects on soil organisms, plants and above-ground wildlife. The main concern, however, is usually for indirect impacts on soil functions, particularly the capacity of soil to act as substrate for plants and those organisms important for proper soil functioning and nutrient cycle conservation.

6.2.2.1 Three Applications of Risk Assessment

To address the problem of contaminated sites, risk assessment methodologies are applied at three different levels (Carlon 2007):

1. Risk-based ranking of sites at a regional scale helps to define priorities of interventions. It is usually based on poor or limited information and utilizes qualitative modelling;

2. Screening risk assessments are typically used for the preliminary identification of both sites and contaminants of concern. They typically use environmental (soil and water) screening values (which may also be referred to as standards, criteria, guidelines, or benchmarks) that are usually derived with quantitative risk assessment methodologies. These values, which are available in several countries, take the form of contaminant concentration thresholds $(mg/kg_{soil-dry\ weight})$, above which certain actions (ranging from the need of further investigations up to the need for immediate or "emergency" remedial actions, depending on national regulatory frameworks) are recommended or endorsed;

3. Site-specific risk assessments are based on environmental and exposure conditions specific to a given contaminated site. The site specific risk assessment aims at estimating the risk for human health or ecological receptors at a specific site, taking into account local or site-specific conditions that may affect exposure or effects. It is clear that a site-specific risk assessment can substantially improve the accuracy of risk estimations for a particular site by reducing the need for the generic and precautionary assumptions underlying screening values. However, the quality of the site-specific risk assessments depends mainly on the proper characterization of the site. The need for further investigations to improve an assessment's reliability must be weighed against the cost of those investigations, as well as the implications (economic, health and environmental) of refined risk estimates (Carlon et al. 2000).

In the U.S., at the federal level, a Natural Resource Damage Assessment (NRDA) is a type of impact assessment that may be required after completion of the risk assessment component (human health and/or ecological) of the Superfund program. Per Section 301(c) of CERCLA, natural resource trustees have the authority to assess [monetary] damages for injury [impact] to, destruction of, or loss of natural resources resulting from a discharge of oil or release of a hazardous substance. An NRDA is used to identify additional actions, beyond any response actions directed at contamination issues, to address injuries to natural resources. Examples could include actions needed to restore the productivity of habitats or the species diversity that were injured by the past releases or to replace them with substitute resources.

6.2.3 The Technological Dimension

Risk assessment results represent the background needed in order to approach risk management solutions. But there are a number of factors that need to be considered in selecting an effective remediation or risk management solution. These include considerations of core objectives such as remedial technology, technical practicability, feasibility, cost/benefit ratio and wider environmental, social and economic impacts (Carlon et al. 2004). In addition, it is also important to consider the manner in which a decision is reached. This should be a balanced and systematic process founded on the principles of transparency and inclusive decision making. Decisions about which technology (or technologies) is (are) most appropriate for a particular site need to be considered in a holistic manner.

6.2.3.1 The Remedial Technology Selection Process

In general, remedial technologies fall into one or more of the following broad categories:

- Excavation and containment – This includes: removal to landfill, i.e., the disposal of material to an engineered commercial void space; deposition within an on-site engineered cell, generally with a view to combining the disposal of waste with the reclamation of land area from the void space; engineered land-raising and land forming, where materials are deposited on the land surface to make a hill or mound above the natural surface level suitably contained.
- Engineered systems – These include: In situ physical containment which is designed to prevent or limit the migration of contaminants left in place or confined to a specific storage area, into the wider environment. Approaches include in-ground barriers, capping and cover systems, hydraulic containment and pump-to-contain approaches.
- Site rehabilitation measures – These are used to bring back some measure of utility to a site whose contamination cannot be treated or contained for technical or economic reasons. Examples include growth of grass cover tolerant of contaminants, covering with soil or soil substitute, raising soil pH by liming and other cultivation measures. Land use controls may be required in order to avoid potential human exposure.
- Treatment based approaches – These are used to destroy, remove or detoxify the contaminants contained in the polluted material (e.g. soil, ground water etc.). Using treatment technologies in contaminated land remediation is encouraged by agencies in many countries, because they are perceived as having added environmental value compared with other approaches to remediation such as excavation and removal, containment or covering/revegetation. The "added" environmental value is associated with the destruction, removal or transformation of contaminants into less toxic forms and with a greater flexibility for site reuse.

Treatment based approaches can be further described as:

- Biological processes: contingent on the use of living organisms.
- Chemical processes: destroy, fix or concentrate toxic compounds by using one or more types of chemical reaction.

- Physical processes: separate contaminants from the soil matrix by exploiting physical differences between the soil and the contaminant (e.g. volatility) or between contaminated and uncontaminated soil particles (e.g. density).
- Solidification and stabilization processes: immobilize contaminants through physical and chemical processes (Solidification processes are those which convert materials into a consolidated mass. Stabilization processes are those in which the chemical form of substances of interest is converted to a form which is less available).
- Thermal processes: exploit physical and chemical processes occurring at elevated temperatures.

Ex situ technologies are applied to excavated soil and/or extracted groundwater. In situ technologies use processes occurring in unexcavated soil, which remains relatively undisturbed. On site techniques are those that take place on the contaminated site. They may be ex situ or in situ. Off site processes treat materials that have been removed from the excavated site (CLARINET 2002a).

The BATNEEC (Best Available Technology Not Entailing Excessive Costs) principle is relevant to make sure that the best technologies are being used for soil remediation, while also taking into account secondary effects and costs of a technology based on cost-effectiveness and cost-benefit analysis. The decision on what is the BAT for remediation of a site is site-specific.

In order to select the "best" remediation technology, decisions are generally carried out on the basis of the following considerations:

- remedial (cleanup) objectives to be reached for contaminated media in order to protect human health and the environment;
- effectiveness of the technology with respect to the contaminant(s) present at the site, with the location/environmental context and contaminated media identified;
- effectiveness and durability in dealing with the contamination with respect to specific risk management approaches (i.e. treating source, breaking the pathway and /or controlling the receptor);
- previous performances of the technology in dealing with particular risk management problems;
- availability of the technology (commercial context);
- verification and long-term monitoring/management requirements;
- regeneration and need to produce land that is suitable for use;
- limitation of potential liabilities;
- practicability, i.e. technical constraints, site constraints, time constraints, regulatory constraints;
- adverse environmental impacts;
- stakeholder demands and perception;
- costs and benefits of remediation;
- combination of different technologies.

The technology selection might then be – especially at large sites affected by a complex contamination pattern – a difficult decision making process that may require a structured decision support procedure. Procedures of this kind have been developed in the UK, Germany and France.

A model procedure for Best Practice in Decision Making (Technology selection), based on existing advanced procedures, has been suggested within the EURODEMO project (EURODEMO 2007). An analysis and review of available Decision Support Tools (DST) in risk assessment and risk management processes has been performed within the CLARINET project (CLARINET 2002b).

Well known technology selection and technology information support tools have been developed in the U.S. The USEPA keeps an on-line updated database and information system of available and innovative treatment and site characterization technologies (www.clu-in.org) and the Federal Remediation Technology Roundtable provides several on-line DSTs and a Treatment Technology Screening Matrix (www.frtr.gov). The Screening Matrix is a user-friendly tool for screening potentially applicable technologies for a remediation project. It allows a user to screen 64 in situ and ex situ technologies for either soil or groundwater remediation.

Documents from a NATO Committee on the Challenges of Modern Society (CCMS) Pilot Study that was designed to share information among countries on innovative treatment technologies are also available for download from the web (www.nato.int/science/pilot-studies/pri/pri-index.htm).

6.2.3.2 Technologies Applied in the European Union

"Today, remediation technology development is evolving differently and at different rates in unconnected, isolated pockets of Europe, without joint sharing of experiences, successes, and lessons learned in technology demonstration. Despite the successful development and demonstration of novel technologies with these features, conventional methods generally still prevail in the market." (EURODEMO 2006).

The national backgrounds for contamination and its management are varied in Europe, related to different cultural and industrial histories, environmental and geological settings, population developments and policy frameworks.

Innovative remediation solutions which have minimal environmental impacts and low resource-consumption have become a tangible priority in most European countries, because of the growing awareness of environmental sustainability issues (CEC 2005). However, the actual implementation of innovative methods that deliver such results may vary very much throughout different countries.

Data from the European Environment Agency (EEA 2007) show that, in the reporting countries, there is a balance in the application of innovative in situ and ex situ techniques. A significantly high percentage of the most-frequently applied techniques can be defined as traditional, such as the so-called "dig and dump" technique and the containment of the contaminated area. This reflects the fact that contaminated soil is frequently treated as waste to be disposed of rather than a valuable resource to be cleaned and reused.

Conventional techniques such as dig and dump are indeed still prevailing through Europe while innovative technologies struggle to emerge. Generally speaking, regulatory bodies still favour conventional techniques compared to innovative techniques, for which they believe there are higher risks of not reaching the clean up goals, and thus, of increasing the potential for their liability.

In most recent years sharing responsibilities between problem holder, service providers and authorities has been indicated as an additional possible way to minimize differences of opinion between problem holders and authorities related to risk management solutions.

Another point is that information on remediation activities is fragmented and isolated across Europe and much information from some countries still remains unavailable. Making national remediation efforts and especially innovative remediation efforts visible and accessible at a European level would support international experience exchange and transnational knowledge transfer. It would also minimize duplication efforts, and, by a faster advancement of innovative technologies, increase effectiveness of remediation activities. Finally, the competitiveness of European technologies could be strengthened in a global market and a European state-of-the-art in remediation could be approached.

As for the technology selection process, individual expertise and know-how play a key role. There is a great amount of well documented reports, guidance and procedures related to technology selection and implementation, even on emerging ones. However there is a gap between the number of available technical documents and the actual use by most of the decision makers. The same observation is valid for the decision support tools which are still used quiet marginally.

Training, communication and sharing experiences about failure and successes are determinant to promote the acceptance of promising technologies (EURODEMO 2005).

Available data show that across Europe the costs for the same technology can vary by several orders of magnitude. This may reflect the different economic basis for environmental work, and particularly the support for implementation of innovative technology. The cost of delivering contaminated soil to landfills varies throughout Europe. Where it is cheap there is limited incentive for alternative approaches. Demonstration projects together with a technology verification system on a European basis could increase the confidence in the reliability of innovative environmental technologies and would support the practical implementation of enhanced solutions (EURODEMO 2005; CLARINET 2002c).

For the sake of example, the results of a survey on the application of remediation technologies in Germany, United Kingdom, and the United States are described below.

6.2.3.3 Technologies Applied in Germany

The source of remediation technologies information in Germany is the Referenzkatalog Altlasten/Schadensfallsanierung (RefAS) (Spira 2006; Umweltbundesamt 2006; RefAS 1995). The RefAS catalogue contains around 1000 remediation projects with different technology applications and this catalogue was published in 1995 in order to enable remediation planners to use experiences from projects with comparable specifications.

The numerous remediation projects completed in Germany result to a great extent from the industrial past of the country and an early start of remediation activities. In Fig. 6.2 it can be seen that Dig and dump (D&D), biological, and physical methods are mostly applied. Pump and treat (P&T) and thermal methods have also been often applied. The high number of completed projects suggests that the treatment methods that have been used until 1995 are rather well developed and used with confidence. Regarding thermal methods, some innovative in situ applications have been reported in more recent years. Additionally, some Permeable Reactive Barrier (PRB) projects have been carried out in Germany, and experience in this technology is being gained (Spira 2006).

6.2.3.4 Technologies Applied in the United Kingdom

The information source for the state of the art in the UK is a remediation status survey undertaken in 2005 with the aim of ascertaining opinions and facts on remediation practices (Spira 2006; Henstock 2006). The number for the Dig and dump (D&D) remedies, derived from the survey, shows that around 41% of completed remediation projects have a D&D component.

The survey reveals that, besides the conventional D&D and Pump and treat (P&T) methods used, there is a high amount of technologies based on biological methods applied in the UK (Fig. 6.3). Moreover, all considered treatment technologies have been applied quite recently to some extent.

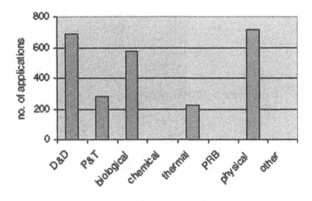

Fig. 6.2 Number of remedial technology applications in Germany (Spira 2006; RefAS 1995). D&D = Dig and Dump; P&T = Pump and Treat; PRB = Permeable Reactive Barrier

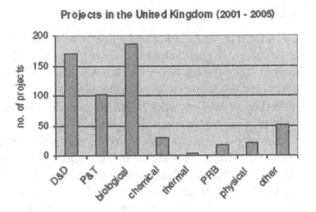

Fig. 6.3 Number of remedial technology applications in United Kingdom (Spira 2006; Henstock 2006). D&D = Dig and Dump; P&T = Pump and Treat; PRB = Permeable Reactive Barrier

6.2.3.5 Technologies Applied in the United States

A considerable progress in the application of innovative and in situ remedial technologies has been experienced in the last 20 years in the United States. The technological advancement is evident when looking at remedy solutions for soil and groundwater contamination completed at National Priority List sites.

Remedy solutions applied at NPL sites from 1982 to 2005 include treatment of groundwater in 50% of the cases, with or without soil source treatment. Soil source treatment alone covers 12% of the cases while containment, off site soil disposal, other source control and no treatment groundwater remedy cover 18% of the cases. No remedy decision and no action/no further action cover 20% of the cases (Fig. 6.4) (USEPA 2007b).

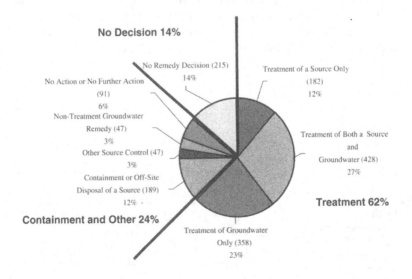

Fig. 6.4 Remedy types at 1557 NPL sites (1982–2005)

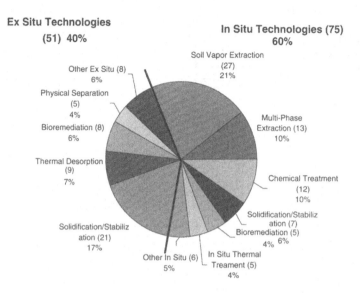

Fig. 6.5 Recent Source Control Treatment Projects – 2002–2005 – total of 126 (Prelim. Data)

According to preliminary data from the USEPA Record of Decision System (RODS), remedial technologies applied in recent (2002–2005) source control and treatment projects at Superfund sites are mostly in situ solutions (60%) (Fig. 6.5) (USEPA 2007b).

Application of in situ source treatment technologies has greatly increased at NPL sites in 20 years, starting from less than 30% in 1985 to over 60% in 2005 (USEPA 2007b).

Progress in groundwater treatment has evolved from pump and treat only to in situ remedies. Application of in situ groundwater treatments has increased from nearly 0% in 1986 to over 30% in 2005 (USEPA 2007b).

6.2.4 The Socio-Economic Dimension

The cleanup and redevelopment of contaminated sites is presently considered as a priority environmental and socioeconomic issue in many countries in Europe and overseas. Policies traditionally often view contaminated land problems from two main perspectives. The first is the perspective of protection – relating to the impact of contamination on human health and environmental quality. The other is the spatial planning perspective – managing the impact of contaminated land on the way land is used, for example regenerating industrial areas, or increasing agriculture use, or for creating a residential or a natural area, or promoting sustainability processes.

In the past, these different perspectives influenced the various legal regimes used in different countries: some countries used environmental legislation as the

Fig. 6.6 Drivers to the
cleanup and redevelopment
of contaminated sites
(modified after CLARINET
2002d)

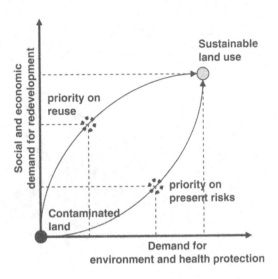

primary means of preventing impacts from land contamination on land use and
the environment, others used spatial planning legislation.

The major trend in policy development today is to address these two aspects
simultaneously (Fig. 6.6). This is increasingly evident in the development of a
more holistic approach to management of urban development. This, in turn,
increasingly links to economic issues, such as changes to land values and use of
the market to drive environmental improvements.

In other words, the socioeconomic implications related to environmental
and human health protection are strictly linked to the socioeconomic implications related to the use and management of the land and to the redevelopment of
contaminated or potentially contaminated sites. Different drivers for solving
contaminated land problems ultimately aim at restoring the capacity to reuse
the land. Defining contaminated land problems as an environmental and spatial
planning problem instead of a general burden for society will assist in finding
sustainable solutions.

Underlying all this is the wider perspective of sustainable development, in
particular the need to consider the timing of any intervention and the future
consequences of any particular solution in relation to environmental, economic,
social and cultural dimensions (CLARINET 2002d; REVIT&CABERNET 2007).

6.2.4.1 The Brownfields Problem

In the wide field of cleanup and rehabilitation of contaminated sites, Brownfield
sites, as previously defined in the introductory section, present particular challenges to national and regional policy makers in terms of bringing the land back
into beneficial use and in terms of cleaning up contaminated land and groundwater. It is in the interest of citizens, local authorities, regional development
agencies and national states that:

a. previously developed lands be regenerated for new uses and
b. any environmental damage or risk associated with prior uses be reduced and future damage mitigated.

Experience in Europe and in the US suggests that brownfields regeneration is less likely to be undertaken solely by the public or private sectors, but rather through collaboration. While there are many real estate market settings in which land values make private investments profitable with no public support, such conditions are exceptions for suspect or confirmed contaminated properties in areas with low property values and in low-income communities. Public-private collaboration should thus characterize much of the regeneration effort (REVIT&CABERNET 2007).

6.2.4.2 The Situation in Europe

There are some general indications of the nature and extent of the problem in Europe. Three main categories of brownfields can be identified:

– brownfields in traditional industrial areas which have declined (especially in the coal, steel and textile areas, but nowadays also in the chemical and power sector);
– brownfields in metropolitan areas (which may include infrastructures such as railways and docks and some of the 19th century smaller industrial uses);
– brownfields in rural areas (mainly associated with agriculture, forestry, mining or military activities).

Across Europe, the presence of derelict land is a subject of concern in many countries. In almost all European countries there are large-scale regional problems, such as those in the Ruhr area, in Catalonia and in South Wales, as well as urban problems, in particular in cities of rapid growth such as Helsinki and Dublin, and rural regional problems, such as those in Lavrion/Attika. Countries which have recently joined the European Union are equally affected, in some cases at a greater scale.

Another aspect of the problem relates to the value of the land. Brownfield land located within urban areas has a high potential value for reuse. Where the land does not have real economic value, the problem is that the land may be abandoned forever. In both cases brownfields may pose exceptional investment risks and opportunities for private investors while carrying the promise of substantial contribution to more sustainable human settlements. These scenarios are common across Europe.

Even though at the European level there is increasing evidence of policies focused on the need to reuse brownfield sites for future urban development, current practice in many industrialized countries still involves a significant level of development on greenfield sites. In Germany, alone, an estimated 129 hectares per day of greenfield land is lost for building purposes. Urban sprawl and the spatial separation of different land uses are ongoing and lead to an increasing need for mobility of the public (i.e., transportation infrastructure). In addition there is ongoing consumption of greenfield sites for housing, retail and industry. Mechanisms for re-integrating brownfields into the property markets may help to shift development back to central urban locations, which are generally considered to provide a more sustainable built environment.

In this respect, successful brownfield redevelopment policies and strategies particularly need the combination of environmental approaches with social/economic needs and spatial and urban planning approaches to be integrated into policy approaches and vice versa (CLARINET 2002c; 2002d).

6.2.4.3 The Situation in the US

Since 1995 the US government has provided technical and financial assistance for brownfields revitalization through an approach based on four main goals: protecting the environment, promoting partnerships, strengthening the marketplace, and sustaining reuse. This approach created a dynamic, flexible program that evolved in response to the needs of state, tribal, and local governments and other stakeholders. Since 1995, the investment in USEPA's Brownfields Program – less than $ 700 million – had leveraged $ 5 billion in cleanup and redevelopment funding from the public and private sectors, and created more than 24,000 jobs, often in economically disadvantaged areas that needed them most. Brownfield grants were used to assess more than 4300 brownfields properties, approximately one third of which were found to have no significant contamination, or levels so low they required no cleanup prior to the property's reuse.

It is estimated that there are more than 450,000 brownfields in the U.S. Cleaning up and reinvesting in these properties increases local tax bases, facilitates job growth, utilizes existing infrastructure, takes development pressures off of undeveloped, open land, and both improves and protects the environment.

The enactment in 2002 of the Small Business Liability Relief and Brownfields Revitalization Act, commonly referred to as the Brownfields Law, marked the beginning of a new era for the USEPA Brownfields Program. The law supports the existing approach of EPA's Brownfields Program, offers additional opportunities for financial assistance to communities, strengthens liability protections for contiguous property owners and prospective purchasers of brownfields properties, and expands assistance to states and tribes for their brownfields response programs. Additionally, the new Brownfields Law included an expanded definition of brownfields: EPA's investment in the Brownfields Program has resulted in many accomplishments, including leveraging more than $6.5 billion in brownfields cleanup and redevelopment funding from the private and public sectors and creating approximately 25,000 new jobs USEPA 2007a.

6.2.5 *The Stakeholder Dimension*

The management of contaminated and potentially contaminated land presents complex problems. The issues that land managers have to address in order to ensure sustainable solutions include involving stakeholders.

The stakeholders at the core of the decision making process for site remediation are typically the site owner and/or polluter, whoever is being affected by contamination, the state and local governments and municipal officials. However, other stakeholders can also be influential, such as:

- Site users, workers (possibly unions), visitors;
- Financial community (developers, investors, banks, founders, lenders, insurers);
- Site neighbors (tenants, dwellers, visitors, local councils);
- Citizen groups, community leaders and non-profit organizations;
- Lawyers;
- Environment agencies;
- Other technical/research institutions and private consultants.

Different stakeholders have a variety of perspectives and needs. Better awareness and common understanding are vital to the successful development of integrated and sustainable solutions.

Contaminated land is the same as any other environmental issue in terms of the range of "stakeholders" – those who have a direct or indirect interest in the outcome of decisions. For some aspects it is more complicated, since it often touches at the heart of a society or individuals in that it affects not only their own immediate environment, but also the value of something precious to them – their health and their land.

Dialogue with stakeholders may affect the choice of certain solutions over others. It will have to deal with "values" which are difficult to express in terms of risk or utilitarian concepts like land use or soil function. The conservation of a pristine underground environment and the conservation of geologically or archaeologically important sites are examples of these "values" (CLARINET 2002a).

Involvement of stakeholders is necessary in order to reach a consensus on the protection and remediation objectives. Stakeholders will have their own perspective, priorities, concerns and ambitions regarding any particular site. The most appropriate choice of risk management approach(es) will offer a balance between meeting as many of the "stakeholder needs" as possible, in particular risk reduction and achieving sustainable development, without unfairly disadvantaging any individual stakeholder. It is possible that a proposed solution may be appropriate in risk reduction terms, technically suitable and cost effective, but not accepted by some or all stakeholders because of more subjective or non-technical factors. These include:

- fixed preferences of stakeholders (for example a view that contaminated materials must always be dealt with off site);

- fear of exposure and lack of trust in the government;
- lack of confidence in the technique, which may be affected by concerns about its previous performance, the adequacy of its validation, the expertise of the provider or the ability to verify its actual performance;
- the absolute cost, which may be more than the project can economically afford;
- the resulting timeframe, which may be critical for certain projects, such as Brownfield site redevelopment to provide a new use quickly.

In the case of Brownfield redevelopment many stakeholders still see protection of the environment as an obstacle rather than an integral part of the decision making process. Political, investor and community involvement and a thorough analysis of financial indicators associated with the project are major factors in successful redevelopment. Communication/Education of people, regulators and all kind of stakeholders and decision makers are major factors in order to align the perceived risk with the assessed risk.

6.3 The Need and Role of Decision Support Systems

From what has been described in the previous sections, it is clear that the management of a contaminated site can be a very complex problem implying the integrated analysis of several aspects, ranging from the human health and ecological risk posed by contaminants to the technological feasibility of interventions and socio-economic interests associated with redevelopment plans at the site. In many cases the risk and the technological assessments require the application of advanced modelling or experimental work, as well as the management and analysis of large pools of data. Very often, socio-economic implications and stakeholder conflicts require proper analysis and intense communication/negotiation. It is also clear that the level of complexity of the problem is not always the same. For the sake of example, the remediation of a gasoline pump station has less socio-economic and environmental implications than the rehabilitation of a large Brownfield in a residential area or the remediation of a large semi-abandoned industrial area or region (also named megasites) (Wolfgang and ter Meer 2003; Bardos 2004). Notwithstanding, the high number of small contaminated sites, e.g. gasoline pump stations, might require strategic planning at a regional scale with consideration of area-wide social, economic and environmental impacts.

In general, the management of contaminated sites appears to be a complex problem that highly benefits from the application of a structured formal analysis, like the one offered by decision support systems (Pollard et al. 2004). The development of a decision support system for the management of contaminated land should allow:

- the identification of the main determinants of the decision within the liability, risk, technological, socio-economic and stakeholder dimensions,

- the analysis of the available data on each determinant, with proper inclusion of time and spatial variabilities,
- the development of management scenarios,
- the integrated analysis of the determinants for each scenario and communication of the results in such terms that can be appreciated by decision makers,
- the involvement of stakeholders in the decision process and the transparent incorporation of their preferences in the system,
- the possibility to support planning and to monitor the changing of determinants over time in order to support adaptive management strategies (Linkov et al. 2006).

Some decision support tools for contaminated sites remediation already exist and a comprehensive overview was provided by Bardos et al. (2001a; 2001b). The review highlighted that, in spite of the high number of software programs supporting specific steps of the site remediation process (i.e. site sampling and characterisation, analysis of contaminant transport, risk assessment, selection of remediation technologies, comparison of remediation options by different stakeholders), at that time (2001) only few software applications were able to combine technical and economic aspects and no one was able to support a more comprehensive approach to the entire site remediation process. Out of the most popular DSS softwares, SMARTSampling (Sandia National Laboratories) and SADA – Spatial Analysis and Decision Assistance (University of Tennessee Research Corporation) mainly focus on site characterisation and risk assessment, while DARTS – Decision Aid for Remediation Technology Selection (UNIDO, Vranes et al. 2001) and RACER – Remedial Action Cost Estimation (Delta Research Corporate) are only concerned with the selection of remediation technologies. RAAS – Remedial Action Assessment System (BATTELLE Pacific Northwest Laboratory) covers various steps, i.e. site characterisation, contaminant transport analysis and selection of remediation activities, but it does not provide tools for the management and analysis of spatial data. The DESYRE DSS (Carlon et al. 2007; 2004) was designed for priority planning of remediation interventions on megasites. It was developed after the review made by Bardos et al. in 2001 and tried to provide a comprehensive and coherent support to various phases of the remediation process, including the decision making of multiple stakeholders.

In spite of their powerful range of functionalities, these DSSs have found poor application in the real world. It may be partly due to differences between the decision making process proposed by the DSS and that occurring in practice. In general, DSSs propose transparent and participatory processes where technical experts can efficiently support educated judgments of decision makers, whereas the current decision process can be different: less coordinated, strictly hierarchical and affected by political factors that can hardly be foreseen and simulated by the DSS. In the view of some decision makers, the DSS can be perceived as a reduction of flexibility in negotiation and not fully acknowledging the hierarchy in current political decision making. For the same reasons, but in an opposite perspective, the use of DSSs can also be seen as the way to facilitate the improvement of coordination, transparency and participation in the current decision making system, i.e. technological support for a political and cultural change.

It follows that the future of DSSs relies in both their capability to feed the current decisional process at contaminated sites, and the evolution towards more structured decisional processes.

References

Bardos P.R. (2004) Sharing experiences in the management of megasites: towards a sustainable approach in land management of industrially contaminated areas. Report of the NICOLE workshop, October 2003, Lille, France, p. 55.

Bardos R.P., Mariotti C., Marot F., Sullivan T. (2001a) Framework for Decision Support used in Contaminated Land Management in Europe and North America. Land Contamination & Reclamation, 9, 149–163.

Bardos P., Lewis A., Nortcliff S., Mariotti C., Marot F., Sullivan T. (2001b) Review of Decision Support Tools and their use in Europe: Report of CLARINET WG2. http://www.clarinet.at/library/WG2_0901.pdf Accessed March 2005.

Carlon C. (Ed.) (2007) Derivation methods of soil screening values in Europe. A review and evaluation of national procedures towards harmonisation opportunities. Report EUR 22805 EN, European Commission Joint Research Centre, Ispra.

Carlon C., Critto A., Nathanail P., Marcomini A. (2000) Risk based characterisation of a contaminated industrial site using multivariate and geostatistical tools. Environmental Pollution, 111(3), 417–427.

Carlon C., Critto A., Ramieri E., Marcomini, A. (2007) DESYRE decision support system for the rehabilitation of contaminated megasites. Integrated Environmental Assessment and Management, 3(2), 211–222.

Carlon C., Giove S., Agostini P., Critto A., Marcomini A. (2004) The role of Multi-Criteria Decision Analysis in a DEcision Support sYstem for REhabilitation of contaminated sites (the DESYRE software). In: Pahl, C., Schmidt, S., Jakeman, T. (eds) iEMSs 2004 International Congress: "Complexity and Integrated Resources Management". International Environmental Modelling and Software Society, Osnabrueck, Germany, June 2004, pp. 1–6

CEC (Commission of the European Communities) (2005) Thematic strategy on the sustainable use of natural resources. COM (2005) 670 final, Brussels.

Christie S., Teeuw, R.M. (2000) Policy and administration of contaminated land within the European Union. European Environment, 10, 24–34.

CLARINET (2002a) Remediation of Contaminated Land Technology Implementation in Europe, A Report from the Contaminated Land Network for Environmental Technologies, Umweltbundesamt, Austria, October 2002.

CLARINET (2002b) Review of Decision Support Tools for Contaminated Land Management and Their Use in Europe, A Report from the Contaminated Land Network for Environmental Technologies, Umweltbundesamt, Austria, November 2002.

CLARINET (2002c) Brownfields and Redevelopment of Urban Areas, A Report from the Contaminated Land Network for Environmental Technologies, Umweltbundesamt, Austria, August 2002.

CLARINET (2002d) Sustainable Management of Contaminated Land: An Overview, A Report from the Contaminated Land Network for Environmental Technologies, Umweltbundesamt, Austria, August 2002.

EEA (2007) Progress in management of contaminated sites, Assessment published Aug 2007, http://themes.eea.europa.eu/IMS/ISpecs/ISpecification20041007131746/IAssessment 1152619898983/ view_content.

EEA, European Environmental Agency (2006) Towards an EEA Europe-wide assessment of areas under risk for soil contamination (http://eea.eionet.europa.eu; accessed October 2006).

EURODEMO (2007) Deliverable Reference Number: D 4-2. Title: Protocols and Guidance for Best Practice in Decision-Making, February 2007.

EURODEMO, Deliverable Reference Number: D 4-1. Title: Status Report on Decision Making Processes and Criteria, November 2005.

EURODEMO Newsletter (2006) 1. 2006. Available from: http://www.eurodemo.info.

European Commission (2006) Impact assessment of the thematic strategy on soil protection. Thematic Strategy for Soil Protection SEC(2006)1165

Henstock J, (2006) "CL:AIRE Contractor-Consultant Remediation Status Survey 2005". In: Brownfield Briefing (Newzeye publication). Issue 4 Remediation Solutions, Issue No. IV, p. 23.

Linkov I., Satterstrom F.K., Kiker G., Bridges T., Benjamin S., Belluck D. (2006) From optimization to adaptation: shifting paradigms in environmental management and their application to remedial decisions. Integrated Environmental Assessment and Management, 2, 92–98.

Ostertag N. (2003) Brownfields: a European perspective. Environmental Practice, 5, 13.

Pollard S.J.T., Brookes A., Earl N., Lowe J., Kearny T., Nathanail C.P. (2004) Integrating decision tools for the sustainable management of land contamination. Science of the Total Environment, 325, 15–28.

REFAS (1995) Referenzkatalog Altlasten/Schadensfallsanierung (RefAS 1.3) (in German), 1995. Available from: http://www.xfaweb.baden-wuerttemberg.de/alfaweb/index.html.

REVIT&CABERNET (2007). 2nd International Conference on Managing Urban Land, Conference Proceedings, Stuttgart, April 2007.

Spira Y. (2006) State-of-the-art of remediation of contaminated sites in Europe, Ecomondo proceedings, Rimini 2006.

Umweltbundesamt (2006) Remediated contaminated sites information (in German), http://www.umweltbundesamt.at/umweltschutz/altlasten/altlasteninfo/German register

USEPA (2007b) Treatment Technologies for Site Cleanup, Annual Status Report, 12th Edition, EPA-542-R-07-012, September 2007.

USEPA (2007a) Brownfields and land revitalization, http://www.epa.gov/brownfields/, (visited December 2007).

Vranes S., Gonzalez-Valencia E., Lodolo A., Miertus S. (2001) Decision support tools: applications in remediation technology evaluation and selection. In proceedings of NATO-CCMS Pilot Study. Decision Support Tools. N. 245, EPA 542-R-01-002.

Wolfgang C., ter Meer J. (2003) Boundary conditions as parts of an integrated management system for contaminated megasites. In: Annokkée, G.J., Arendt, F., Uhlmann, O. (eds) Consoil 2003 Conference Proceedings, pp. 45–49. Forschungszentrum Karlsruhe publishing, Karlsruhe.

Chapter 7
Decision Support Systems for Contaminated Land Management: A Review

Paola Agostini, Andrea Critto, Elena Semenzin and Antonio Marcomini

Abstract The aim of this review is to introduce and compare some of the latest progress in the development of decision support systems (DSSs) for the management of contaminated sites.

Integration of different disciplines, knowledge and decision support are critical in the remediation and management of contaminated sites. Many issues must be considered such as: reduction of risk, socio-economic impacts on the area, technical suitability and feasibility, time and cost perspectives, possible re-use options, stakeholders' points of view.

Therefore, many efforts have recently been devoted to the definition and development of DSSs that, separately or jointly, can assist the responsible authorities in planning rehabilitation and management activities at the site of interest. Some existing DSSs are here presented. Differences and common aspects are discussed in relation to main assessment and management options.

7.1 Introduction

Decision-making is a very critical and difficult task when applied to the remediation and revitalization of large contaminated sites. The main challenge is to find sustainable solutions by integrating different disciplinary knowledge and expertise with multiple views and values (Siller et al. 2004; Kiker et al. 2005).

Technical tools are therefore needed in order to integrate the wide range of decisions related to contaminated land management and re-use, including environmental, technological and economic issues (CLARINET 2002a).

P. Agostini (✉)
IDEAS (Interdepartmental Centre for Dynamic Interactions between Economy, Environment and Society), University of Ca' Foscari, San Giobbe 873, I-30121, Venice, Italy; CVR (Venice Research Consortium), via della Libertà 5-12, I-30175, Venice, Italy

A. Marcomini et al. (eds.), *Decision Support Systems for Risk-Based Management of Contaminated Sites*, DOI 10.1007/978-0-387-09722-0_7,
© Springer Science+Business Media, LLC 2009

Decision support systems (DSSs) are proven to be an effective support for any kind of decision-making process including contaminated sites management (CLARINET 2002b). Decision Support Systems can be generally defined as tools that can be used by decision makers in order to have a more structured analysis of a problem at hand and define possible options of intervention to solve the problem (Jensen et al. 2002; Loucks 1995, Simonovic 1996; Salewicz and Nakayama 2003). Nevertheless, as explained in the chapter, there is no unique definition of Decision Support Systems.

This chapter aims at providing a review of Decision Support Systems devoted to complex contaminated sites management problems. Some European and US systems are analyzed and compared: DESYRE (Carlon et al. 2007; Chapter 8), ERA-MANIA (Semenzin et al. 2005; Chapter 10), SADA (Stewart 2001; Chapter 11), RAAS (Johnson et al. 1994), REC (Vrije Universitat Amsterdam 2005), NORISC (Morvai et al. 2005), WELCOME IMS (WELCOME project 2005) and SMARTe (SMARTe 2007; Chapter 9).

In the next section, a description of the process for the management of complex contaminated sites will be presented in order to highlight critical elements in each step. Then the technical characteristics of a DSS will be presented, followed by the discussion on the main benefits that the use of a DSS can provide with reference to contaminated site redevelopment. Finally, some already existing systems for complex contaminated site management will be described and compared.

7.2 The Process of Complex Contaminated Sites Management

The complexity of the management process for contaminated sites derives from the need to choose the possible remediation intervention on the site of interest by balancing the environmental concerns caused by the contamination with the socio-economic constraints and benefits, the technological limitations, the social acceptance of redevelopment alternatives and the active participation of concerned stakeholders.

The European network CLARINET has established a framework for contaminated land management by defining the risk-based land management concept (CLARINET 2002a). Risk-Based Land Management (RBLM) has three main components: fitness for use, which considers proper use of land on the basis of acceptance by the concerned people; protection for the environment, which aims at preventing or reducing impact on the environment; and long-term care, which takes into account the intergenerational effects of present choices (CLARINET 2002a). RBLM combines within the management process the principles of sustainability, whose main goal is the integration of social, environmental and economic aspects (Bruntland 1987).

The management process becomes even more complex when the contaminated site has a large extension. In fact, two main categories of contaminated sites may be identified: brownfields and megasites. A Brownfield is a real property, the expansion, redevelopment or reuse of which may be complicated by the presence or potential presence of a hazardous waste substance, pollutant, or contaminant (US-EPA 1999). A megasite is instead "a large (km^2 scale) contaminated area or impacted area (current or potential), characterized by multiple site owners, multiple stakeholders, multiple end-users, unacceptable costs for complete clean-up (due to political, economical, social or technical constraints) within currently used regulatory timeframes, and need for an integrated risk-based approach at a regional scale" (WELCOME project 2005).

As shown particularly in the definition of the megasites, the remediation process for contaminated sites may need to address different issues (environmental, economic, and technological), to involve several stakeholders and to take into consideration many different data.

It is possible to divide the decisional process into two main phases: assessment and management, as reported in Fig. 7.1, with the reuse of the site being the final objective. Within each of the two phases, the decision-makers should address specific questions and decisions.

ASSESSMENT

- What is the site contamination? What is the environmental situation at the site?
- Which analytical procedures and technologies can be used to detect this contamination?
- Which risk does this contamination pose, to humans and/or to the environment?
- Which are the regulatory and legislative frameworks that should have to be complied with?
- Are there planning requirements that should be taken into account when defining the remediation plans?
- What would be the most beneficial land use of the site after remediation? Which is the socio-economic context the site is influenced by? Which are the economic benefits derived by the site remediation?
- Which remediation technologies may be applied? It is possible or suitable to apply more than one technology (i.e. a train of technologies)?
- How effective will these technologies be? How costly? What will be their impact?

MANAGEMENT

- How many and which different redevelopment solutions for the site can be considered?
- How can these different management options be efficiently evaluated?
- Which other decision makers and/or stakeholders should be considered in order to take a widely consensus-based decision? How can these stakeholders be effectively involved in the process and how can their values be included in the final decision? How can all the required technical experts be effectively involved?

REUSE

Fig. 7.1 The phases of the process of contaminated sites redevelopment and the related questions that decision-makers should address

In the first phase (the assessment), collection and evaluation of all the information about the site of interest and definition of remediation activities are the necessary tasks to be carried out.

After those questions, in the management phase, the definition of remediation options and solutions for site redevelopment need to be performed. Also the definition of the future use of the site is a specific decision of this phase.

Once all the assessment and management questions have been answered and clean-up activities have been performed, the process ends with the actual reuse of the site.

The list of possible questions that a decision maker can face during the redevelopment process of a contaminated site, as reported in Fig. 7.1, shows how many problems must be solved and how many different perspectives and issues must be taken into account. Specifically, in the assessment and remediation phases, decision-makers and experts are called to collect different information and take into consideration different factors of the decision: economic drivers and funding opportunities; stakeholder involvement and responsibility; access to information, legislation and policy; estimation of human health and ecological risk; finding of reliable remediation techniques; costs and benefits evaluation (Bardos et al. 2001; Siller et al. 2004).

Many other considerations may go together with the primary ones identified in Fig. 7.1: for example the economic valuations should also include financing aspects; sustainability principles should guide the entire process in order to produce a reuse that is also beneficial for future generations; effects on public health should be deeply investigated. With reference to legislative and regulatory aspects, institutional controls (e.g. for sites which are not completely cleaned up but must be monitored over time) should also be guaranteed when suitable. Moreover, reuse of the sites is the goal of the process and therefore should be carefully planned in the assessment and management phases in order to be effective and sustainable.

The diversity of questions and problems also produces a variety of methodologies and tools that the decision-maker can use.

For instance, many sampling activities can be conducted on the site: therefore the decision maker should consider which are the available sampling techniques and which ones can be effectively applied in the considered site, taking into consideration the possible presence of contaminants in different media (e.g., soil or water). Or equally, many remediation technologies may be applied, e.g. in-situ or ex-situ treatments (see also Chapter 6). In sites of large dimensions, the choice of a sampling or a remediation technology is not trivial or straightforward, due to the economic and logistical restrictions.

Another problem is the quantification of risk posed by the identified contaminants. A risk assessment procedure must be applied, following the required steps from site characterization to the assessment of risk for human health and the environment. The site characterization includes the integrated elaboration of environmental data previously collected and the definition of a conceptual model for the site (US-EPA 1989; 1998). Risk assessment is the examination of risks resulting from stressors that threaten ecosystems and people, and it involves determination of both release and resulting consequence (exposure and effect) (NRC-NAS 1983). Problems usually associated with this assessment concern the correlation of screening and site-specific procedures, uncertainties analysis and results, and decisions communication (Bridges 2003).

Even if common to the whole process, the identification of the future use of the site is an important step of public participation, particularly in cases of urban areas, and not only for communication objectives, but mostly for active involvement.

Within the assessment phase, the economic analysis must also be performed, mainly concerning the land use identification and selection (Njikamp et al. 2002), but also the assessment of costs caused by the remediation activities. Once more, different methodologies may be applied in this case to have an economic valuation of the site redevelopment, from pricing techniques (such as the cost-benefit analysis) to valuation techniques (such as the contingent valuation) (Kontoleon et al. 2001). These tools should also include the consideration of costs and values perceived by communities concerned with the redevelopment process, and with respect to impacted ecosystems.

The last aspect to consider is the linking and integration of the different involved issues. The consideration of all the aspects is essential in order to produce long-term sustainable redevelopment solutions. For example the definition of a remediation plan should be based on the reduction of risk according to the defined land-use, but also on economic constraints and drivers, and on preferences of stakeholders, mainly people directly concerned with the contaminated site as workers or inhabitants. The balance of economic (costs) and technological (treatment efficiency) aspects is critical in the selection of the remediation technologies in the second phase (i.e., management) of the process.

In summary, due to the many involved aspects, the process of contaminated sites management seems too complex to be solved only with a summation of different traditional tools and methodologies. A more structured, complete, easy to use and supportive instrument is needed. A Decision Support System may be considered a suitable solution.

7.3 Decision Support Systems for Contaminated Sites Management

Many definitions have been proposed for Decision Support Systems (DSSs) applied in different management fields (Pereira and Quintana 2002). As a general characterization, DSSs are computer technology solutions that can be used to support complex decision making and problem solving. Conventional DSSs consist of components for database management, powerful modeling functions and powerful (but simple) user interface designs (Shim et al. 2002; Ascough et al. 2002). These computer-based systems aid site managers and interested parties in gathering and integrating information, selecting and applying analytical procedures and defining management options (Shim et al. 2002; Ascough et al. 2002; Jensen et al. 2002).

In fact, a generic DSS is generally composed of the following components (Jensen et al. 2002; Loucks 1995, Simonovic 1996; Georgakakos 2004; Salewicz and Nakayama 2003):

- Database(s) and data retrieval system
- Analytical models or algorithms
- Spatial analysis, usually performed through GIS
- Graphic and visualization tools, through Graphic User Interface
- Simulation and optimisation models.

The common objective of all Decision Support Systems, according to Loucks (1995), is to "provide timely information that supports human decision makers – at whatever level of decision making." However, a DSS can be alternatively used in different manners, as an information tool, as a learning tool, as a communication tool and as a management tool (Lahmer 2004).

The main function of the DSS is usually to provide appraisal of planning variants, to simulate scenarios by numerical modelling and then to aid the stakeholders judgment and the achievement of consensus on a management solution. The DSS is expected to create and compare different scenarios by combining different modelling tools and participatory processes (Jensen et al. 2002).

In order to handle different temporal and spatial scales, the majority of DSSs include GIS (Geographic Information System) tools (see also Chapter 4). These specific DSSs are often referred to as Spatial Decision Support Systems (SDSSs) (Dietrich et al. 2004).

Moreover, some DSSs are Web-based, in order to reach as many users as possible and allow information integration and sharing among different users. In these cases, in addition to the abovementioned basic elements, the components for user application (such as the web server and the client browser) are also included.

Due to the abovementioned functionalities of DSSs (i.e. the ability of elaborating and evaluating different data sets, presenting results in understandable formats and providing a common platform for consensus-based decisions), these instruments may be proposed to answer questions of complex contaminated sites management. In fact, DSSs can guide the users in the two phases of the redevelopment process and can provide the appropriate tools to the decisions that are posed in each phase.

A significant benefit that a DSS can provide to contaminated sites managers is the possibility to have a structured analysis of the process, in the best case from the problem formulation to the reuse of the site. The decision process can be guided by the DSS in each or in the majority of its decisional steps, by providing related information, suitable tools to address problems and possible optimal solutions, by encouraging discussion of necessary tradeoffs.

Another general characteristic of DSSs, essential for a sustainable management of contaminated sites, is the technical feature of providing a combination of models and data and of considering different aspects in an integrated way. This quality is helpful for the consideration of the economical, technological and environmental aspects with respect to stakeholders' preferences. The management of numerous and differentiated data to be presented in a condensed and directly understandable format to decision-makers is extremely supportive in the overall process.

But there are functionalities and features of DSSs that are important in answering specific questions identified in the two process phases (assessment and management) of Fig. 7.1. In the assessment phase, risk assessment represents the core of the phase. A DSS may provide a platform that supports the user in the characterization of the environmental situation and in the automatic calculation of risks. The spatial functionalities that some systems include can be used to visualize, for example, the conceptual model of the site, the contaminants distribution and the extension of risk. Therefore, the system can allow the user to get answers about the first questions he is confronted with in the assessment phase (Fig. 7.1).

In the case of selection of sampling or remediation technologies, a DSS may aid the user by offering several criteria for an accurate definition and evaluation of the effects of the choices in terms of performance, costs, wider environmental impacts and so on (Bonano et al. 2000; Khadam and Kaluarachchi 2003; Khan et al. 2004). The decision-makers are facilitated in their selection because the system can automatically show advantages and disadvantages of each option and may provide a ranking or evaluation of tradeoffs based on decision-makers' preferences.

In the management phase, another critical question is the involvement of experts and stakeholders. A DSS can facilitate the achievement of a shared vision for the redevelopment of the site of interest between both experts and stakeholders (Pollard et al. 2004). For example, Multi-Criteria Decision Analysis (MCDA) tools can be included into DSSs in order to allow consideration of different stakeholders' priorities and objectives (Linkov et al. 2004; Giove et al. 2006; Kiker et al. 2005) and also integration of different issues (see also Chapter 3).

This last feature of the DSS is also helpful for the definition of alternative management options and for the analysis of multiple redevelopment scenarios. The functionalities of the modeling components of a DSS (i.e. simulation and optimization) allow the users to explore different lines of actions to reach the management objectives. The computational capabilities of a computerized system allow the user to play with different solutions and to reiterate the process in order to evaluate differences of approaches and results of actions. Different reductions of risk may be tested, different remediation plans may be proposed, and different uses (e.g. residential, recreational and industrial) can be defined. The computerized system may in this case provide cost-benefit analysis of the different options, may construct the alternatives in consideration of stakeholders' preferences, or may simulate their economic, ecological, or social effects. The system can also link an economic perspective with the planning opportunities and the legislative requirements for site redevelopment. The possibility to analyze multiple scenarios and therefore to optimize contaminated land management is also very often connected with the incorporation of process uncertainties.

The capability of DSSs to integrate different analysis by merging results of models or simulations allows the user(s) to evaluate the most effective reuse option for the site under consideration, by taking into account both the present conditions and the possible remediation achievements with respect to ecological, economic and societal impacts.

Moreover, the use of a DSS can provide transparency and openness to the process, since all decisions can be traced back by the system and can be accurately justified. The application of DSSs by management authorities may avoid the risk that decisions are taken only in consideration of partial information or with a disagreement of preferences.

Nevertheless, some disadvantages may also be faced in the use of DSSs: difficulty in gaining acceptability; limitations in providing highly integrated information without the rationale behind clearly explained; need to be continuously updated; and the necessity to have reliable input data and clear assumptions (Sullivan et al. 1997). Moreover, as a general feature, a DSS should create a balance between the complexity needed to address the wide range of site conditions, and ease of use (Sullivan et al. 1997). This is not simple to provide, because the system should be complete and helpful but at the same time easy-to-use, flexible in site-specific evaluations, manageable in terms of data collection and input, and reliable in the correctness of results.

The general characteristics of DSSs for contaminated sites presented here should nevertheless be valued with reference to the specific objectives for which a single system is constructed. In fact, when designing and building a DSS, the choice of functionalities and software components depends on the management objectives for which the system is created. The management objectives are inevitably linked to the management questions concerned with the process phases. As a consequence, there is a great variety of Decision Support Systems and tools that can be proposed to answer specific questions in the process of contaminated sites management. In the next section, some of these systems will be presented and compared in their main functionalities.

7.4 Comparison of Selected Decision Support Systems

As mentioned in the Introduction, eight systems were selected for this review. These systems were chosen because they were developed and presented to the scientific community in the last 10 years and they were validated by the application to real cases within internationally approved projects. The selection of the

DSSs was primarily driven by their capability to answer complex management questions and therefore to be helpful in defining general considerations about structure and functionalities of a suitable Decision Support System for complex contaminated site redevelopment.

DESYRE (DEcision Support sYstem for REhabilitation of contaminated sites) is designed particularly to manage large contaminated sites. DESYRE is a GIS-based software composed of six interconnected modules that provide site characterization, socio-economic analysis, risk assessment before and after the technologies selection, technological aspects and alternative remediation scenarios development (see Chapter 8).

The ERA-MANIA DSS was developed to support the site-specific phase of the Ecological Risk Assessment (ERA) for contaminated soils. In particular, it is based on the Triad approach (Rutgers and Den Besten 2005), where the results provided by three Lines of Evidence (LoEs) (i.e., chemistry/bioavailability, ecology and ecotoxicology) are gathered and compared to support the assessment and evaluation of the ecosystem impairment caused by the stressor(s) of concern (see Chapter 10).

RAAS (Remedial Action Assessment System) is a decision support system designed to assist remediation professionals at each stage of the investigation and feasibility study process. RAAS is based on two main components: ReOpt (which provides descriptive information about technologies, contaminants, or regulations) and MEPAS (which is a human health risk model). RAAS has the objective to identify remedial technologies for the specific site conditions, remediation strategy, and cleanup objectives, and to estimate the effects of applying those technologies.

SADA (Spatial Analysis and Decision Assistance) is a software that incorporates tools from environmental assessment fields into an effective problem solving environment. The capabilities of SADA can be used independently or collectively to address site specific concerns when characterizing a contaminated site, assessing risk, determining the location of future samples, and when designing remedial action (see Chapter 11).

The REC system includes three tools developed to evaluate risk reduction, environmental merit and costs of remediation alternatives. The use of the tools can be modular, which means that the three tools may be used independently, but the main aim is the integration of the risk, environmental impact and cost aspects.

The NORISC Decision Support System basically guides the development of a methodology for investigating and assessing a contaminated site, in particular, for determining the pollution occurrence in soil and groundwater, as well as the risks involved and the potential site reuse.

The Web-based WELCOME IMS (Integrated Management Strategy) is a stepwise approach to establish integrated risk based management plans for large contaminated sites, from the initial screening to the final definition of the remediation scenarios and long-term site management plan. In fact WELCOME differs from the other systems because it provides an operational framework within which different tools are proposed for application to address specific concerns (for example, risk assessment or scenarios creation), but the outputs of the different applications are not linked together and elaborated by the system.

SMARTe (Sustainable Management Approaches and Revitalization Tools – electronic) is a free, open-source, web-based, decision-support system to help revitalize communities and restore the environment. It is primarily intended to help bring potentially contaminated land back into productive use. It contains resources and analysis tools for all aspects of the revitalization process from definition of future land use and stakeholder involvement to economic analysis of financing, market costs and benefits, environmental issues and liability aspects. It is a holistic decision analysis system that integrates these aspects of revitalization while facilitating communication and discussion among all stakeholders (see Chapter 9).

More information about the different systems may be found in the references list, as well as in the related chapters.

Technically, all the systems are constructed in connected modules that perform separated analyses. For example, the software DESYRE is characterized by six interconnected modules (Characterization, Socio-economic, Risk Assessment, Residual Risk Assessment, Technological and Decision Modules), while the RAAS system has two main components (ReOpt for information on technologies, regulation and contaminants; and MEPAS for human health risk assessment), and the REC system is organized into three separate steps (Risk reduction, Environmental merit and Cost components). Equally, in SMARTe, the different sections of the system address specific aspects of the site revitalization process, from site description and identification of future land use, to community involvement and environmental management, as well as liability and financial assessments.

Nevertheless, in the different systems, a connection is established within the modules or single tools, in order to ensure that a subsequent analysis may be performed. A flow of information is maintained from input data to the definition of remediation or revitalization options, as in DESYRE, SMARTe or WELCOME systems, where the user is guided step-by-step during the application in the different phases (see Fig. 7.1).

The connected components of the DSS correspond to database systems and analytical models previously identified in Section 7.3 about DSS generalities. These components allow users to perform the different analyses required during the assessment phase of the contaminated sites redevelopment. The other components of the DSS, such as the GIS platform and the tools for visualization and graphic modeling, are instead those elements that allow the system to link the results of the analytical steps and to present them to the user in a friendly interface.

Operatively, the DSSs perform different functionalities. With reference to the decision questions identified in the previous paragraphs for the assessment and management phases of the redevelopment process for contaminated sites, the main supporting functionalities of the different systems are reported in the Text Boxes 7.1–7.7. The content of the Text Boxes supports the following discussion and comparison of tools.

Text Box 7.1.   Decision maker question: What is the site contamination? What is the environmental situation at the site?

DESYRE: Users can perform environmental characterization, through geostatistical techniques which allow visualization on GIS maps (Characterization module). The module is fed by chemical and idrogeological data. The Module also allows the selection of priority substances.

RAAS: User can input the conceptual model of the site, considering contaminants and receptors, while the system uses information about the site to identify potential contaminant transport mechanisms.

SADA: Data exploration tools include two- and three-dimensional data visualization options. In order to allow data visualization with respect to site characteristics, SADA can accept map layers from a Geographic Information System (GIS).

WELCOME: The PriCon tool (Priority Contaminants) allows for a systematic and comparable assessment of a large number of contamination hot spots. It allows detection of all relevant contaminants; assessment of these contaminants with regard to their toxicological potential for different exposure scenarios; and combination of respective physiochemical substance characteristics to quantify their mobility, retardation and accumulation.

NORISC: GIS spatial information from different investigation methods can be visualized allowing for a more comprehensive definition of the contamination.

SMARTe: There is a site description tool, which allows users to input the description of the existing conditions at the site. Moreover, the site characterization tool provides some statistical analysis capabilities that can be used to support data analysis and risk assessment. The monitoring data analysis tool offers standard statistical methods for displaying temporal/spatial data.

Text Box 7.2.   Decision maker question: Which can be the analytical procedures and technologies that can be used to detect the contamination?

ERAMANIA: MCDA based comparison procedure is used for the selection of the most suitable set of bioavalaibility, eco-toxicological and ecological tests to be performed at the site of interest.

SADA: It provides different strategies to determine future sample locations, depending on the geospatial interpolation.

NORISC: Geochemical, (hydro-) geological, geophysical and biological investigation methods are described in detail and their combinations are evaluated based on suitability, cost and time criteria. A ranking system defines the suitability of each method.

SMARTe: The Risk Assessment section discusses the preparation of an investigation of hazardous substances on the site.

Text Box 7.3.   Decision maker question: Which risk does this contamination pose, to humans and/or to the environment?

DESYRE: Risk calculation is deterministic and probabilistic. It follows ASTM-RBCA guidelines. It provides indices for risk, including risk magnitude and extension reduction and risk uncertainty. There is an uncertainty analysis associated. Considered receptors are adults and children. Exposure scenarios are: residential, industrial and recreational. Spatial analysis obtained by GIS techniques allows visualization of risk extension on the site for contaminants and class of contaminants in soil and groundwater. The system provides calculation of residual risk.

ERAMANIA: Ecological risk assessment, with definition of integrated ecological risk indexes and analysis of impairment on soil biodiversity and functional diversity.

RAAS: Human health risk model, based on EPA risk assessment guidance (MEPAS module). Four potential exposure media: airborne contamination, groundwater, surface water, direct exposure to external radiation. To this, media are associated corresponding to the potential exposure routes: ingestion, dermal contact, inhalation, external radiation.

SADA: Risk calculation is deterministic. It considers soil and water exposure pathways. Exposure scenarios are: future unrestricted industrial, residential, recreational, excavation, and agricultural exposures. The ecological and human risk assessment is integrated with spatial analysis and interpolation (Nearest Neighbor, Natural Neighbor, Inverse Distance, Ordinary Kriging, and Indicator Kriging, 2 and 3D visualization).

WELCOME: Human health risk assessment, through the HIRET tool. Risk calculation is deterministic. Exposure scenarios are: Residential, Recreational, Excavation, Industrial and Agricultural. Due to the incorporation of a geographic information system (GIS) environment, contamination maps may be produced from transport models external to the tool (Sesoil, AT123D, Modflow, others) and GIS interpolation functionalities.

NORISC: Risk calculation is deterministic. Exposure scenarios are: residential, industrial and recreational. 1D, 2D and 3D spatial visualization of contamination. Risk zones can be obtained by point grouping (clustering) from single points with specific risk level (below and above Target Risk levels).

REC: The calculation of the risk reduction leads to a certain score per remediation option, which correspond to a risk index. Considered exposure scenarios are: residential, commercial, Natural area, Rural area, Urban and Industrial area. It calculates the risk associated to the exposure of both people and ecosystem, prior to, during and after a remediation operation. The time profile of exposure is compared to that in absence of remediation (the do-nothing option).

SMARTe: A Human health risk calculator is included in the system. Risks are calculated for various exposure pathways for a selected chemical and its concentration. Considered exposure scenarios include ingestion, dermal contact, and inhalation.

Text Box 7.4. Decision maker question: Which are the regulatory and legislative frameworks that should have to be complied with? Are there planning requirements that should be taken into account when defining the remediation plans?

DESYRE: The socio-economic Module takes into consideration planning variables (e.g. distances from main roads, economic attractiveness for specific land use) when defining the most suitable land use for the site.

RAAS: The ReOpt Module about technologies contains also regulatory information.

WELCOME: The system provides guidelines for the analysis of the boundary conditions for site redevelopment, which include legislative frameworks review.

SMARTe: The system provides extensive resources for regulatory and liability aspects, with reference to mainly US legislations, in forms of information sources, tools and checklists.

Decision maker question: What would be the most beneficial land use of the site after remediation? Which is the socio-economic context the site is influenced by? Which are the economic benefits derived by the site remediation?

DESYRE: The Socio-economic Module of the DESYRE software allows to study, based on a Fuzzy Logic analysis, the constraints and benefit of the site in relation to different land uses.

SMARTe: Many economic tools, included in the Future Land Use section address the contaminated site revitalization, identifying, among others, common vision for the future use, main drivers for revitalization, community and regional needs, and opportunities for innovation.

Text Box 7.5. Decision maker question: Which remediation technologies may be applied? It is possible or suitable to apply more than one technology (i.e. a train of technologies)?

DESYRE: The DESYRE system provides the user with a wide database of technologies from which the user can select the most suitable for the specific case according to criteria such as suitability for contaminants, applicability, and so on. The technological Module of DESYRE allows the creation of several technology sets, composed of different remediation technologies, i.e. train of technologies.

RAAS: The system contains an extensive database on remediation technologies. It helps the user identify remedial technologies for the specified site conditions based on the clean up strategy, the contaminated medium and the contaminants.

WELCOME: PRESTO (PRESelection of Treatment Options) is a user-friendly program that allows the user to conduct a first evaluation on the technical applicability of available remediation technologies to risk clusters. The applicability check is done on the basis of site-specific characteristics such as the geological conditions, the geochemistry of the groundwater and the contaminants present in the area.

REC: The Environmental merit part of REC highlights clean-up strategies which do not put a particular burden on environmental resources.

SMARTe: The Managing Risk section includes approaches to identification of alternative remediation actions.

Decision maker question: How effective will the technologies be? How costly? What will be their impact?

DESYRE: The technological Module of the software allows a user to compare and evaluate by means of MCDA methodologies the technological choices. Several criteria are used for this evaluation (e.g. efficiency with specific contaminants, public acceptability, costs, and commercial availability).

RAAS: A feasibility study can be performed and the effects of application of technologies can be performed based on any special constraints related to site, medium and contaminant, or technology characteristics. The system also calculates costs for the applied technologies.

SADA: it can estimate costs associated with various clean-up activities, and can estimate clean-up volumes.

REC: The system considers the overall costs for remediation options, including operational investment and maintenance costs.

SMARTe: The Net Benefit Calculator compares the cost elements and revenues of the different revitalization options.

Text Box 7.6.   Decision maker question: How many and which different redevelopment solutions for the site can be considered?

DESYRE: The DESYRE software can create several remediation scenarios, which can be evaluated by stakeholders. A scenario represents a suitable solution for the site in consideration of technological choices, environmental impacts, cost and time issues, risk reduction (both for the environment and for humans) and socio-economic benefits related to a specific land-use.

RAAS: The system provides redevelopment options linked to the technological choices and risk reduction functions.

NORISC: The Revitalisation Module allows a user to select the optimal "site-option", which is a remediation option for a specific site. It is also possible to create combination(s) from a set of sites and a set of options. The selection is based on the social benefits (avoided risks) and the financial benefits (change in land value) achieved by the site-option combinations, and on the costs of the options.

SMARTe: In the My project section there is the possibility to identify a revitalization project, evaluating information and elaborations from other sections of the system, and comparing different reuse options.

Decision maker question: How can the different management options be efficiently evaluated?

DESYRE: The Decision Module provides the user with the comparison of remediation scenarios. Each scenario is identified by indices related to the risk, technological and socio-economic aspects.

RAAS: The system evaluates the different options by considering costs, implementability, extent of treatment, long and short-term effectiveness and objectives compliance.

SADA: SADA can produce site-specific cost-benefit curves that demonstrate the specific relationship between a given remedial cleanup goal and the corresponding cost.

WELCOME: CARO (Cost Analysis of Remediation Options) simulates effects (reduction of mass and mass flow) and costs over time. In doing so, essential information is provided that will help site managers identify the most cost-efficient remediation setup, i.e. the setup that meets the prefixed goals at a minimum of costs.

NORISC: Different options can be characterized and evaluated in consideration of: the clean-up methods, risk reduction, costs (sum of investigation, clean-up and monitoring costs), and a simplified cost-benefit analysis.

REC: The REC model is based on a weighting methodology that considers cost, risk reduction and environmental merit, in order to support the choice of the most cost-effective remediation option.

SMARTe: All the investigations on the site are brought together in an MCDA in the My Project part of SMARTe, that helps the user to sort through the many different options considering costs and benefits primarily. There is also an opportunity for users to include information regarding community values. Uncertainty of the various options can also be calculated.

Text Box 7.7. Decision maker question: Which other decision makers and/or stakeholders should be considered in order to take a widely consensus-based decision? How can these stakeholders be effectively involved in the process and how can their values be included in the final decision? How can all the required technical experts be effectively involved?

DESYRE: The inclusion of MCDA methodologies in the system allows for the active participation of both experts (specifically in the analytical phases) and stakeholders (in the decision phase).

ERAMANIA: It contains an MCDA-based procedure in which the experts assign specific evaluations to the tests to be compared and provide crucial information for the results normalization.

WELCOME: The system provides guidelines for the involvement of relevant stakeholders NORISC: GIS spatial information form different investigation methods can be visualized allowing for a more comperhensive definition of the contamination

NORISC: The Revitalization Module allows stakeholders to select one or more contaminated sites and to rank the suitable remediation options, by filling out a questionnaire and input sheets.

SMARTe: The system includes a tool for stakeholders' identification and another tool for selecting the most effective community involvement approach taking into account different criteria (audience, target, costs, time). Stakeholders are actively involved in the definition of the vision for the reuse of the site. My Project encourages all stakeholders to come together to discuss and evaluate reuse options and determine where trade-offs will be made.

It is evident that the systems can be used by decision-makers according to the assessment and management questions they need to answer and according to the methodologies they include to better address the same questions.

First of all, the systems may be divided into two main categories: the ones that follow the majority of the assessment and management steps, such as DESYRE, NORISC, SMARTe or WELCOME, and those systems that are concerned with more specific issues, such as ERA-MANIA or REC.

One important aspect is represented by the site characterization and the risk assessment, whose related questions are addressed by the majority of the systems. Taking into consideration the risk assessment functionality, which is one of the most critical parts of the assessment process, the systems present common features but also some exclusive features.

All the systems are concerned with human health risk assessment and consider mainly the same exposure scenarios, but some of them provide additional functionalities. For example DESYRE, SADA, HIRET in WELCOME, and NORISC provide spatial analysis of risk, through GIS or other interpolation methods. These visualization functionalities are particularly interesting in consideration of the support provided to the decision-makers. In fact, visualization of risk levels and distribution on the site of interest may be a more effective way to communicate risk assessment results. Therefore, those systems may be more respondent to managers who want to answer the questions about the risk to humans in a more spatially resolved way, especially when the site is very large.

At the same time, some other functionalities may be provided by the systems. For example, the risk assessment procedure included in the DESYRE software allows users to calculate residual risk, to obtain risk indices and to have a probabilistic risk assessment. In SMARTe, human health risk assessment inputs may be entered as constants but also, if the user has an understanding of the uncertainty associated with their inputs, as distributions, allowing the performance of probabilistic risk assessment. The SADA software is the only one that can spatially evaluate both human and ecological risk.

There are some questions that are less considered by the systems: the regulatory and legislative frameworks, the planning requirements and the socio-economic aspects not connected to remediation costs are aspects which are not dealt with by the majority of the systems. A remarkable exception is represented for instance by SMARTe, where the many tools and information are devoted to the consideration of regulatory, political and marketing aspects, as well as the assessment of motivators or drivers for site revitalization.

For the questions related to the technical aspects of the remediation activities, DESYRE, RAAS, and WELCOME provide databases of technologies and methods to choose the most suitable technologies, based on site-specific conditions and on each technology's characteristics.

The characterization of redevelopment scenarios is connected to an integrated approach, by which environmental, technical, economic and other broad aspects are taken into consideration altogether. Some of the DSSs allow this integration by providing integrated indices, as in the REC system with the value of costs, environmental impact and risk reduction; or in DESYRE with indices derived from the five analytical modules.

The definition and evaluation of redevelopment options for the site is supported by several systems, but with some distinguishing characteristics. In fact, a common approach is to consider the risk reduction in relation to cost evaluation. The REC system also incorporates the environmental impact issue, the RAAS system also considers the actions' effectiveness and to what degree the management options comply with the objectives. DESYRE integrates risk, technological and socio-economic aspects in a time perspective, and SMARTe dedicates a specific section to the evaluation of different revitalization options, by calculating costs and revenues.

The involvement of stakeholders is ensured in the ERA-MANIA, DESYRE, WELCOME, SMARTe and NORISC systems. SMARTe gives emphasis to this aspect through tools that allow identification of relevant stakeholders and selection of the most effective public participation techniques. MCDA methodologies are most commonly included in the systems to support integration of aspects and inclusion of stakeholders' and experts' perceptions.

Even if some common functionalities and included methodologies may be equally present in the reviewed systems, the fact that some of them have been constructed and are specifically devoted to address some assessment and management questions should guide the selection of the most suitable tool for the application in a contaminated site redevelopment. For example, there are instruments that analyze ecological risk, systems that have developed complex methodologies for risk characterization and spatial visualization, or systems that are mostly concerned with the procedures for the detection of contamination more than with the selection of remediation techniques.

The differences in approaches also result in different levels of complexity of these systems. For example, the GIS and spatial functionalities may require a user that has familiarity with these methodologies and therefore can better evaluate the potentialities of these analyses.

The selection of a system is also connected to a general objective of the application. If the user is interested in considering all the phases of the process, he will choose those systems that follow the process from the initial phases until the reuse of the site. Otherwise, he will just select the system that provides an in-depth analysis of a specific aspect or question.

Therefore, the multiplicity of systems can represent an advantage for a generic user, because these systems can be constructed differently and be adapted to different viewpoints.

7.5 Conclusions

The presented review discussed the relevant aspects and decision questions involved in a complex contaminated site management process and identified the most important functionalities and technical characteristics for a suitable Decision Support System. The review was based on the examination of some DSSs currently developed or under development, which proved that the end-users could choose among several systems, according to the questions they help to address.

This comparison may provide important lessons learned for the design and building of new instruments. For example, the more comprehensive those systems will be, the more useful and probably up-to-date they will be. These systems should take into consideration risk assessment not only as a calculation but also as a mapping exercise, in order to facilitate understanding of problem significance by non-experts. The systems should provide more clear management of uncertainties in the analytical steps and they should take the legal and planning issues more into consideration with respect to management solutions. The communication and involvement features should be enhanced to take into account that the real protagonists of the remediation process are mainly the people living and working on the contaminated land.

In general, the recent developments in Decision Support Systems show that the developers are more and more concerned with the needs of decision makers that require simple but comprehensive tools for approaching contaminated site problems.

However, much more information and experience for improving DSSs development may be gained from a real comparison between different DSSs accomplishing the same goals.

In fact, at the present stage of DSS development, it is not yet possible to evaluate comparatively the performances of each system and the efficiency in achieving the management objectives. Therefore, for future developments in this field, a comparison of different DSSs through the application to the same problem may be beneficial in order to reveal advantages and disadvantages of each system and derive useful suggestions for new implementations.

References

Ascough, J C, Rector H, Hoag D L, McMaster G S, Vanderberg B C, Shaffer M J, Weltz M A and Ahjua L R. 2002. Multicriteria spatial decision support systems: overview, applications and future research directions. In: Rizzoli, A E and Jakeman, A J (eds.), Integrated Assessment and Decision Support, Proceedings of the First Biennial Meeting of the International Environmental Modelling and Software Society, iEMSs, 2002 3: 175–180.

Bardos, R P, Mariotti C, Marot F and Sullivan T. 2001. Framework for decision support used in contaminated land management in Europe and North America. Land Contamination and Reclamation 9 (1): 149–162.

Bonano, E J, Apostolakis G E, Salter P F, Ghassemi A and Jennings S. 2000. Application of risk assessment and decision analysis to the evaluation, ranking and selection of environmental remediation alternatives. Journal of Hazardous Materials 71: 35–57.

Bridges, J. 2003. Human health and environmental risk assessment: the need for a more harmonised and integrated approach. Chemosphere 52: 1347–1351.

Bruntland, G. (ed.). 1987. Our common future: The World Commission on Environment and Development. Oxford, Oxford University Press.

Carlon, C, Critto, A, Ramieri, E and Marcomini, A. 2007. DESYRE decision support system for the rehabilitation of contaminated mega-sites. Integrated Environmental Assessment and Management 3: 211–222.

(CLARINET) Contaminated Land Rehabilitation Network for Environmental Technologies. 2002a. Sustainable Management of Contaminated Land: An Overview. A report from the Contaminated Land Rehabilitation Network for Environmental Technologies, p. 128.

(CLARINET) Contaminated Land Rehabilitation Network for Environmental Technologies. 2002b. Review of Decision Support Tools for Contaminated Land Management, and Their Use in Europe. A Report from the Contaminated Land Rehabilitation Network for Environmental Technologies, p. 180.

Dietrich, J, Schumann, A and Lotov, A. 2004. Workflow oriented participatory decision support for the integrated river basin planning. Proceeding of the IFAC Workshop Modelling and Control for Participatory Planning and Managing Water Systems, Venice September 29th–October 1st, 2004.

Georgekakos, A. (2004). Decision support systems for integrated water resources management with an application to the Nile basin. In Proceeding of the IFAC Workshop Modelling and Control for Participatory Planning and Managing Water Systems, Venice September 29th —October 1st, 2004.

Giove, S, Agostini, P, Critto, A, Semenzin, E and Marcomini, A. 2006. Decision Support System for the management of contaminated sites: a multi-criteria approach. In: Linkov, I, Kiker, G A and Wenning, R J (eds.), Environmental Security in Harbors and Coastal Areas: Management Using Comparative Risk Assessment and Multi-Criteria Decision Analysis. Amsterdam, The Netherlands: Springer, pp. 267–273.

Jensen, R A, Krejcik J, Malmgren-Hansen, A, Vanecek, S, Havnoe, K and Knudsen, J. 2002. River Basin Modelling in the Czech Republic to Optmise Interventions Necessary to Meet the EU Environmental Standards. In Proceedings of the International Conference of Basin Organizations, Madrid 4–6 November 2002, DH ref 39/02.

Johnson, C D, Bagaasen, L M, Chan, T C, Lamar, D A, Buelt, J L, Freeman, C J and Skeen, R S. 1994. Overview of technology modeling in the Remedial Action Assessment System (RAAS). In: Proceedings Spectrum '94: Nuclear and Hazardous Waste Management International Topical Meeting, Volume 2. American Nuclear Society: La Grange Park, IL, pp. 1078–1085.

Khadam, I and Kaluarachchi, J J. 2003. Applicability of risk-based management and the need for risk-based economic decision analysis at hazardous waste contaminated sites. Environment International 29: 503–519.

Khan, F I, Husain, T and Hejazi, R. 2004. An overview and analysis of site remediation technologies. Journal of Environmental Management 71: 95–122.

Kiker, G A, Bridges, T S, Varghese, A, Seager, T P and Linkov, I. 2005. Application of Multicriteria Decision Analysis in Environmental Decision Making. Integrated Environmental Assessment and Management 1 (2): 95–108.

Kontoleon, A., Macrory, R., Swanson, T. 2001. Individual preferences, expert opinions and environmental decision making: an overview of the issues. Paper presented at the Symposium on Law and economics of environmental policy, UCL, 6-7 September 2001.

Lahmer, W. 2004. Multi-disciplinary approaches in River Basin Management – an example. Presentation at the International Conference on Water Observation and Information System for Decision Support, Ohrid, Macedonia, 25–29 May 2004.

Linkov, I, Varghese, A, Jamil, S, Seager, T P, Kiker, G, and Bridges, T. 2004. Multi-criteria decision analysis: a framework for structuring remedial decisions at contaminated sites. In: Linkov, I and Ramadan, A (eds.), Comparative Risk Assessment and Environmental Decision Making. Kluwer, pp. 15–54.

Loucks, D P. 1995. Developing and implementing decision support systems: a critique and a challenge. Water Resources Bulletin 31 (4): 571–582.

Morvai, B, Kriszt, B and Szoboslay, S. 2005. NORISC revitalization strategy for contaminated sites In: Proceeding of CABERNET 2005: The International Conference on Managing Urban Land. Compiled by Oliver, L, Millar, K, Grimski, D, Ferber, U and Nathanail, C P. Land Quality Press, Nottingham. ISBN 0-9547474-1-0.

(NRC-NAS) National Research Council of the National Academy of Science. 1983. Risk Assessment in the Federal Government: Managing the Process. National Academy Press, Washington, DC, USA.

Njikamp, L, Rodenburg, C A and Wagtendonk, A J. 2002. Success factors for sustainable urban brownfield redevelopment. A comparative case study approach to polluted sites. Ecological Economics 40: 235–252.

Pereira, A G and Quintana, S C. 2002. From technocratic to participatory decision support systems: Responding to the new governance initiatives. Journal of Geographic Information and Decision Analysis 6 (2): 95–107.

Pollard, S J T, Brookes, A, Earl, N, Lowe, J, Kearny, T and Nathanail, C P. 2004. Integrating decision tools for the sustainable management of land contamination. Science of the Total Environment 325: 15–28.

Rutgers, M and Den Besten, P. 2005. Approach to legislation in a global context. The Netherlands perspective – soils and sediments. In: Thompson, K C, Wadhia, K and Loibner, A P (eds.), Environmental Toxicity Testing. Blakwell Publishing CRC Press, Oxford, pp. 269–289.

Salewicz, K A and Nakayama, M. 2003. Development of a web-based decision support system (DSS) for managing large international rivers. Global Environmental Change 14: 25–37.

Semenzin, E, Critto, A, Carlon, C, Mesman, M, Schouten, T, Rutgers, M, Giove, S and Marcomini, A. 2005. ERA-MANIA DSS: a decision support system for site-specific Ecological Risk Assessment (ERA) for contaminated sites. Extended abstract, 9th

International conference on soil-water systems "ConSoil 2005", 3–7 October 2005, Bordeaux, France, pp. 627–630.

Shim, J P, Warkentin, M, Courtney, J F, Power, D J, Sharda, R and Carlsson, C. 2002. Past present and future of decision support technology. Decision Support Systems 33: 111–126.

Siller, D, Blodgett, C, Cziganyik, N and Omeroglu, G. 2004. Factors involved in urban regeneration: remediation and redevelopment of contaminated urban sites – a comparison of France and the Netherlands. In: Proceeding of CABERNET 2005: The International Conference on Managing Urban Land. Compiled by Oliver, L, Millar, K, Grimski, D, Ferber, U and Nathanail, C P. Land Quality Press, Nottingham. ISBN 0-9547474-1-0.

Simonovic, S P. 1996. Decision Support Systems for sustainable management of water resources: 1. General principles. Water International 21 (4): 223–232.

SMARTe project. 2007. www.smarte.org, visited on August 2007.

Stewart, R. 2001. Geospatial decision frameworks for remedial design and secondary sampling. In NATO/CCMS Pilot Study Evaluation of demonstrated and emerging technologies for the treatment of contaminated land and groundwater. Special Session Decision Support Tools. Committee on the challenges of modern society, p. 139.

Sullivan, T M, Gitten, M and Moskowitz, P D. 1997. Evaluation of selected environmental decision support software. BN-64613, Brookhaven National Laboratory, p. 46.

(US-EPA) US Environmental Protection Agency. 1989. Risk assessment guidance for Superfund (RAGS): volume 1 – Human health evaluation manual (HHEM) (Part A, baseline risk assessment). Office of Emergency and Remedial Response. EPA/540/1-89/002. Washington, DC.

(US-EPA) US Environmental Protection Agency. 1998. Guidelines for ecological risk Assessment. EPA630-R-95-002F. Risk Assessment Forum. Washington DC.

(US-EPA) US Environmental Protection Agency. 1999. A sustainable Brownfields model framework, Sustainable redevelopment linking the community and business for a brighter future, EPA500-R-99-001. Washington DC.

Vrije Universitat Amsterdam. 2005. REC decision support system for comparing soil remediation alternatives, http://www.falw.vu.nl/Onderzoeksinstituten/index.cfm/home_file. cfm/fileid/EA9 454E9-7E6F-4DBF-A34D63D78A93D022/subsectionid/602C4835-C246-41FA-8DD706E7084B0D06

WELCOME project. 2005. http://euwelcome.nl/kims/index.php, visited on March 2005.

Chapter 8
A Spatial Decision Support System for the Risk-Based Management of Contaminated Sites: The DESYRE DSS

Lisa Pizzol, Andrea Critto and Antonio Marcomini

Abstract A GIS-based Decision Support System (DSS) called DESYRE (Decision Support sYstem for the REqualification of contaminated sites) was developed in order to address the integrated management and remediation of contaminated megasites. The DESYRE DSS supports decision makers during the main phases of a remediation process, i.e. analysis of social and economic benefits and constraints, site characterization, risk assessment, selection of best available technologies, creation of sets of technologies to be applied, analysis of residual risk and comparison of different remediation scenarios. Within the DESYRE DSS these functionalities were implemented in six interconnected modules. In the characterization module, chemical and hydrogeological data are organized in a relational database in order to support the definition of the conceptual model of the site and to provide information regarding contaminant distribution and transport through the different environmental media. Georeferenced information system tools are used for handling spatial data. The socio-economic module addresses the socio-economical constraints though a Fuzzy Logic analysis, in order to provide decision makers with a tool that compares the different land use options outlining possible scenarios linked to alternative uses of the considered site, on the basis of socio-economic considerations and local characteristics. In the pre-remediation phase, an original risk assessment methodology (risk assessment module) allows the evaluation and estimation of the spatial distribution of risks posed by contaminants in soil and groundwater, providing a risk-based zoning of the site in support of the definition of the remediation technologies plan. The latter phase is performed in the technology assessment module, where a selection of suitable technologies and the creation of different technology sets are carried out by experts supported by Multi-Criteria Decision Analysis tools. The selection takes into account both technical features and requirements of available technologies, as well as site-specific environmental conditions of the site of concern, such as chemical

L. Pizzol (✉)
Consorzio Venezia Ricerche, Via della Libertà 5-12, I-30175, Marghera, Venice, Italy
e-mail: lisa.pizzol@unive.it

A. Marcomini et al. (eds.), *Decision Support Systems for Risk-Based Management of Contaminated Sites*, DOI 10.1007/978-0-387-09722-0_8,
© Springer Science+Business Media, LLC 2009

contamination levels and remediation objectives. A simulation of the selected technologies application is performed in the post-remediation risk assessment (residual risk assessment module) in order to provide residual risk maps with related uncertainty maps. Finally, the decision module provides a methodology for the description and comparison of alternative remediation scenarios by defining a set of aggregated indices. In this chapter, the structure and the functionalities of the software DESYRE are presented. The application of the whole system to a case study gives an explanation of the flow of information within the system and the presentation of the outputs of all the above mentioned modules.

8.1 General Introduction

The management of contaminated sites and the assessment of the potential risks they pose to human health and ecosystems is a worldwide issue. In order to achieve a proper environmental, economical, social and industrial rehabilitation of these sites, the risk-based prioritization of the remediation interventions becomes a valuable step within the contaminated sites management process. This is especially relevant in the case of contaminated "megasites". The term "megasites" is used to indicate large (km² scale) contaminated areas or impacted areas, like industrial harbours, petrochemical districts and mining areas, characterized by unacceptable costs for complete clean-up due to political, economic, social or technical constraints since they usually involve multiple owners and stakeholders (Wolfgang and ter Meer, 2003; Rijnaarts and Wolfgang, 2003; WELCOME, 2004). Within this complex context, it is even more pressing for decision-makers and managers to find sustainable remediation solutions, by integrating all the involved aspects and by taking into consideration stakeholders' points of view. The development of technical tools, able to integrate the wide range of decisions related to contaminated land management and re-use, including environmental, technological and economic issues, is therefore critical (CLARINET 2002a).

An important category of these technical tools is represented by Decision Support Systems (DSSs), already proposed by the network CLARINET in 2002 (2002b). As illustrated in several parts of this book, Decision Support Systems can be generally defined as tools that support decision makers in structuring a decision problem and in defining options of intervention to solve the same problem (Jensen et al., 2002; Georgakakos, 2004; Salewicz and Nakayama, 2003).

A decision support system called DESYRE (DEcision Support sYstem for the REqualification of contaminated sites) was specifically developed to address the integrated management and remediation of contaminated megasites encompassing site characterization, risk assessment, intervention selection, risk-based prioritization of the remediation interventions and stakeholder involvement. The DSS objective is the creation of spatially resolved remediation scenarios and their comparison in terms of the residual risk, technological choices and socioeconomical benefits. DESYRE can be defined as a Spatial Decision Support System since it is able to: (1) integrate large volumes of geo-referenced heterogeneous information; (2) perform a spatially resolved environmental risk assessment; (3) allow the spatial evaluation of the technological aspects of the remediation process; and (4) allow decision makers, and more generally multiple stakeholders, to compare potential alternative scenarios in terms of spatially resolved environmental, socioeconomic, and technological benefits and constraints.

DESYRE was structured into six modules (*Socio-economic, Characterization, Risk Assessment, Technological Assessment, Residual Risk Assessment* and *Decision*) integrated into a GIS software platform. Since the six modules represent the main phases of contaminated sites management, which involve different expertise, the software users should be a multi-disciplinary team of experts, including risk assessors, socio-economists and technology engineers. DESYRE provides support to the experts along the several steps of the contaminated sites rehabilitation process allowing integration, spatial evaluation and analysis of available information by using GIS functions and a number of decision support tools implemented in the system. Finally, the DESYRE output enables decision makers to compare benefits, costs and impacts of different strategies and technical solutions.

The whole methodological background and the mathematical algorithms at the basis of each module are explained in the dedicated papers (Facchinetti et al., 2005; Critto et al., 2006; Carlon et al., 2007, 2008), while the focus of this Chapter is the presentation of a complete application of the DESYRE software. Therefore, this chapter discusses how the DESYRE DSS addresses the described objectives and presents how the information flows within the system. Moreover, the complete system application allows the understanding of the entire remediation process by the analysis of the input-output of all the modules and the connections between them.

8.2 Description of Framework and Functionalities

DESYRE is a totally new DSS, composed by original assessment modules fully integrated in the popular GIS platform ArcGIS 9.2 (ESRI) by means of friendly Visual Basic interfaces which also makes DESYRE an easy tool for users who are not GIS-trained. The software is structured into six modules (Fig. 8.1): *Socio-economic, Characterization, Risk Assessment, Technological Assessment, Residual Risk assessment* and *Decision* (Carlon et al., 2007) which can be accessed directly by the ArcGIS interface which activates the different software packages implemented within the system.

The *Socio-economic assessment module* supports the comparison of the possible future land uses of the site and the subsequent identification of the optimal future land use on the basis of socio-economical variables. These socio-economic variables are inputs of a fuzzy expert system which generates an index of suitability for six alternative land uses: (1) residential, (2) recreational,

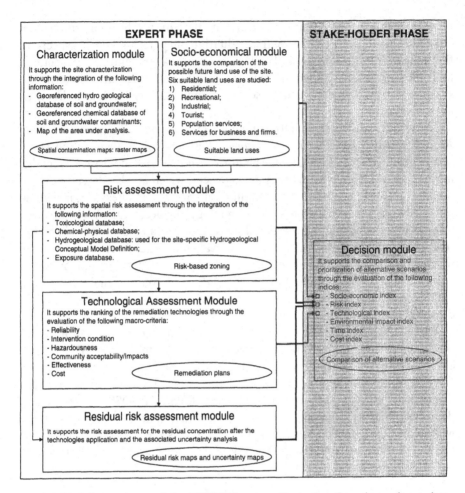

Fig. 8.1 Modular structure of the DESYRE system illustrating the primary input data requested, functionalities and, in *circles*, the main outputs of each module

(3) industrial, (4) tourist, (5) population services, (6) services for business and firms. The socio-economic variables include, among others, number of factories, traffic connections, distances in time from principal connections (i.e. 30 min. isochrones), traffic impacts, land value, population density, presence of alternative sites, etc. On the basis of a fuzzy tree structure, the socioeconomic variables are aggregated into four indicators for each land use: (1) the use demand at the local level, (2) the attraction of alternative sites for the same use, (3) the site attraction, and (4) the consistency of the use with the surrounding territory (vocation). These indicators are aggregated into a final numerical index, also termed land use suitability index, which is the first input of the decision module (Fig. 8.1). A detailed description of the fuzzy expert system is provided by Facchinetti et al. (2005). The fuzzy system was built by using the commercial software fuzzyTECH® (INFORM GmbH), while the calculation

sheet is in EXCEL Microsoft® and can be accessed directly by the ArcGIS interface.

The *Characterisation module* combines a GIS database of site maps and administrative data with a relational geo-referenced database (based on ACCESS Microsoft®) of contaminant concentrations in soil and groundwater, chemical-physical characteristics of contaminants, as well as geological and hydrogeological data. The module includes a system for selecting the contaminants of concern to be considered in the risk assessment. The spatial distribution of contaminants is analyzed by means of geostatistical methods, including variography and Kriging (Carlon et al., 2000).

The *Risk Assessment module* provides tools for human risk analysis of soil and groundwater contaminants and allows the risk-based zoning of the site accordingly. Input data are the future use of the site (from the Socio-economic module), maps of contaminants distribution in soil and groundwater, hydrogeological data and contaminant properties (from the Characterisation module) and exposure parameters. The latter are defined on the basis of the choices made in the definition of the exposure diagram, which is the first phase of the risk assessment module. In DESYRE, a procedure for the spatial resolution of the risk-based corrective action (RBCA) risk estimation algorithms (ASTM, 1998) was developed. According to this procedure, chemicals are gathered into six classes referring to the US-EPA classification system (FRTR, 2002), i.e.: Nonhalogenated Volatile Organic Compounds, Halogenated Volatile Organic Compounds, Nonhalogenated Semivolatile Organic Compounds, Halogenated Semivolatile Organic Compounds, Fuels, and Inorganics. These categories mainly depend on basic physico-chemical properties (e.g. water solubility, vapour pressure, bio-degradability) that heavily affect suitability and performance of various remediation interventions. Two contamination sources are considered in the risk analysis, i.e. soil and groundwater. Considered exposure pathways are the ingestion and dermal contact with soil and groundwater, and the inhalation of vapour and particulate emissions. Considered receptors are humans (on site) and groundwater/surface water quality (on site and/or off site). Risk based Acceptable Concentrations (ACs) are estimated by means of ASTM risk estimation algorithms for the different exposure routes related to a specific medium (i.e. soil or groundwater). The integration of the different risk based ACs is performed in order to estimate the Multi-pathway Acceptable Concentrations (MACs) in the soil medium and in the groundwater medium.

The contaminant concentration is interpolated by means of Kriging in the characterization module and, for each node of the interpolation grid, a Risk Factor (RF) is calculated as the quotient between the interpolated concentration and the estimated MAC. The result of this methodology is the spatial distribution of the Risk Factor and the risk based zoning of the contaminated site (Carlon et al., 2008). In DESYRE, the risk mapping is performed using the 6 main contaminant categories (following FRTR, 2002) instead of individual substances, in order to significantly reduce the number of risk maps and to make them more helpful for the formulation of suitable remediation plans.

The *Technological Assessment module* supports the selection of the best available technologies for the site and the formulation of remediation plans. It is based on a database of available technologies and requires risk maps from the Risk module as input. The Technology Selection sub-module provides the user with a database (ACCESS Microsoft®) of more than 60 types of treatment technologies described in terms of target contaminants, application features, range of applicability under different site-specific parameters and commercial availability. Out of the entire dataset, a restricted group of suitable technologies can be selected by applying criteria of applicability in the site specific conditions, such as types of contaminants to be treated, hydrogeological conditions, soil texture or commercial availability. The set of suitable technologies are ranked by means of Multi-Criteria Analysis which considers a set of macro-criteria, including cost, duration, performance, reliability, flexibility and public acceptability of technologies (Critto et al., 2006). On the basis of the technologies ranking, the user can formulate different technological scenarios, which will be evaluated through the use of appropriate indices in the Decision module. One of the outputs of the Technological Assessment module is the simulation of the remediation technologies application, which leads to the reduction of contaminant concentration in the analyzed media on the basis of the performance of the remediation plans. The maps of the residual concentration after the remediation application is an input of the *Residual Risk Assessment module* which provides the estimation of the residual risk after the remediation and the estimation of the associated uncertainty through the use of Monte Carlo probabilistic analysis (Carlon et al., 2008).

Finally, in the *Decision Module* the scenarios characterized by different remediation plans are described by means of a set of indices that enable decision makers to compare and prioritize alternative solutions.

While the first five assessment modules can be used by a team of technical assessors, the decision module is specifically addressed to stakeholders, such as public decision makers, planners, site owners, investors and associations.

8.3 Application of DESYRE to a Case Study

The DESYRE software was applied to a 150 ha sub area of a contaminated megasite of national interest: the Porto Marghera megasite. This megasite covers an area of 3595 ha, out of which 479 ha are occupied by canals. The site was originally a mudflat (so-called *barene*) that has been raised up 2 m above sea level by filling with material from the dredging of lagoon canals (Gatto and Carbognin, 1981) and waste production residues, including industrial toxic waste.

Below the topsoil, a 1st impermeable layer is found, consisting of Holocene deposits of lagoon mudflats and an overconsolidated silt clay layer called *caranto*. The 1st impermeable layer is followed by a semiconfined aquifer, delimited at the bottom by clayey Pleistocene sediments that constitute the deepest impermeable layer (Carbognin et al., 1972).

Chemical contamination was characterized extensively on a 100-m sampling grid. In the topsoil (i.e. the filled material layer), several classes of pollutants were found: amines, chlorobenzenes, chloronitrobenzenes, chlorophenols, dioxins, aliphatic hydrocarbons, polynuclear aromatic hydrocarbons, metals, metalloids, and inorganic anions (Venice City Council, 2001).

Metals and metalloids showed the highest concentration levels and the widest spread of contamination. The soil samples displayed high concentrations of arsenic, chromium, cadmium copper, mercury and lead.

The application of the DESYRE DSS aims at identifying the risks posed by the characterized contamination and at defining and comparing different remediation scenarios. The application follows the information flow described in Fig. 8.1 and is divided according to the six modules describe above. Since the spatial risk assessment and the following remediation technologies selection and allocation are complex modules which require a huge number of input data, in order to simplify the information flow description, the whole application will consider only one contaminants category (the Inorganics) of the six FRTR contaminants categories encoded in the system. Moreover, in order to further simplify the application, only two potential future land uses are considered: residential and industrial.

8.3.1 The Socio-Economic Module

In order to compare the two possible future land uses of the study area (i.e. residential and industrial) and to identify which one is the most suitable, the socio-economic module was applied. The socio-economic variables required by the methodology were collected for each land use scenario and the indicators and the final index were calculated by the use of the fuzzy expert system, described in Facchinetti and colleagues (2005). All the input and output data are reported in Table 8.1.

As illustrated by Table 8.1, according to the Socio-Economic module application, the most appropriate future land use of the site is the industrial one, having achieved the higher socio-economical score. According to this result, the application of the other modules to the study area, and in particular the subsequent risk assessment, will be based only on the industrial land use related exposure scenario.

8.3.2 Characterization Module

In the characterization module all the site-specific information concerning the hydrogeology of the site, the spatial distribution of the contamination and the chemical characteristics of the analyzed substances are collected and organized. In order to identify the substances, which have concentrations in soil above

Table 8.1 Application of the socio-economic module to the selected scenarios (residential and industrial): input variables, indicators and final index (n.a. = not applicable). For the explanation of the following table refer to Facchinetti et al., 2005

Variable	Unit of measure	Associated land use[a]	Residential scenario values	Industrial scenario values	Indicator	Residential scenario values	Industrial scenario values	Residential scenario final index	Industrial scenario final index
Number of factories	(–)	3,6	n.a	86192					
Born/death factories	(–)	6	n.a	n.a					
Number of inhabitants in the municipality	(–)	1, 2, 3, 5	270639	270639					
Number of inhabitants in the municipality at the age 25–40	(–)	1	41308	n.a	Demand for the use	n.a	1.00		
Number of inhabitants in the province	(–)	1, 2, 3, 5	814581	814581					
Saturation of the Hotels	(%)	4	n.a	n.a					
Mega stores already working	(m²)	5	n.a	n.a				0.36	0.67
Attractivity of alternative sites	(€)	1, 3, 4, 5, 6	5165	37.88	Attractivity of alternative sites	0.50	0.75		
mq recreational areas elsewhere	(m³)	2	n.a	n.a					
Isochrones 30 min	(–)	1, 2, 3, 4, 5, 6	81000	81000					
Land value after remediation	(€/m²)	1, 2, 3, 4, 5, 6	51.65	51.65	Site attractivity	0.92	1.00		
Distance from doors	(m)	3, 4, 5, 6	n.a	24.7					
Local priorities	(–)	1, 2, 3, 4, 5, 6	4	6	Context index (vocation)	0.00	0.33		
Impacts on traffic	(–)	3, 5, 6	n.a	130000					

[a]The associated land use column indicates the land use scenario to which the variable is associated. Land use scenarios are: (1) residential area; (2) recreational green area;

Table 8.2 Inorganic substances with concentration in soil above Italian regulatory threshold limits and the corresponding minimum, mean, maximum and regulatory concentration in soil

Analyzed substances	Min concentration in soil (mg/kg$_{dw}$)	Mean concentration in soil (mg/kg$_{dw}$)	Max concentration in soil (mg/kg$_{dw}$)	Regulatory limits for industrial land use (mg/kg$_{dw}$)
Arsenic	9.6	53.7	350	50
Cadmium	0.4	6.6	996	15
Chromium	11.1	46.5	4200	800
Manganese	234.1	359.8	3036	–
Mercury	0.1	7.9	130	5
Lead	14.9	190.5	1348	1000
Copper	12.7	83.5	3040	600
Vanadium	11.3	46.7	860	250
Zinc	19.3	630.7	3997	1500

regulatory threshold limits, and to obtain basic inference statistics, appropriate statistical tools were implemented. According to the statistical results, the inorganic substances above regulatory threshold limits are reported in Table 8.2 with the corresponding minimum, mean, maximum and regulatory concentration in soil.

For each of the substances reported in Table 8.2, the geostatistic analysis was performed in order to define the spatial distribution of the contamination in soil for the estimation of the spatial distribution of the risks posed by these inorganic substances.

8.3.3 Risk Assessment Module

On the basis of the most suitable future land use identified in the socio-economical module (i.e. industrial), in the Risk Assessment module, a spatial risk assessment methodology is applied to the substances which are present in soil concentrations above regulatory threshold limits (Table 8.2) and for which the spatial distribution of the contamination was performed in the Characterization module. The DESYRE interface of the risk assessment module is shown in Fig. 8.2.

The first step of the spatial risk assessment approach is the calculation of acceptable concentrations in soil which is based on the standard algorithms elaborated by the American Society for Testing and Material (ASTM, 1998). In accordance with these equations, the software requires many data inputs concerning the toxicity and chemical-physical properties of the analyzed substances, the target exposure parameters and the hydrogeological parameters needed for the fate and transport models. The application of the ASTM algorithms leads to the estimation of the Acceptable Concentrations in Soil (ACSs) for each of the analyzed receptor exposure routes checked in Fig. 8.2 for the soil pathway (i.e. ingestion of soil, dermal contact with soil, inhalation of vapor and

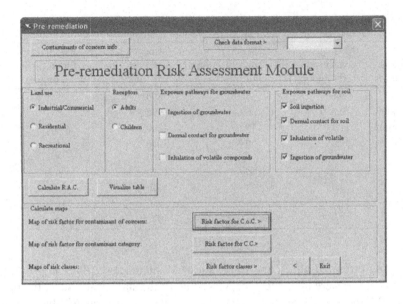

Fig. 8.2 DESYRE interface of the risk assessment module. R.A.C = Risk-based Acceptable Concentration. C.o.C = Contaminant of Concern. C.C. = Contaminants Category

particulate emissions from soil and ingestion of groundwater contaminated by the soil). According to the most suitable socio-economical scenario identified by the socio-economical module, the considered targets are humans employed in the industries.

The second step of the spatial risk assessment approach is the estimation of the Multi-pathway Acceptable Concentrations in the Soil medium (MACSs) which integrates the acceptable concentrations calculated for the different exposure routes related to a specific medium, as reported in the paper Carlon et al., 2008.

Table 8.3 reports the acceptable concentrations in soil for the considered exposure routes and the multi-pathway acceptable concentrations in soil. They are estimated for a reasonable worst case scenario, i.e. establishing the characteristics of the release and the exposure that provide the upper bound of a potential intake.

Finally, the proposed spatial risk methodology is based on the estimation of the Risk Factor (RF) for a specific medium (i.e. soil or groundwater). RF is defined as the ratio between the measured concentration in the analyzed medium and the acceptable concentration in that medium (Carlon et al., 2008).

The ratio between the measured concentration in soil and the acceptable concentration in soil is calculated for each cell of the raster concentration maps of the selected substances in order to obtain the raster maps of the spatial distribution of the Risk Factor. In Table 8.3 the site area (m^2 and %) where the contaminant concentration is above the acceptable level (i.e. > MACS) is reported for each of the analyzed contaminants. The percentage of the area

Table 8.3 The acceptable concentrations in soil for the considered exposure routes and the multi-pathway acceptable concentrations in soil for each of the analyzed substances. ACS_IS = Acceptable Concentration in Soil medium related to the ingestion pathway; ACS_DCS = Acceptable Concentration in Soil medium related to the dermal contact pathway; ACS_IVS = Acceptable Concentration in Soil medium related to the inhalation pathway; ACS_IGW = Acceptable Concentration in Soil medium related to leaching and water ingestion pathway; MACS = Multi-pathway Acceptable Concentration in Soil medium; EXCEEDING AREA = site area where the contaminant concentration is above the acceptable level (i.e. > MACS); % EXCEEDING AREA = % of the site where contaminant concentration is above the acceptable level. The acronyms C, NC and CNC are referred to Carcinogenic, Non Carcinogenic and both Carcinogenic – Non Carcinogenic contaminant toxicological effects, respectively

Contaminants belonging to the inorganic category	Toxicity	ACS_IS mg/kg_{dw}	ACS_DCS mg/kg_{dw}	ACS_IVS mg/kg_{dw}	ACS_IGW mg/kg_{dw}	MACS mg/kg_{dw}	Exceeding area m^2	% Exceeding area %
Arsenic	CNC	3.82E+00	2.47E+01	3.98E-01	9.29E-01	2.57E-01	1.438.125	100
Cadmium	CNC	1.02E+03	1.62E+02	3.26E+00	9.34E+01	3.08E+00	6.61.250	46
Chromium	CNC	7.84E+02	3.10E+02	4.74E-01	1.83E+01	4.61E-01	1.438.125	100
Manganese	NC	9.40E+04	5.95E+04	1.02E+02	7.45E+03	1.00E+02	1.438.125	100
Mercury	NC	6.13E+02	6.79E+02	1.50E+02	3.95E+01	2.85E+01	24.375	1.7
Lead	NC	7.15E+03	1.13E+05	2.49E+04	7.83E+03	3.16E+03	0	0
Copper	NC	1.02E+06	1.62E+07	3.55E+06	5.32E+05	3.12E+05	0	0
Vanadium	NC	1.43E+04	2.26E+03	4.97E+04	1.74E+04	1.70E+03	0	0
Zinc	NC	7.15E+03	1.94E+06	2.13E+06	4.64E+04	4.13E+04	0	0

where the contaminant concentration is above the acceptable level is estimated as the ratio between the number of cells with a concentration above the acceptable concentration and the total number of cells in the raster map.

As highlighted by Table 8.3, the contaminants which pose a significant risk to human health are the first five contaminants, since they present a high percentage of the site area where the concentration in soil is above the Multi-pathway Acceptable Concentration in Soil, while the concentration in soil related to the last 4 contaminants reported in Table 8.3 is below the Multi-pathway Acceptable Concentration in Soil in the entire site. Likewise, Table 8.3 shows that the concentration in soil for Arsenic, Chromium and Manganese is above the Multi-pathway Acceptable Concentration in Soil for the entire site under investigation. In fact in Table 8.2, the minimum concentration in soil of Arsenic, Chromium and Manganese is higher than the MACS for these substances as reported in Table 8.3.

With the intention of obtaining a single representation of the risks posed by the inorganic contaminants and finalizing the risk assessment results for the identification of suitable remediation interventions, a summarizing map of the Inorganic category is produced by assigning to each cell of the grid the maximum value of RF (RFmax) as calculated from the analyzed substances.

Although the raster map of RFmax represents the spatial distribution of the risk factor for the Inorganics, it does not provide a discrete zoning of the site and, more importantly, it does not retain relevant information about the risk characterisation, like the identification of most relevant contaminants, exposure pathways and impacted receptors which complement spatial features for the planning of interventions. In DESYRE, a process of vector transformation overcomes this limitation. RFmax values are divided into five classes: $RF \leq 1$, $1 < RF \leq 3$, $3 < RF \leq 10$, $10 < RF \leq 100$, $RF > 100$, respectively. In Fig. 8.3a, the spatial distribution of the RFmax classes for the Inorganics is reported, while in Fig. 8.3b, the substances, within the Inorganics category, which produce the RFmax are identified.

As shown by Fig. 8.3a, in almost the entire site area, the RFmax belongs to the higher class ($RF > 100$), while the remaining areas are characterized by an RFmax belonging to the class with RF included in the range $10 < RF \leq 100$. According to Fig. 8.3b the substance which produces the RFmax in almost the entire site is arsenic, replaced by chromium in small parts of the site area (the black areas). Combining the information reported in Fig. 8.3a and b with the information reported in Table 8.3 it is possible to underline that the most critical issue for the site is determined by the inhalation of particulate from arsenic and chromium contaminated soil followed by the ingestion of drinking water contaminated by arsenic. In fact the most critical substances are arsenic and chromium and for both of them the Acceptable Concentration in Soil medium related to the inhalation pathway (ACS_IVS) is the most significant exposure route (i.e. the lower AC in soil). Similar considerations can be made for the arsenic Acceptable Concentration in Soil medium related to leaching and water ingestion pathway (ACS_IGW). This information is useful for the experts who

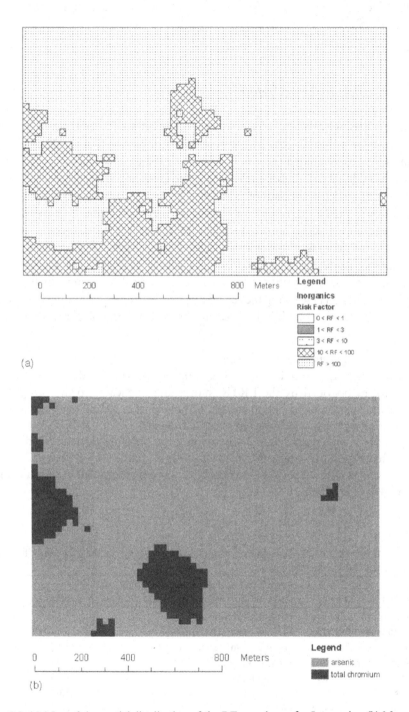

Fig. 8.3 (a) Map of the spatial distribution of the RFmax classes for Inorganics. **(b)** Map of the spatial distribution of the substances which produce the RFmax

will decide the remediation intervention and the risk based definition of the future use of the site.

8.3.4 Technological Module

In the technological module the definition of the different "remediation scenarios" is performed. Each scenario represents a suitable solution for the remediation of the contaminated site and is defined on the basis of the socio-economic module results, the risk assessment results, the identified technological solutions and the residual risk estimation. The first two aspects are already described in Sections 8.3.1 and 8.3.3, respectively, while the identified technological solutions and the residual risk estimation are explained in this section. All the steps for the selection and ranking of the suitable technologies, which can be applied to the Inorganics category, are fully described in Critto et al. (2006). The implemented methodology for the identification of the technological solution is divided into three main steps: (1) selecting the remediation technologies, (2) setting comparative criteria, and (3) ranking the selected remediation technologies with a comparative procedure.

(1) *Selection of remediation technologies:* to select the remediation technologies, two subsequent selection filters were applied to the input database of technologies provided by the FRTR matrix and applicable to the Inorganics category in order to obtain the list of remediation technologies applicable to the case study which are reported in Table 8.4.

(2) *Setting comparative criteria:* to compare the pool of selected remediation technologies, six comparative macrocriteria and a set of the evaluation criteria associated with each macrocriterion were identified as described in Critto et al., 2006.

Table 8.4 Ranking of the remediation technologies applicable to the case study for soil medium and inorganic contaminants

Typology of remediation treatment	Pool of technologies applicable to the case study for inorganic category and soil medium	Scoring	Rank
Treatment for soil sediments and sludge in situ	Phytoremediation	2,57	6
	Electrokinetic separation	3,64	2
	Solidification/stabilisation	3,58	3
Treatment for soil sediments and sludge ex situ	Separation	3,23	4
	Soil washing	3,85	1
	Solidification/stabilisation	3,07	5
Containment and other treatments	Landfill cap	2,02	7
	Landfill cap alternatives	1,96	8
	Excavation, retrieval and off-site disposal	1,43	9

(3) *Ranking the selected remediation technologies with a comparative procedure*:
the comparative process uses a Multi-Criteria Decision Analysis (MCDA)
algorithm that is based on the weighted averaging operator associated with
the absolute AHP to structure the problem into a suitable hierarchy and to
determine the criteria weights. The MCDA procedure is divided into the
following steps described in detail by Critto et al. (2006):

1. the definition of the qualitative and quantitative rating of each evaluation criterion;
2. the definition of the evaluation matrix for the Inorganics category applied
   by the expert for the macro-criteria judgement;
3. the evaluation of the matrix of pairwise comparisons applied to the
   calculation of macrocriteria weights;
4. the definition of the judgment matrix for the evaluation of the remediation technologies selected for inorganic contaminants.

The integration of all these steps finally leads to the ranking of the remediation technologies applicable to the case study which is reported in Table 8.4.

Once the remediation technologies are ranked, the system allows the user to
define different remediation scenarios which are characterised by the application
of one or a chain of remediation technologies to the analysed site. In this case
study, two different remediation scenarios are compared and reported in Table 8.5.

The simulated application of the selected technologies leads to the creation
of the residual risk maps which are reported in Fig. 8.4a and b.

The remediation technologies' performances used for the simulation of the
remediation technologies' application are those reported in Table 8.5. Based on
the estimation of post-remediation contaminant concentration, the residual risk
factor maps show that the residual risk associated with Inorganics is significantly reduced but not to acceptable levels since many areas in Fig. 8.4a and b
belong to the third RF class ($3 < FR < 10$) and no area belongs to the first RF
class ($FR < 1$). Focusing on the comparison of the two residual risk factor
maps, Fig. 8.4a shows some areas in the North East part of the site where
the Risk Factor falls in the fourth class ($10 < RF < 100$), while in Fig. 8.4b
the highest Risk Factor class is represented by the third class ($3 < FR < 10$) and
the West part of the site belongs to the second RF class ($1 < RF < 3$). As a
consequence, it is possible to state that the application of the train technology
defined in the Scenario 2 leads to a better result as far as risk reduction is
concerned. Finally, due to the limitation of any remediation intervention on
Inorganics, the risk management of the area would likely include the control of
the exposure route which also influences the risk estimation. As emphasized in
the Risk Assessment module (Section 8.3.3), the exposure route of most concern
is the inhalation of suspended soil particles which can be limited by an adequate
impermeable clay layer or paving. The relevance of this route of exposure also
suggests that further investigations and modelling refinement of fine particles
contamination, suspension and inhalation should be performed in order to
avoid excessive risk overestimations.

Table 8.5 Remediation scenarios created for the case study area. For each remediation scenario, the applied technologies with associated substance performance are reported. NAV = not available

Remediation scenario name	Applied technologies	Performance (%)									
		Arsenic	Cadmium	Chromium	Manganese	Mercury	Lead	Copper	Vanadium	Zinc
Scenario 1	Soil washing	96	NAV	93	NAV	NAV	91	87	NAV	98
Scenario 2	Electrokinetic separation	88	93	90	NAV	NAV	89	85	NAV	73
	Solidification/ Stabilization ex situ	99	98	99	NAV	98	98	97	NAV	NAV

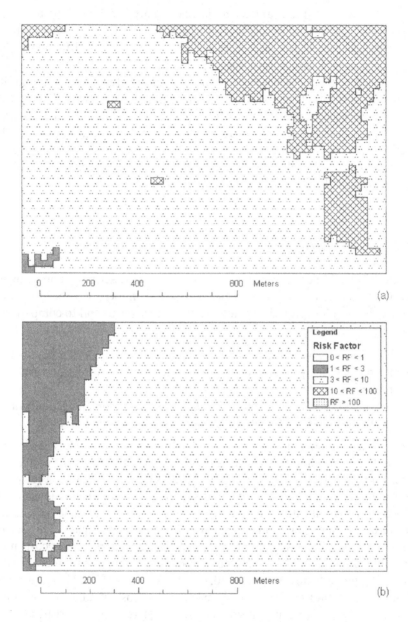

Fig. 8.4 Residual risk map obtained through the simulated application of (**a**) soil washing remediation technology (i.e. Scenario 1) and (**b**) a technology train composed of the Electrokinetic separation and the solidification/stabilisation ex situ (i.e. Scenario 2)

In the residual risk estimation module, the DESYRE software supports the probabilistic estimation of the risk factor, which also provides an indication of the propagation of the uncertainty of the input values into the risk estimates. The residual risk probabilistic assessment is based on the Monte Carlo analysis.

This technique is used to characterize the uncertainty and variability in the risk estimates by the estimation of the 50th and 5th percentiles of the acceptable concentrations in soil for the considered exposure routes and for the multi-pathway acceptable concentrations in soil for each of the analyzed substances. An uncertainty index is calculated as the difference between the 50th and 5th percentiles of the multi-pathway acceptable concentrations in soil. The system can generate raster and vector maps of the risk factor spatial distribution as well as the uncertainty indices spatial distribution. For simplification reasons, the description of the DESYRE probabilistic risk assessment module is not given in this chapter. Further details of the probabilistic risk assessment methodology applied in DESYRE are provided by Carlon et al. (2008).

8.3.5 Decision Module

The Decision module provides a methodology for the description and comparison of alternative remediation scenarios by the definition of a set of indices. The decision module procedure allows the user to describe and to compare the following scenario aspects:

- the post remediation site use and related socio-economic benefits;
- the remediation plan and related costs, time of interventions, performance reliabilities and environmental impacts;
- the reduction of the risk posed by contaminants in soil and groundwater, resulting from the simulated application of the remediation plan.

The socio-economic, technological and risk assessment modules generate a set of indices describing relevant features of each scenario (in terms of average values for the whole site): socioeconomic benefits for the selected post remediation land use, technological and logistical quality of the technological set, residual risk (spatial extension, average magnitude and magnitude reduction), total cost and duration of interventions, environmental impact. This set of indices identifies advantages and drawbacks of each scenario: lower costs may be combined with longer intervention periods for the rehabilitation of the site; high treatment performances may lead, especially in case of large contaminated volumes, to relevant environmental impacts; the most suitable site use may require very strict risk minimization targets and high remediation costs. Specific mathematical algorithms are used to calculate the indices reported in Table 8.6. The table presents (in the second and third columns) the indices name and description, in the fourth the reference scale, and in the fifth and sixth the values of each index for Scenario 1 and Scenario 2, respectively.

The reference scales are usually between 0 and 1, where 1 represents the best index value. For the residual risk magnitude, the scale varies instead from a worse value of 5 (i.e. highest class of residual risk) to a best value of 0 (i.e. lowest class of residual risk). The environmental impact index scale varies from a worse

Table 8.6 Decision Module Indices description and related values estimation for the case study area

Index number	Indices name	Indices description	Indices scale	Indices value	
				Scenario 1 — Application of the Soil washing remediation technology	Scenario 2 — Application of the Electrokinetic separation and Solidification/Stabilization ex situ remediation technologies
1	Residual risk extension	represents the areas of the site where risk is unacceptable after the remediation technology application	1–0	1	1
2	Residual risk magnitude	represents the averaged non-acceptable residual risk class at the site after remediation	5–0	3.22	2.88
3	Risk magnitude reduction	represents the benefit of reduction of risk magnitude	0–1	0.33	0.40
4	Technological set quality	measures the technological set value on the basis of experts ranking and the area of technologies application	0–1	1.00	0.67
5	Logistical set quality	measures the advantage derived by the application of a low number of technologies	0–1	1.00	0.50
6	Socio-economical	expresses the attractivity of a specific land use	0–1	0.67 (industrial land use)	0.67 (industrial land use)
7	Environmental impact	represents the expert judgment which expresses the community acceptability of the impacts caused by the analyzed technological set, in consideration of the areas of application of each technology	5–1	2.00	2.00
8	Cost	expresses the costs of the technologies in the considered set	million Euro	238	320
9	Time	expresses the time of the remediation process, provided by expert judgment	years	5	5

value of 5 (i.e. the maximum technological impact derived by the assignment to a considered technology of the worse judgement by experts) to a best value of 1 (i.e. minumum technological impact derived by the assignment to a considered technology of the best judgement by experts). Finally, the residual risk extension index scale varies from a worse value of 1 (i.e. all the areas have non acceptable residual risk) to a best value of 0 (i.e. no areas with non acceptable risk remains after remediation).

As shown in Table 8.6, the two scenarios do not differ for the residual risk extension, scored 1 in both of them, meaning that there is still unacceptable risk in all the cells composing the site map, and for the socio-economic index, since the industrial land use is the most suitable one for the area. As far as the environmental impact index is concerned, the index values reveal an equivalence between the application to the whole area of the soil washing technology, which has a community acceptability/impacts score defined by expert judgment equal to 2, in scenario 1, and the application to the same area of two remediation technologies (Electrokinetic separation and Solidification/Stabilization ex situ).

On the other hand, scenario 2 performs better in terms of risk magnitude (2.88 versus 3.22 of scenario 1), meaning that there is a lower average unacceptable residual risk in the area. In fact the residual risk magnitude index can range between 0 and 5, where 0 represents the class with an acceptable risk factor ($RF<1$) and 5 represents the class with the higher residual risk factor ($RF > 100$). The risk magnitude reduction is again in favor of scenario 2 (0.40 versus 0.33 of scenario 1), since scenario 2 is more effective in reducing risk levels in the considered area of remediation technologies application.

As far as the technological indices are concerned, scenario 1 performs better in terms of technological set quality and logistical set quality. In fact, the technological set quality values (1 versus 0.67 of scenario 2) are derived by the application to the whole site area of a remediation technology which has a high remediation technologies ranking score (first position in Table 8.4), versus the application to the whole site area of two technologies which have a lower remediation technologies ranking scores (second and fifth positions in Table 8.4). The logistical set quality values (1 versus 0.5 of scenario 2) are justified by the inclusion of only one technology in the set of scenario 1 versus the two technologies of scenario 2.

In conclusion, the indices results for the analyzed scenarios show that the two scenarios have comparable performances, characterized, in a relative sense, by both positive and negative issues, so that neither scenario is significantly better that the other. For instance, scenario 1 performs better when considering the indices of the remediation technologies option and the lower costs of intervention (238 million Euro versus 320 of scenario 2), while scenario 2 performs better when considering the indices of the residual risk evaluation. Therefore, the choice of the best remediation technologies solution depends on the decision makers' goals (i.e. the residual risk reduction, the technological quality, the costs etc.) and should be discussed by decision-makers and stakeholders on the basis of the advantages and limits of each scenario, weighted with preferences and objectives of those involved.

8.4 Conclusion

With respect to other existing DSS software, DESYRE offers a comprehensive consideration of almost all the assessment and decisional phases implied in contaminated site remediation. From this perspective, the development of comprehensive and integrated systems like DESYRE allows for a more coherent formalization of the problem with respect to the combination of various tools originally designed for different scopes and phases of the remediation process. Moreover, some methodologies included in DESYRE have strong aspects of originality, e.g. the socio-economic analysis of benefits based on fuzzy analysis, the residual risk simulation in support of the spatial allocation of technologies and the probabilistic analysis of risk distribution.

Among the future developments of the software, the adoption of an MCDA methodology that allows users to integrate the results of the indices calculation into one single index for each scenario is under evaluation. This solution would allow users to technically take into consideration, in the assessment of scenarios scores, the stakeholders' preferences and weights of the different indices, as well as to provide one condensed value per scenario, which should facilitate comparisons.

Acknowledgments The DESYRE project was funded by the Italian Ministry of University and Scientific Research and developed by the Venice Research Consortium with the support of University Cà Foscari of Venice in 2003. The authors are grateful to the wide group of scientists involved in the development of specific modules and functions, in particular a special thank you goes to Claudio Carlon (project management), Manuela Samiolo and Gian Antonio Petruzzelli (technology selection), Nadia Nadal (risk assessment), Stefano Foramiti and Luca Dentone (GIS application), Stefano Soriani, Ilda Mannino, Gisella Facchinetti, Antonio Mastroleo, Silvio Giove and Stefano Silvoni (multicriteria analysis and fuzzy analysis). The authors are also grateful to INSIEL spa which provided a valuable technical contribution to the software implementation, allowing the transformation of the initial research prototype into a consolidated and commercial product.

References

ASTM, American Society for Testing and Materials Standard (1998) Provisional Guide for Risk-Based Corrective Action, Philadelphia (PA), USA: ASTM. PS 104–98.

Carbognin L, Gatto P, Mozzi G (1972) Hydrogeologic aspects of Venice Lagoon Venice (Italy): National Research Council (CNR. Lab St.Din.Gr.- Ma.), TR32 Final Report, (In Italian.)

Carlon C, Critto A, Nathanail P, Marcomini A (2000) Risk based characterization of a contaminated industrial site using multivariate and geostatistical tools. Environ Poll 111, 3, pp 417–427.

Carlon C, Critto A, Ramieri E, Marcomini A (2007) DESYRE: Decision support system for the rehabilitation of contaminated megasites. Integr Environ Assess Manag, 3, pp 211–222.

Carlon C, Pizzol L, Critto A, Marcomini A (2008) A spatial risk assessment methodology to support the remediation of contaminated land. Environment International 34, 3, pp 397–411.

CLARINET, Contaminated Land Rehabilitation Network for Environmental Technologies. (2002a) Sustainable Management of Contaminated Land: An Overview. A report from the Contaminated Land Rehabilitation Network for Environmental Technologies, p 128.

CLARINET, Contaminated Land Rehabilitation Network for Environmental Technologies. (2002b) Review of Decision Support tools for Contaminated Land Management, and Their Use in Europe. A Report from the Contaminated Land Rehabilitation Network for Environmental Technologies, p 180.

Critto A, Cantarella L, Carlon C, Giove S, Petruzzelli G, Marcomini A (2006) Decision support-oriented selection of remediation technologies to rehabilitate contaminated sites. Integr Environ Assess Manag 3, pp 273–285.

Facchinetti G, Mastroleo G, Mannino I, Soriani S (2005) A fuzzy expert system for the socio-economic analysis of contaminated sites. In: Pejas J., Piegat A., editors. Enhanced Methods in Computer Security, Biometric and Artificial Intelligence System. New York (USA): Springer-Verlag, p 396.

FRTR, Federal Remediation Technology Roundtable. Remediation Technologies Screening Matrix and Reference Guide, 4th ed; 2002. (www.frtr.gov/matrix2/, visited March 2006).

Gatto P, Carbognin L (1981) The lagoon of Venice: Natural environmental trend and man-induced modification. Hydrol Sci Bull 26, pp 379–391.

Georgakakos A (2004) Decision support systems for integrated water resources management with an application to the Nile basin. In Proceeding of the IFAC workshop Modelling and Control for participatory planning and managing water systems, Venice, September 29th–October 1st, 2004.

Jensen R A, Krejcik J, Malmgren-Hansen A, Vanecek S, Havnoe K and Knudsen J (2002) River Basin Modelling in the Czech Republic to optimize interventions necessary to meet the EU environmental standards. In Proceedings of the International Conference of Basin Organizations, Madrid 4–6 November 2002, DH ref 39/02.

Rijnaarts H, Wolfgang C (2003) Development of an Integrated Management System for Prevention and Reduction of Pollution of Water bodies at Contaminated Industrial Mega-sites. Annokkée, G.J, Arendt, F., Uhlmann, O. (Ed.) Consoil 2003 Conference Proceedings, pp. 45–49. Forschungszentrum Karlsruhe publishing, Karlsruhe.

Salewicz K A, Nakayama M (2003) Development of a web-based decision support system (DSS) for managing large international rivers. Global Environ Change 14, 25–37.

Venice City Council (2001) Knowledge framework to write out the master plan for the rehabilitation of Porto Marghera contaminated site, according to the DPCM 12.02.99. Venice (IT): Council of Venice, Environmental Planning Department. (In Italian.) p 20.

WELCOME, Water, Ecology, Landscape at contaminated Megasites (2004) Project website, http://www.euwelcome.nl/

Wolfgang C, ter Meer J (2003) Boundary conditions as parts of an integrated management system for contaminated megasites. In Annokkée G J, Arendt F, Uhlmann O (Eds.) Consoil 2003 Conference Proceedings, pp 45–49. Forschungszentrum Karlsruhe publishing, Karlsruhe.

Chapter 9
SMARTe: An MCDA Approach to Revitalize Communities and Restore the Environment

Ann Vega, Roger Argus, Tom Stockton, Paul Black, Kelly Black and Neil Stiber

Abstract SMARTe (Sustainable Management Approaches and Revitalization Tools – electronic) is a free, open-source, web-based, decision-support system that helps revitalization stakeholders (communities, developers, regulators, etc.) overcome obstacles to revitalization and allows them to develop and evaluate future reuse scenarios for potentially contaminated sites (e.g., brownfields). SMARTe currently has four primary components:

- eDocument: Through a series of menus (i.e., table of contents), SMARTe provides information, resources, links and case studies for all aspects of revitalization including visioning/planning, environmental risk management, social acceptance, and economic viability.
- Tool Box: SMARTe contains stand-alone analysis tools and checklists for many aspects of revitalization including human health risk, site characterization and monitoring data analysis, net revenues, public involvement, and selecting a lawyer, environmental consultant, and developer.
- My Project: A project-specific, password protected portion of SMARTe contains an integrated decision support system that helps users evaluate and assess both market and non-market costs and benefits of potential reuse options (i.e., future alternatives). This part of SMARTe also enables users to develop a revitalization plan.
- Search Engine: A search bar available on the SMARTe banner enables users to directly search for specific information within the system.

SMARTe ultimately helps communities preserve greenspace and natural systems since it facilitates revitalization of previously used sites, thus preserving previously undeveloped land.

SMARTe is primarily intended for community members who are trying to learn about the revitalization process, improve their communities and

A. Vega (✉)
US Environmental Protection Agency, 26 W. Martin Luther King Dr., Cincinnati,
OH 45268, USA
e-mail: Vega.ann@epa.gov

A. Marcomini et al. (eds.), *Decision Support Systems for Risk-Based Management of Contaminated Sites*, DOI 10.1007/978-0-387-09722-0_9,
© Springer Science+Business Media, LLC 2009

encourage change. Other stakeholders, however (such as regulators, redevelopment agencies, local governments, developers, property owners, etc.), also benefit from SMARTe and are the primary users of the Tool Box and My Project components. SMARTe is available at smarte.org.

This chapter will describe the framework, functionalities, and structure of the system, the design aspects including stakeholder involvement, and a case study to demonstrate the application of SMARTe to a specific site.

9.1 Introduction

Revitalization of sites potentially contaminated with environmentally toxic or hazardous materials (e.g., brownfields) is a global concern requiring a multi-disciplinary approach to mitigate the risks to human health and the environment. Many countries have committed extensive resources to address environmental, social, and economic issues related to the cleanup and redevelopment of contaminated sites. The challenge is to determine how to capitalize on the available resources, expertise, and knowledge to effectively share and transfer that information to the organizations and individuals responsible for making decisions and implementing redevelopment.

In 1990, the United States Environmental Protection Agency (USEPA) entered into a cooperation with the German Federal Ministry of Education and Research (Bundesministerium Für Bildung und Forschung [BMBF]). The original purpose of the cooperation was to understand the other's approach to the cleanup of chemical contamination in order to protect human health and the environment. Over the years, the cooperation expanded to include other organizations in the US and Germany and became known as the US-German Bilateral Working Group (BWG). The purpose of the cooperation also expanded to include revitalization (and in 2006, the beginnings of sustainable land use planning) in addition to remediation. Phase 1 and Phase 2 activities under the BWG were completed between 1990 and 2000 and included demonstrations of 20 innovative remedial technologies. The U.S.-German partnership completed Phase 3 (2000–2005) with a focus on providing a variety of tools, approaches, and technologies that could facilitate streamlined, cost-effective cleanup and revitalization of potentially contaminated sites (for example, Brownfield sites). The Interstate Technology and Regulatory Council (ITRC), a U.S. state-led organization, was also a significant partner in the third phase activities. Phase 4 (2006–2010) will continue the Phase 3 activities with an emphasis on providing tools, approaches and technologies to facilitate planning land use sustainably. The primary tool being developed in the U.S., begun in Phase 3 and continuing in Phase 4, is called Sustainable Management Approaches and Revitalization Tools – electronic (SMARTe) and is the subject of this chapter.

SMARTe is a free, open-source, web-based, decision-support system that is being developed to help revitalize communities and restore the environment. SMARTe contains resources and analysis tools for all aspects of the revitalization process: visioning (future land use), stakeholder involvement (including communities), economic viability (financing, market costs and benefits), environmental issues (site assessment, risk assessment and risk management), liability, and community benefits. It is a holistic decision analysis system that integrates these aspects of revitalization while facilitating communication and discussion among all stakeholders. It can be directly accessed at smarte.org.

The objectives of SMARTe are to:

- Educate the general public about the revitalization process;
- Help revitalization stakeholders (communities, developers, regulators, etc.) overcome obstacles to revitalization;
- Allow revitalization stakeholders to jointly develop and evaluate future reuse scenarios for potentially contaminated sites (e.g., brownfields) and develop a revitalization plan;
- Promote sustainable revitalization;
- Provide a complete project management system for revitalization to enable stakeholders to interactively analyze their situation at the same time they are given guidance, interpretation, and explanation of the analysis – promoting communication and understanding between and among stakeholders;
- Use open-source software so that is available to the general public at no cost.

In order to receive direct feedback from users in a timely manner, SMARTe is built in "levels" and released to the public as they are built. The first level of development, for each topic in SMARTe, consists of relatively simple textual information and access to resources and databases. The second level of development consists of stand-alone analysis tools (both technical and non-technical) that support each topic in SMARTe. Finally, the third level consists of integrating all information, databases and analysis tools using a Multi-Criteria Decision Analysis engine.

The first level of SMARTe development attempts to bring content, information and resources from multiple sources (EPA and non-EPA) together in one place so that: all aspects of sustainable revitalization are identified and considered, information is made available for practitioners and stakeholders on each aspect of the sustainable revitalization process, and the decision support system is structured to accommodate adding analysis tools that will enhance the capabilities of SMARTe.

The second level involves building stand-alone analysis tools to support each of the topics that are included in the SMARTe structure in the first level. For example, environmental topics are supported by analysis tools for sampling design, site characterization and monitoring, human health risk assessment, and fate and transport modeling. The social topic is supported by a community involvement tool that enables the user to select different community involvement techniques for his/her specific situation. The economic topic is supported by an economic model that tracks market (and eventually non-market) costs and benefits. In this second level, the economic evaluation consists of a spreadsheet algorithm that compiles the market costs and benefits of a revitalization option.

In the third level, all the tools are brought together in an integrated Multi-Criteria Decision Analysis (MCDA). This includes an expert system that helps the user sort through the many different options, costs and benefits that could be considered, and an MCDA integrator that captures the costs and benefits of a completed user-supplied application.

Some other elements also contribute to SMARTe's functionality. These include Quality Assurance (QA) aspects such as review and feedback, navigation, access to supporting information, an interface to each SMARTe topic, search capabilities, and a complex database structure to house input and information provided during a SMARTe application.

Feedback from the user community and new information received through workshops, the open literature, participation in national and international conferences, experts, and review comments are incorporated into SMARTe on an annual basis. SMARTe also is peer reviewed and reviewed by quality assurance personnel on an annual basis. A new version of SMARTe is released each October at smarte.org in order to keep information current. For example, in October 2007, SMARTe 2008 was released.

SMARTe 2008 is complete at Level 1. Level 2 contains multiple stand-alone analysis tools and checklists for many aspects of sustainable land revitalization. It is desirable to add more tools and checklists to assist stakeholders with additional aspects of sustainable land revitalization. Level 3 is available in a simplistic form. It is desirable to add greater complexity in the future to further enhance SMARTe's decision analysis capability.

By combining access to information and data with environmental risk, social acceptance and economic analysis tools, SMARTe enhances the decision making process and helps stakeholders overcome obstacles to revitalization. By providing potential solutions for sites where many obstacles and few benefits are perceived (that is, facilitating the sustainable reuse of contaminated land), SMARTe promotes sustainable revitalization. This will help preserve precious greenspace and revitalize blighted areas.

9.2 Description of Framework and Functionalities

The SMARTe framework is structured as dynamic, interactive on-line documents, tools, checklists and an expert system that serve as interfaces to and integrators of information, database searches and technical analysis tools. The on-line documents and checklists provide detailed information on specific topics. The tools provide both technical and non-technical analysis of data and information. Implementing and integrating the tools provides stakeholders with a decision analysis system that allows them to develop various revitalization alternatives. Using an iterative approach with extensive communication and discussion, stakeholders can use SMARTe to review and evaluate optimal revitalization options and develop a revitalization plan.

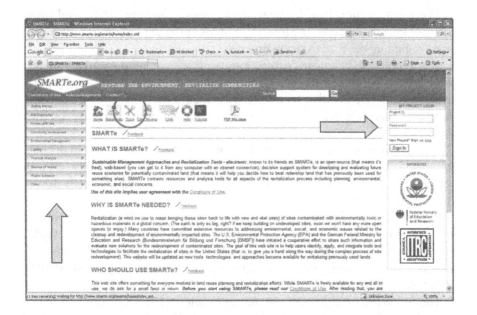

Fig. 9.1 SMARTe 2008 Home Page (http://smarte.org/smarte/home/index.xml). The *left-most arrow* points to the content "table of contents." Content is accessible by clicking on menu items. The *red circle* surrounds the "tools" icon. Click on the tools icon to access analysis tools and checklists. The *right-most arrow* points to the My Project Login area. Users can develop and access their project-specific options for revitalization and use SMARTe's decision analysis capabilities

The framework is comprised of content, tools and checklists, and the My Project decision analysis system. These are shown in Fig. 9.1 and briefly described in the following subsections.

9.2.1 Content

The content of SMARTe consists of a series of topics and subtopics accessed through a side-bar menu (Fig. 9.1).

9.2.1.1 Getting Started

When developing a revitalization plan strategy it is important to address, early in the revitalization effort, the variety of complex issues that may arise. The Getting Started section of SMARTe provides information for beginning the site revitalization process, including: Revitalization Plan Strategy, Project Stake-holders, Timeline of Events, Case Studies, Previous Site Use Scenarios, and Environmental Stigma.

The Revitalization Plan Strategy section simply identifies the broad categories of information (i.e., the topics in SMARTe) which should be considered when beginning the revitalization process.

The Project Stakeholders subsection identifies potential members of the project team who need to be a cohesive group that can work as a group, as well as independently. In addition to identifying the individual team members, this subsection outlines their roles and responsibilities on the project.

The Timeline of Events subsection summarizes the processes for managing the revitalization project schedule. It identifies key project milestones, critical path activities, and support activities required in order to prevent costly delays and project failure.

Several Case Studies are available through this section (as well as from the icon across the top of every page) enabling users to review what other revitalization practitioners have done in the broad areas of social aspects, economic aspects, environmental aspects and cultural heritage preservation. Case studies are available from the US as well as Germany.

The following previous site use scenarios within SMARTe provide background site information for specific types of sites:

- Rural Community Sites
- Mine-Scarred Land Sites
- Methamphetamine Sites
- Landfills and Junkyards
- Gas Stations
- Federal Facility Sites
- Superfund Sites
- Resource Conservation and Recovery Act (RCRA) Sites
- Railfields
- Dry Cleaner Sites
- Pulp and Paper Mill Sites
- Automotive Recycling Sites
- Iron and Steel Mill Sites
- Metal Finishing Sites

The Environmental Stigma subsection includes site-specific and regional approaches to removing environmental stigma through a variety of means.

9.2.1.2 Site Description

Developing, understanding, and maintaining a good description of the site is an important step in the revitalization process. A site description consists not only of a legal description, site background, history, and existing conditions, but also a detailed description of the proposed revitalization project.

9.2.1.3 Future Land Use

The Future Land Use section of SMARTe addresses the following future land use considerations in the revitalization process: Project Vision, Previous Plans, Revitalization Motivation, Regional and Local Needs, Marketing the Project Benefits, Sustainable Practices, Keys to Success, Innovative Project Features, and Construction and Demolition.

An attractive project vision of the future land use should be developed by the community and other stakeholders. A shared project vision will ultimately lead to a successful reuse of the site in terms of environmental, economic, and social benefits and is key to obtaining support from all stakeholders. The development of the project vision is critical to the success of the revitalization project.

In considering revitalization options the stakeholders should review previous plans for the site, such as previous environmental work plans or Phase I and Phase II environmental site assessments.

The Revitalization Motivation subsection identifies the motivators or drivers for site revitalization that could determine the course of the revitalization process. Drivers that should be addressed include:

- Ecological and human health
- Social
- Economic
- Political

In the Regional and Local Needs subsection, local community and regional needs that should be considered are presented. Examples include:

- Local Political Considerations
- Land Reuse Issues in the Urban and Local Context
- Environmental Justice
- Quality of Life
- Community Infrastructure Improvement
- Cultural Heritage and Historical Preservation

Effective marketing of the project to obtain support from critical stakeholders is important to the project's success. The stakeholders must present a focused program to market the advantages of the project. Marketing tools can include public advertising and individual meetings with stakeholder groups both to garner their input and support for a planned project.

Sustainable revitalization is a holistic approach that considers more efficient use of social, economic, and environmental resources. The Sustainable Practices subsection describes how identifying sustainable practices, applicable to the revitalization effort, may increase the probability of social, economic, and ecological sustainability.

The Keys to Success subsection provides guidance to the developer on essential stakeholder involvement, barriers that need to be addressed, and the schedule and timing required to overcome obstacles.

Site revitalization provides an opportunity for innovation. The Innovative Project Features subsection describes innovative approaches to cleanup and reuse of former industrial/commercial properties.

Construction and Demolition describes how the construction planning process may be complicated and can quickly become prohibitive to success when revitalization initiatives are added to environmental contamination, infrastructure, community opinion, financing, and other factors associated with site revitalization. The subsection provides information to help overcome these and other complications.

9.2.1.4 Community Involvement

The Community Involvement section discusses performing a Community Assessment to better understand community needs, problems, population, and potential impacts of revitalization. Strategies for setting up effective community participation programs to improve the public's access to information on revitalization are discussed in the Community Involvement Techniques section. The Community Involvement Techniques tool is described in the Tools and Checklists summary below.

9.2.1.5 Environmental Management

The Environmental Management section covers the following topics associated with site cleanup and regulatory negotiation: site assessment, assessing risk, communicating risk, managing risk, and public health.

The Site Assessment subsection covers the EPA Brownfields Roadmap, infrastructure considerations, All Appropriate Inquiry, and the environmental schedule for the revitalization effort. It emphasizes the importance of the environmental site assessment prior to the revitalization effort.

The Assessing Risk subsection discusses the preparation of an investigation of on-site hazardous substances as well as an assessment of the human health and ecological risks that may result from those substances, and the potential broader health impacts of concern that a revitalization effort might present to the community. In addition to discussing the preparation of an investigation, this subsection addresses the following four processes of quantitative risk assessment:

- Data Collection and Evaluation
- Exposure Assessment
- Toxicity Assessment
- Risk Characterization

The Communicating Risk subsection provides some helpful information for explaining, to potentially affected parties, the risk of exposure to a contamination source.

The Managing Risk subsection includes approaches to identification of alternative remediation actions at a site, risk-based corrective action, potential remediation options for site revitalization, and long term stewardship solutions to prevent exposure to contamination left in place at the site.

The Public Heath subsection gives a very limited overview of, and links for, public health concerns related to a potentially contaminated site – beyond the obvious contamination concerns.

9.2.1.6 Liability

This section introduces liability risk concerns, regulatory liability, third party liability, and environmental insurance as they relate to the revitalization effort.

A property owner faces liability risks in two broad categories: regulatory liability (federal and state agencies) and third party liability (cost recovery by private parties). The Regulatory Liability subsection presents various tools to address regulatory liability related to the purchase, environmental condition, and financial status of revitalization properties. The Third Party Liability section describes how a redeveloper (and perhaps other associated financial stakeholders) may obtain appropriate releases from possible liability through federal and state government officials. Such releases or clarifications may be incorporated into prospective purchase agreements, a covenant not to sue, no further action letters, and comfort letters.

The Environmental Insurance subsection describes how insurance can be used as a tool to overcome potential environmental liability problems in transactions. Policies such as cost cap, pollution liability, secured creditor, and insurance improvements are discussed.

9.2.1.7 Financial Analysis

The Financial Analysis section discusses the economics associated with site revitalization including: financial management and controls, market analysis, economic risk analysis, estimating economic viability, lender issues, investor issues, key financial indicators, information and advisory services, and long term economic impacts.

Revitalization projects are, by nature, unpredictable; therefore, financial management tools are used to increase the potential for success. There are a variety of tools available to the project team to monitor and control costs during the project to prevent cost overruns.

The Market Analysis subsection describes perceived or real environmental impairment costs to be considered in addition to the typical real estate costs.

The Economic Risk Analysis subsection presents the six main variables associated with income-producing properties: net operating income, debt coverage ratio, cap rate, break even ratio, cash on cash return, and loan to value.

There are many tools for estimating the economic viability of public and private real estate revitalization opportunities, most of which can be used in the case of potentially contaminated sites; however, there are few tools for specifically evaluating the economic viability of site revitalization. The Estimating

Economic Viability subsection provides some example tools that can be useful in evaluating the economic viability.

The Lender and Investor Issues subsections discuss the issues for both parties related to revitalizing a potentially contaminated property, including elements to include in a project finance plan and using an analytical market-profiling approach to determine whether a property is positioned for its highest and best use.

The Key Financial Indicators subsection covers items that should be considered to complete a thorough analysis of financial indicators associated with the project and for the property to maximize the potential for success.

Resources for lessons learned from the efforts of previous project teams in revitalizing contaminated properties are presented in the Information and Advisory subsection.

Sustainable economic revitalization strives to achieve a level of economic viability that enables local ownership to effectively use financial resources to meet local needs. Returns on sustainable economic revitalization should be used to maintain income and growth in a community, as well as to further evaluate ecological and sociocultural impacts of the revitalization. The Long Term Economic Impacts subsection suggests information to be ascertained and evaluated for a sustainable economic revitalization effort.

The Net Benefit Calculator Tool is described in the Tools and Checklists below.

9.2.1.8 Sources of Money

The Sources of Money section introduces a variety of financing strategies currently used in site revitalization, including: public financing (e.g., subsidized low interest loans, bonds, revolving loan funds, grants, and taxes and special assessments), private financing, and foundation funding. The Financial Resources Tools are described in the Tools and Checklists below.

The Public Financing subsection describes how the public sector plays a valuable role in assisting developers in overcoming some of the hurdles that can be encountered in the revitalization of potentially contaminated sites. The tools available to the public sector to catalyze revitalization expand each year, and are becoming more creative and sophisticated. A number of federal, state, and regional financial assistance programs, grants, tax incentives, loans, bonds, and more are introduced.

Private companies and individuals are finding the revitalization of potentially contaminated sites profitable, and are currently enjoying a surge in demand for property in formerly neglected urban areas. Some revitalization projects may require a combination of lenders to obtain the full financing needed. The Private Financing subsection presents revolving loan funds, grants, real estate investment trusts, and trust funds.

The Foundation Funding subsection contains information and resources for foundations that provide funds and technical assistance to non-profit organizations for a range of activities including education, public health, science, the arts, the environment, and social services.

9.2.1.9 Project Schedule

Detailed planning and scheduling should be conducted in order to prevent costly delays and project failure. The project team should work diligently during the planning phase to identify key project milestones and the stakeholders and support activities required to achieve the milestones on schedule. A detailed timeline of events and environmental schedule is essential when working with projects involving the complex mix of issues and activities in site revitalization. The Project Schedule section describes the key project phases requiring scrutiny during the development of the project schedule including: development of the project team, encouragement of stakeholder involvement, completion of a project objective plan and goals, development and implementation of a community involvement plan, and potential decontamination, construction, and demolition activities.

9.2.2 Tools and Checklists

SMARTe combines the power of the Internet with analysis and presentation tools that can be used interactively to build decision models for solving revitalization problems.

- Community Involvement Techniques – This tool helps the user find approaches to community involvement that meet selected criteria (e.g., number of participants, time required, cost, etc.). Users select the desired criteria for each variable and click "Search". Approaches to community involvement that match the selected criteria are returned in a document layout that includes a description of each approach, potential advantages and disadvantages (strengths and weaknesses), and some further recommendations for their use.
- Potential Stakeholders – This tool provides a list of potential stakeholder groups who could be involved in a land revitalization project. The intent is that the selected list of stakeholders will form the basis for the project team that implements a specific land revitalization project.
- Visioning – A project vision drawing tool provides users an interface to visualize the reuse options.
- Human Health Risk Calculator – This calculator supports a human health risk assessment by allowing the user to define receptor scenarios. Risks are calculated for various exposure pathways for a selected chemical and its concentration. The output pages show a table of risks by exposure pathway, and the parameters used in the risk equations for the selected risk scenario.
- Site Characterization – This SMARTe tool provides some statistical analysis capabilities that can be used to support human health or ecological risk

assessment. These include exploratory data analysis, spatial plots, confidence limits, and some Classical one- and two-sample statistical hypothesis tests. Both parametric and non-parametric tests are offered. Subsets of the data may be obtained using the panel and group variables, so that individual chemicals or sites can be evaluated.

- Monitoring Data Analysis – This tool offers standard statistical methods for displaying temporal and spatial data, and for performing temporal trend tests. The statistical tools offered to support monitoring programs include some exploratory data analysis methods (summary statistics and boxplots), time plots, trend estimation, and spatial plots. Subsets of the data may be obtained using the panel and group variables, so that individual chemicals or sites can be evaluated.
- Fate and Transport – This part of SMARTe offers a discussion of environmental modeling as it pertains to spatial and temporal contaminant concentrations.

  o Hydraulic Gradient Calculator – This tool provides gradient calculation from fitting a plane to as many as fifteen points. Inputs include coordinates of the well and head.
  o Plume Diving Calculator – The Plume Diving calculator can estimate the prospects for plume diving assuming simplified flow in a water table aquifer. Inputs to the calculator are the hydraulic conductivity, recharge rate, and head at two points in the aquifer. Taken together all of these parameters determine flow in the aquifer, so a means of calculating the flow is needed.
  o Vapor Intrusion Modeling – This on-line calculator implements the Johnson and Ettinger (J&E) (Johnson and Ettinger 1991) simplified model to evaluate the vapor intrusion pathway into buildings. This J&E model replicates the implementation that the US EPA Office of Solid Waste and Emergency Response (OSWER) used in developing its draft vapor intrusion guidance, but includes a number of enhancements that are facilitated by web implementation: temperature dependence of Henry's Law Constants, automatic sensitivity analysis of certain parameters, and others.

- Net Benefit Calculator – This calculator compares the cost elements (development costs, environmental risk management costs, etc.) and revenues of different revitalization options.
- Financing Resources – This tool helps the user evaluate possible financing resources that are available to support land revitalization projects. Upon the selection of Public or Private financing, the user can select more specific options.
- Financing Resources: Locating Potential Sources – This SMARTe tool is under development and not available for use. In the future, the tool will help the user through the maze of possible financing resources that are available to support land revitalization projects. The user will enter project characteristics and SMARTe will return the financing options that best fit

those characteristics. It is planned to have additional tools that will further help the user determine if a specific financing option is right for their project.

- Financing Resource: Evaluation of Project Criteria for BEDI Program – This SMARTe tool is under development and is not currently functional. In the future, the tool will help the user through the maze of eligibility criteria for Housing and Urban Development's (HUD's) Brownfields Economic Development Initiative (BEDI), providing a means of determining at a high level whether the BEDI program is appropriate for the user's project. BEDI grant funds are primarily targeted for use with a particular emphasis upon the redevelopment of Brownfields sites in economic development projects and the increase of economic opportunities for low-and moderate-income persons as part of the creation or retention of businesses, jobs and increases in the local tax base.

The following SMARTe checklists and fact sheets are available:

- Land Reuse Options – a list and description of land reuse options that can be selected for a land revitalization project.
- Select a Consultant – helpful information for hiring an environmental consultant for a revitalization project.
- Select a Lawyer – helpful information for hiring a lawyer for a brownfields redevelopment.
- Purchasing a Brownfields Property – a checklist of considerations when looking to purchase a brownfields property.
- Finding an Insurance Broker – helpful information for finding environmental insurance coverage.
- Do I Need a Permit – a list providing considerations for planning, building, and site development permits.
- Selecting a Developer – things to consider when selecting a brownfields developer.
- Understanding Units of Measurement – explanation of units of measure that may be seen in technical reports.
- Writing a Request for Proposal – a guide to help those who need to solicit proposals for environmental site assessments or cleanup of brownfields.
- What are Quality Assurance Project Plans? – information regarding the purpose of a Quality Assurance Project Plan (QAPP) and what should be included.

9.2.3 Creating a Revitalization Plan with SMARTe Using My Project

Stakeholders who wish to make use of SMARTe's full capabilities can do so by creating one or more online projects. A SMARTe project combines environmental modeling, risk assessment, statistics, economic modeling, and decision analysis with documentation and presentation capabilities to guide users through the process of creating a revitalization plan. Project information is stored in a "project workspace" which is accessed through a "My Project" sidebar available when the user is logged in to their project (Fig. 9.1). A project is created by registering a project name and password. This initializes the project workspace with a set of links to a revitalization plan outline, as well as links to analysis tools that aid stakeholders in completing the revitalization plan. All stakeholders involved in a project can connect to their project by logging in using the project name and password.

SMARTe's approach to creating a revitalization plan is grounded in the technical framework of probabilistic decision analysis (Berger 1985, Bernardo and Smith 1994). Probabilistic modeling requires specification of probability distributions for each input component. These distributions capture both what is thought to be known (for example, the best guess is the average value) and the associated uncertainty. Decision-making is best supported by a full characterization of uncertainty so that the quality of the final decisions that are made can be measured, the need to collect additional information can be evaluated, and sensitivity analysis can be performed to identify the most important factors in the model. Decisions are nearly always made in the face of uncertainty, and this probabilistic framework allows the uncertainty to be specified, evaluated and managed. This describes the essence of MCDA, which forms the foundation of the SMARTe decision analysis approach.

As a user follows the steps to create a revitalization plan, SMARTe elicits information from the user which is fed into the decision framework. Many inputs allow the specification of a probability distribution. Multiple potential revitalizations options can be built and stored under "My Project". A summary of the benefits, costs, and relative non-monetary values for each revitalization option is provided under "My Project", as well as a comparison of the cost, benefits, and non-monetary values, and their associated uncertainties, of the different revitalization options. This forms an iterative process in which stakeholders can gather information and update the option inputs in attempt to reduce the uncertainty to a level in which stakeholders are comfortable.

The technical objective of "My Project" is to provide an interactive technical guidance program with analysis capabilities that is developed solely with Open Source Software (OSS, www.opensource.org). SMARTe is a complete decision management system for revitalization with the following features:

- Guidance for each aspect and function of SMARTe. This includes interpretation of results and explanation of technical terms and methods.
- Access to and integration of project-specific knowledge bases with further access to the wealth of information available on the internet.
- Environmental modeling, risk assessment, statistics and decision analysis tools that use R as the analytical engine (R is an open-source statistical programming language [www.r-project.org], sometimes embedded in Java).

- An Expert system that helps the user navigate the technical choices available within the analysis tools (for example, risk assessment, financial, and social options, or statistical and decision analysis options).
- A decision analysis engine.
- A document publishing system which can produce revitalization plan documents in a variety of formats (in development).
- An easy-to-use feedback mechanism for user comments.
- Interactive training in each aspect of the SMARTe system.

SMARTe's systems approach reflects the OSS paradigm for sharing information and involving users in development of SMARTe applications.

9.3 Technical Structure of the System

SMARTe is designed to provide the user access to increasing levels of technical details and analysis tools based on the needs of the user. The top or simplest level of SMARTe provides basic revitalization information and resources for revitalization. The next level provides more detailed revitalization guidance and analysis tools. The most complex level integrates detailed technical guidance with the analysis tools needed to implement an MCDA approach to revitalization. Each level of SMARTe will potentially (based on the level of technical complexity) have one or more of the following technical components:

- Content
- Technical documentation
- A database that is queried by the user or that feeds into technical calculations
- A computational engine for performing technical calculations
- An expert system for guiding and providing interpretation of technical results
- A user interface linking technical information, databases, the computational engine, and the expert system

9.3.1 Content

SMARTe is designed to be free OSS based on World Wide Web Consortium (W3C) web specifications. All SMARTe content, including the user interface, has been developed using the eXtensible Markup Language (XML). XML is software independent (that is, the content can be edited with any text editor) and platform independent (that is, Windows or Linux). Presentation is handled using the eXtensible Stylesheet Language (XSL) and Cascading Style Sheets (CSS). This separation facilitates providing web accessibility of SMARTe to the broadest possible audience using the W3C Web Accessibility Initiative specifications. By using XML as a content management tool, SMARTe content is separated from, and thus is independent of, presentation style, technique, and technology. This separation allows great flexibility and variety in the choice of presentation. SMARTe content, for instance, can be seamlessly transformed into HTML (for web browsing), and Adobe Acrobat PDF (for reporting). SMARTe is currently being served using Apache Cocoon.

9.3.2 Technical Documentation

Each of the technical components is presented with explanations of relevant theoretical concepts and technical terms, instructions for using the tool, and instructions for interpreting the results. Technical documentation is integrated with the software using a process to ensure that changes in the software are synchronized with the appropriate changes in the documentation. Technical documentation also includes information regarding software and model evaluation.

9.3.3 Databases

Databases for SMARTe are managed either in PostgreSQL or XML. PostgreSQL is a highly-scalable, Structured Query Language (SQL) compliant, Open Source object-relational database management system. User queries of PostgreSQL databases are performed by SQL queries generated based on user interaction with a database search user interface. The user interface is presented in the user's browser as an HTML form. Form parameters are translated into an SQL query using Java and Javascript (AJAX). User queries of XML-based databases also are based upon user interaction with an HTML form. However, for XML-based databases the form parameters are used directly in Java or Javascript to search the database. Database queries for use in technical calculations are performed using pre-defined SQL queries from R or Java.

9.3.4 Technical Analysis

Technical analysis, computations, and algorithms are programmed either in custom software or in existing free OSS. Custom software will be developed in the common web programming tools Java, Javascript (AJAX), or XSL. An example of OSS that will be relied upon is R, a statistical computing, programming, and graphics language environment (www.r-project.org). Other modeling and computational tools are also used for specific applications. Only those software components used in development of SMARTe and resulting in 508 compliant products will be used.

9.3.5 Expert Systems

In general, expert systems include a set of rules and a knowledge base. A decision logic can provide a visual depiction of the steps implemented in the expert system. In SMARTe, the decision logic is built in XML (in the form of Scalar Vector Graphics, SVG). This allows a rule to be embedded in the graphic (the SVG). When the analysis is run, each rule is run sequentially. Results are then stored back in the decision logic SVG. The result of this application of the rules and knowledge base is a coded decision logic depicting the SMARTe recommended path, and a textual interpretation of the analysis results with language targeted to a non-technical audience.

9.3.6 User Interface

The user interface (UI) for SMARTe includes a system of pull-down menus and a hyperlinked table of contents sidebar. The pull-down menus are implemented through an XML menu specification that is translated to meet the W3C specification for HTML forms. The UI is served dynamically, using Javascript (AJAX), to provide an interactive environment for the user.

9.4 SMARTe Decision Structure Including Stakeholder Involvement

SMARTe is a framework program for building site-specific analyses for the revitalization of potentially contaminated sites. The ultimate goal is for these analyses to coalesce in a decision analysis framework, so that all uncertainties, costs, and value judgments can be brought together in a holistic model. The intent is to bring together the economic, environmental and socio-political components of a problem so that each component contributes to the overall decision problem. Traditionally the focus of revitalization has been economic. This focus is, arguably, shortsighted considering the increased recognition of the importance or value of ecological systems and natural resources, and the increasing need and desire to develop and use green technologies and renewable energy in light of the impact of fossil fuel use on our environment. In addition, improving the quality of life for individuals and communities often leads to better long-term financial rewards. Finding sustainable solutions that encompass economic, environmental and socio-political concerns is becoming more critical.

The technical philosophy behind SMARTe follows the Scientific Method. The appropriate normative paradigm for such a framework is Bayesian statistical decision theory[1] (Berger 1985, Bernardo and Smith 1994). Bayesian

---

[1] The authors recognize that this point is arguable on many fronts, including competing theories of uncertainty and utility using point-valued calculus, and expansion of Bayesian

statistical decision analysis requires specification of probability distributions for uncertain parameters, and specification of costs and value judgments. SMARTe provides simple user interfaces to specify these inputs to a Bayesian decision analysis. These inputs are propagated through the SMARTe system to produce a cost-benefit analysis of the land use and remediation options. That is, the decision analysis approach leads to a cost-benefit analysis across economic, environmental and socio-political factors. This approach requires applying costs to environmental factors such as human health risk and ecological risk, and to community benefits. Methods for applying costs to these types of factors do not currently exist in the mainstream literature. If cost-benefit analysis is not ultimately supportable, then multi-attribute utility approaches can be tried instead. However, cost is the metric by which business decisions are made, and it seems unlikely that SMARTe can be taken seriously in a business context if cost and return on investment is not the ultimate decision metric. This requires finding ways to value environmental factors, ecological systems and community benefits.

SMARTe 2008 allows inputs to be captured for many aspects of a complete decision analysis for land revitalization problems, but, under the current schedule, the system will not be complete until 2011. Some of the topics that require further development include valuing environmental and community benefits factors. Others include uncertain factors such as fate and transport, sustainability, and likelihood of tax credits.

Figure 9.2 shows a schematic of the high-level conceptual structure of SMARTe. Note that this figure is clickable in SMARTe, which opens each high-level topic to reveal the next level of detail.

Implementation of SMARTe begins with building a vision for the specific revitalization project. The vision consists of different options for land use as well as assembling a project team to complete the work. The different land use options can include single use options, which might be applicable at small sites, or multiple use options, which might be more applicable at large sites. In SMARTe 2008, there is a primitive visioning tool that allows a project team to sketch possible land use options. Future versions of SMARTe are envisioned to provide more options and will provide greater flexibility for sketching options. Furthermore, land use options will be linked to receptor scenarios, which will directly feed the risk assessment in projects that involve contamination and human health or ecological risk. In so doing, the decision analysis will be fed directly by the possible decision endpoints.

---

statistical theory to an interval-valued calculus (e.g., lower probability theory). In addition, it has been argued that application of an appropriate normative paradigm does not necessarily translate into an appropriate model of behavior. However, at this point in time, we believe that a standard Bayesian approach is sufficient and reasonable, and through repeated application is likely to improve, over time, the effectiveness of this approach.

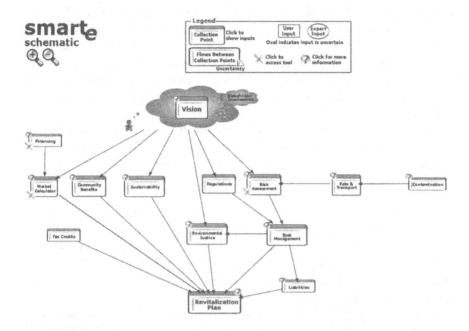

Fig. 9.2 SMARTe schematic (www.smarte.org/smarte/structure/getsmart.xml?layout = no-sidebar-no-mainnav)

At a simple level, there are 3 major components to SMARTe following specification of land use options. These can be labeled environmental, economic and socio-political. The environmental component pertains to the potential need to remediate chemical or biological contamination to cost-effectively reduce human health and ecological risk. When applicable, this begins with a conceptual model for source release, followed by fate and transport of the contaminants, and then, risk assessment and risk management. Source release corresponds to establishing contaminant concentrations. Statistical tools are available in SMARTe to analyze the contaminant data. SMARTe 2008 includes some standalone independent fate and transport models. In future versions of SMARTe it is envisioned that fate and transport models will be coupled with the contaminant data on the front end, and the risk assessment on the back end. A fairly comprehensive human health risk assessment tool is included in SMARTe 2008, with refinement and improvements anticipated so that the risk tool is coupled with advice on some of the idiosyncrasies of risk assessment. The risk assessment tool will be linked to the contaminant database. An ecological risk assessment tool is also envisioned.

Under the Bayesian Decision Theory framework, the vision for future versions of SMARTe is to couple these components that contribute to a risk assessment. The contaminant database will feed fate and transport models or the risk assessment (as appropriate). The fate and transport models will use the contaminant database and will generate concentration terms for the various media that are needed to support the risk assessment. That is, the contaminant concentrations, fate and transport and risk assessment will be fully coupled.

All factors included in the current human health risk assessment tool can be input probabilistically. Uncertainty can, therefore, be included directly and numerically.

The risk assessment itself, including concentration terms and fate and transport modeling, results in estimates of risk, including uncertainty. However, to complete the ensuing decision analysis, a value needs to be placed on risk. This is one step down the path of valuing impacts to the community in the decision analysis. The intent of future releases of SMARTe is that risk management options will also be evaluated, and that risk reduction will be traded-off against the cost of performing different risk management approaches. In so doing, a SMARTe decision analysis is intended to optimize over land use options and risk management options. SMARTe will offer more support for selection and costing of remediation options. At this time, costs related to risk management can be entered as part of the revitalization effort.

Economic components of SMARTe include a tool for market costs and benefits of redevelopment, financing options, liabilities (insurance options) and tax credits or benefits that might result from effective revitalization. The net benefits calculator in SMARTe collects costs for the redevelopment of the site. The impact of financing options is also included through the cost of borrowing money. Insurance costs are similarly included. Part of the SMARTe decision structure includes costs or value judgments for ecological and community benefits factors.

In SMARTe 2008, community values are specified in relative terms for each land use option. The intent of future versions of SMARTe is to provide tools that will allow users to value various types of community benefits, such as parkland as opposed to industrial buildings, street lighting, green buildings, saving ecological systems, jobs, reduction in crime rate, and many others. In order to achieve this goal for SMARTe, methods for valuing community benefits need to be developed. Note also that the potential success of some community benefits might depend on factors that are uncertain. For example, how the crime rate will be reduced in response to street lighting might be best modeled by addressing uncertainty directly.

There are some other cost factors that are part of the SMARTe decision model. For example, SMARTe provides some help and advice on community involvement techniques. In reality, each one of these techniques has associated costs. These costs are not currently included in SMARTe 2008, but this is planned for future releases.

Overall, the decision analysis that is performed in SMARTe addresses the needs of all possible stakeholders individually. Its decision objective can be termed "maximizing societal welfare". That is, determining which land use option has the greatest overall benefit. This might prove insufficient if the minimal needs of any one stakeholder group are not satisfied. Each stakeholder group might need to realize a positive return. If so, the optimization will be constrained.

Stakeholder groups are varied. A list of potential stakeholders is provided in SMARTe. This list is not intended to be complete, and it is not intended to be exhaustive for a specific application. However, the list covers the main stakeholder groups that are likely to be needed as part of a revitalization project. The list includes regulators, community groups, environmental engineers and consultants, property owners, developers, financial institutions, insurers, and lawyers. Once a list of stakeholders is identified, as noted above, the decision analysis might need to address each stakeholder separately, as well as attempting to maximize societal welfare.

SMARTe covers a lot of ground, but there are many refinements that are planned and will be made in each future release of SMARTe to improve the land use and risk management options analysis that is the primary objective. Environmental factors (remediation), economic factors and community benefits are included directly, and they will be improved or refined annually. The basic framework exists; further development is a matter of refining what is there now. The ability to perform a comprehensive decision analysis in an interactive Open Source web-based environment is quite an achievement, and the decision analysis component will only continue to improve. It is possible, even in its current state, to demonstrate SMARTe's functionality through a hypothetical case study.

9.5 Case Study

In Greenville, Nebraska, the site of a former manufacturing facility was being considered for a variety of potential uses. The 3.2 ha site is located right in the heart of this small city. The City of Greenville is the current owner of the property where the former manufacturing facility was located. The City has a variety of ideas for more productive uses for this land.[2]

A Committee for the Reclamation of the Acme Site ("the Committee") was established. The Committee includes the mayor, two interested community members, and three potential developers. One of the community members found SMARTe when looking for resources on the internet. Based on the information they found under the Project Stakeholders topic in SMARTe, the Committee invited their TAB (Technical Assistance to Brownfields) representative to join them. They also decided to hire an environmental consultant, and

---

[2] This case study is based on a combination of several sites and projects that have benefited from aspects of SMARTe. Some liberties are taken with the blending of projects, and with application of the technical tools in SMARTe that were not yet available at the time these projects were conducted. The names are fictitious because they actually represent the combination of several sites and projects.

using the checklist in SMARTe, identified a small environmental firm with one key member who joined the committee, and a variety of other environmental experts who were then made available to them for specific questions.

The City and the developers on the Committee had two main options in mind for this site as they proceeded. The first option, given its prime location, was to move the tourist train depot (the main touristic draw for the area) to this site to stimulate tourist visits to the contiguous areas of the downtown district. The second option was to put in some moderate, but market-priced rental housing units around a new park. Several of the other park areas in the city have been taken over by developers through the years, and this would be an appealing location for a downtown park. Using the Community Involvement tool in SMARTe, the Committee decided to hold some field trips and brainstorming sessions to get community input on their ideas for the site. They followed this up with sketch interviews where interviewees either hand-sketched or used the Visioning tool in SMARTe to draw their ideas for the site. All of this input was collapsed into the two main options (train depot or park with housing), and the original plans for both were vastly improved and given more detail based on the community input.

The manufacturing, done by Wextell Corporation, primarily involved chrome plating of metal instruments. The City was very aware that there was a serious possibility of chromium contamination at the site. Wextell used a large amount of hexavalent chromium in their plating, and although for the most part it was carefully handled and disposed of properly, there were known to have been numerous spills over the years of facility operations. According to the Sampling Plans tool in SMARTe, which the Committee used with the help of their environmental consultants, it was determined that they should collect 25 samples of surface soil at the site. This large number of samples was the result of a very small willingness on the City's part to make an error and misjudge any site contamination. Funding for site characterization was identified, through the listings of financial resources in SMARTe, and applications were sent. The site characterization was conducted with 50% City funds, and 50% federal grants.

When the data from site characterization came back from the chemical laboratory, the Committee chose to use the Site Characterization data analysis and Risk Assessment tools in SMARTe to determine if they had a contamination problem at this site. First summary statistics were computed, and the data were plotted using SMARTe's data analysis capabilities. The average hexavalent chromium concentration in surface soil at the site was 51 mg/kg, with the minimum observed concentration at 31 mg/kg and the maximum observed concentration at 69 mg/kg, as reported in Table 9.1.

Several statistical tests were performed (system results not shown) to determine if the concentrations at the site exceeded the screening level for hexavalent chromium. There were two different screening levels that they considered: the residential scenario screening level of 30 mg/kg was clearly exceeded at the site; the commercial/industrial screening level of 64 mg/kg was questionable – two of the statistical tests (Wilcoxon Sign Rank Test and One Sample t-test) implied

Table 9.1 Summary statistics of site characterization data

Result	N	Num detect	Min ND	Max ND	Min detect	Median	Mean	Max detect	Std dev	Shapiro-Wilk p-value	log Shapiro-Wilk p-value	t UCL	Bootstrap BCa UCL
Result	25	25	NA	NA	30.6	49.6	50.99	69.4	13.67	0.006567	0.007918	55.67	55.11

that the site concentrations were below this threshold, but one test (Sign Test) implied that it was not safely below this level.

To better understand the potential environmental concerns at the site, a risk assessment was then performed. One of the community members on the Committee conducted the risk assessment using the SMARTe tool, and then had the environmental consultant review the assessment to be sure it was conducted appropriately. The findings of the risk assessment, with the consultant's interpretations, are shown in Table 9.2.

Based on the risk assessment, it is clear that some remediation of the contaminated soil would be necessary before locating a park and housing on this site. It is also fairly clear that no remediation at all would be necessary to locate the tourist train depot on the site.

The Committee worked with their TAB representative to estimate the costs for environmental remediation. The developers formed a sub-committee to determine cost estimates for building the park, houses, and train depot with parking facilities. This sub-committee also estimated annual revenues from each option. They were able to find $ 100,000 in federal grants for the revitalization, and an additional $ 100,000 in grants from a foundation that would be available only if they chose to build a park. The Committee used the My Projects capabilities in SMARTe to input these cost and revenue estimates. They also used the input gleaned from the community interviews and brainstorming sessions to estimate the relative value to the community of each of the two options on a scale of 0–10 (5 out of 10 for the train depot, 9 out of 10 for the park and rental housing units). Upon presentation of this comparison of options in a public meeting, they revised the input on relative community value to be 6 out of 10 for the train depot, and 8 out of 10 for the park and housing units option. The train depot was the more financially viable option, but the community expressed a preference for the park

Table 9.2 Risk assessment results

	Park with housing	Train Depot	Interpretation
Systemic Risk for Adult	1.7	0.0015	If this number is above 1, then the site may pose an unacceptable risk.
Systemic Risk for Child	4.3	0.0026	If this number is above 1, then the site may pose an unacceptable risk.
Carcinogenic Risk	1.0×10^{-3}	7.7×10^{-7}	Risks below 1×10^{-6} are generally considered insignificant, whereas risks greater than 1×10^{-4} are often considered unacceptable. Acceptability of risks between these levels is usually negotiated with regulators.

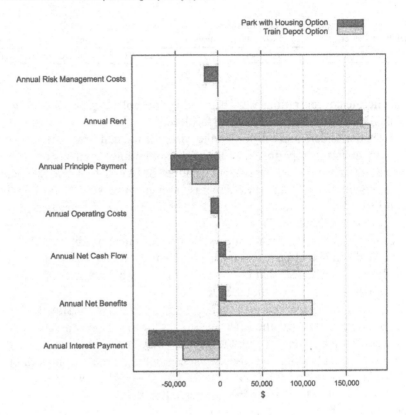

Annual Net benefits components grouped by option

Fig. 9.3 Options comparison

and housing option. The comparison of options was summarized in SMARTe as shown in Fig. 9.3.

The Committee presented this summary to the City Council who considered the community input, but decided ultimately to approve plans for moving the tourist train depot to this location. The City Council had some concerns regarding the consequences of maintaining housing on the site in light of the existing contamination and requirement for long-term monitoring. They also viewed the stated relative community values as extremely important, but not very disparate for the two options, so that wasn't a major swaying concern. The primary deciding factors for the City Council were the clearly better short-term financial outlook for the train station, and their strong expectation that placing the depot on this site would stimulate other commercial development nearby that would increase the tax base and invigorate the downtown area.

SMARTe was used from beginning to end in this project. It provided assistance in creating a project team, involving the community, identifying sources for grants for site characterization and site revitalization, planning for data collection and assessing the data to determine the level of environmental concerns at the site, and finally a system for entering and viewing the economic costs and expected revenues such that they could be readily compared. Although each aspect helped the City of Greenville reach a successful conclusion to this project, the most valuable aspect for them was to be able to present their plans and calculations to the City Council via SMARTe so that the process for arriving at their recommendations was fully documented and supported. The City Council members were shown the SMARTe input tables at a meeting where they were able to alter the inputs and consider what different values, or changes in uncertainty levels, would do to the expected results for each revitalization option. The ability for all to view the project and do real-time adjustments made the basis of the Committee's recommendations very transparent to the City Council, and greatly eased the process of getting Council approval.

9.6 Conclusions

SMARTe provides a comprehensive source of information, tools, and analysis to support the revitalization of potentially contaminated sites. Through its DSS, SMARTe integrates a wealth of information from informative narratives and case studies, functional risk assessment and statistical tools, and an easy-to-use decision interface that assesses market and non-market costs and benefits of revitalization. SMARTe is the product of 5 years of development by a team of U.S. EPA employees and their contractors, the German Federal Ministry of Education and Research in Germany and the Interstate Technology and Regulatory Council. SMARTe is an important resource for revitalizing sites and hence protecting greenfields from future development.

The DSS engine in SMARTe is a sophisticated web-based MCDA tool. It currently allows users to evaluate market costs and benefits of different reuse options. A future version of this tool will integrate a full spectrum of costs and benefits (both market and non-market costs) to provide a holistic picture of the potential financial, social, and health benefits from redevelopment decisions. Included in SMARTe's MCDA system are parameters that calculate values for construction costs, remediation costs, maintenance costs, and future releases will include benefits of reduced ecological and human health risk, social investment costs, and social benefits of improved public recreation.

One of the most exciting and promising aspects of SMARTe is that its development and testing focused on its use to help community members who wish to better understand the revitalization process. SMARTe provides the information, resources, checklists, and calculators to enable citizens to provide educated informed opinions on redevelopments in their communities. Ultimately, by using the DSS features of SMARTe, community members will be able to work remotely, but yet collaboratively, to identify and evaluate redevelopment options for sites of interest. Because SMARTe is a publicly-available web site, it can, and has, assisted people worldwide with their revitalization projects.

SMARTe has been "beta tested" on actual communities around the U.S., including a rural community in Missouri that lacked other information resources. Through the rich content of SMARTe, this community received pertinent knowledge and was empowered to hire environmental professionals and attorneys, negotiate with developers, and pursue a vision of what they want for their revitalization. Now that SMARTe includes its DSS engine, it is hoped that it can provide additional services for many other communities worldwide.

References

Berger, J.O. (1985) Statistical Decision Theory and Bayesian Analysis. Springer-Verlag, New York.

Bernardo, J.M. and A.F.M. Smith (1994) Bayesian Theory. John Wiley & Sons Ltd., New York.

Johnson, P.C. and R.A. Ettinger (1991) Heuristic model for predicting the intrusion rate of contaminant vapors into buildings. Environmental Science and Technology 25: 1445–1452.

Chapter 10
DSS-ERAMANIA: Decision Support System for Site-Specific Ecological Risk Assessment of Contaminated Sites

Elena Semenzin, Andrea Critto, Michiel Rutgers and Antonio Marcomini

Abstract The DSS-ERAMANIA is a decision support system implementing a site-specific Ecological Risk Assessment (ERA) procedure and supporting the experts and the decision makers in the assessment of contaminated soils. It was developed according to a Triad approach, where the results provided by a set of measurement endpoints are evaluated and integrated to support the assessment and characterization of ecosystem impairment (i.e. assessment endpoint) caused by the soil contamination. In the Triad approach, the measurement endpoints refer to three Lines Of Evidence (LOE): environmental chemistry, ecotoxicology and ecology, whose integration should pragmatically reduce the uncertainty in the risk estimation. For the DSS, a framework including three subsequent investigation levels (i.e. tiers) was implemented, that enables the completion of the risk assessment once the provided answer is unequivocal as characterized by a relatively small uncertainty, ensuring at the same time an adequate financial investment. The DSS includes two modules: "Comparative Tables" and "Integrated Ecological Risk Indexes", both based on Multi Criteria Decision Analysis (MCDA). Module 1 ("Comparative Tables") aims at comparing the different measurement endpoints belonging to each LOE (i.e. bioavailability tools, toxicity tests and ecological observations) to guide the expert/decision maker in the choice of the suitable set of tests to be applied to the case study for each tier of investigation. Module 2 ("Integrated Ecological Risk Indexes") provides qualitative and quantitative tools allowing the assessment of terrestrial ecosystem impairment (i.e. the impairment occurring on biodiversity and functional diversity of the terrestrial ecosystem) by integrating complementary information obtained by the application of the measurement endpoints selected in the first module. The two modules were implemented in a software application and validated using the Acna di Cengio contaminated site (located in Savona province, Italy). The objectives, functionalities

E. Semenzin (✉)
Department of Environmental Sciences and Centre IDEAS – University Ca' Foscari,
Venice, Italy
e-mail: semenzin@unive.it

A. Marcomini et al. (eds.), *Decision Support Systems for Risk-Based Management of Contaminated Sites*, DOI 10.1007/978-0-387-09722-0_10,
© Springer Science+Business Media, LLC 2009

and structure of the DSS-ERAMANIA will be presented and the main outcomes of the DSS application to the case study will be discussed.

10.1 Introduction

Criteria and methodologies for the assessment and rehabilitation of contaminated sites are urgently needed worldwide because of the huge number of contaminated sites and the financial implications representing significant constraints for site redevelopment.

In order to address the rehabilitation of contaminated ecosystems, Ecological Risk Assessment (ERA) is the appropriate process for identifying environmental quality objectives and the ecological aspects of major concern (US EPA, 1998; Suter et al., 2000). At the international level, the principles and procedures that have been established (UK-EA, 2003; CLARINET, 2001; INIA, 2000; ECOFRAME, 1999; CARACAS, 1998), suggest an ERA framework based on a tiered approach, including (1) a screening phase that allows the definition of land use based soil screening values (SSVs), and (2) a site-specific phase in order to accomplish a more comprehensive risk characterization. Within the risk characterization, the application of Weight of Evidence methods (WoE) (Burton et al., 2002; Chapman et al., 2002) are used to determine possible ecological impacts based on multiple Lines Of Evidence (LOE) (US EPA, 1998). In the Netherlands, an up-date of the soil protection act is foreseen (VROM, 2003), clarifying the way for a site-specific risk assessment based on one of the WoE methods: the Triad approach (Rutgers and Den Besten, 2005). The Triad approach requires three major LOE for an accurate assessment of contamination: chemical contaminant characterization, laboratory-based toxicity to surrogate organisms, and indigenous biota community characterization (Long and Chapman, 1985). Moreover, it is performed according to subsequent tiers of investigation, from a general to a specific one. The results obtained in each tier have to support the decision making process, whose main challenge is to find sustainable solutions by integrating different disciplinary knowledge, expertise and views (Siller et al., 2004; Kiker et al., 2005). The concept of decision support has evolved from highly technocratic systems aimed at improving understanding of technical issues by individual decision makers to a platform for helping all parties involved in a decision process engage in meaningful debate (Pereira and Quintana, 2002). Therefore, simple technical tools are increasingly needed in order to integrate the wide range of decisions related to contaminated land management and re-use (CLARINET, 2002a). Instruments, proved to be an effective support for any kind of decision making process including contaminated sites management, are Decision Support Systems (DSSs) (CLARINET, 2002b). Although some specific tools for human health risk assessment are included within the existing DSSs, relevant development should be made in order to implement suitable support systems for site-specific ecological risk assessment (CLARINET, 2002b).

For this reason, within the ERA-MANIA project ("Ecological Risk Assessment: development of a Methodology and Application to the contaminated site of National Interest Acna di Cengio (Italy)") the DSS-ERAMANIA was developed, as a Multi Criteria Decision Analysis (MCDA)-based system capable of supporting experts/decision makers in the site-specific phase of ERA of contaminated sites. As far as the ERA-MANIA project is concerned, the Acna di Cengio contaminated site was selected as a case study for the DSS design and application. The complexity of the site contamination was especially useful for the DSS design that resulted from a close collaboration among different experts within a multidisciplinary team (including risk assessors, ecotoxicologists, ecologists, environmental chemists, mathematicians and informatics) and thanks to the valuable support of an international steering committee, including decision makers and stakeholders.

This Chapter is intended to provide an overview of the DSS-ERAMANIA and of the application to the Acna site. The detailed presentation of each tool composing the DSS is the subject of four dedicated manuscripts (Critto et al., 2007; Semenzin et al., 2007, 2008a, b), where specific analyses of the adopted methodological approaches are discussed.

10.2 DSS-ERAMANIA Framework and Functionalities

In order to support experts and decision makers in the application of site-specific Ecological Risk Assessment (ERA) for contaminated soil, a specific framework was developed and implemented in the decision support system DSS-ERAMANIA. The proposed framework is mainly based on the Triad approach (Rutgers and Den Besten, 2005). Moreover, it includes three subsequent investigation levels, named tiers, each one allowing the assessment to stop depending on the acceptability of the estimated risk and related uncertainty.

The first one (i.e. Tier 1) can be regarded as a preliminary investigation level, in which the analysis can stop only if the estimated risk is considered to be acceptable by the experts and the decision makers and affected by a minimum level of uncertainty. The unacceptable risks found in Tier 1 can be investigated in the second tier (i.e. Tier 2), to reduce the uncertainty in the risk estimation and to achieve a more accurate risk assessment. Because of economic and time reasons, it is preferable to stop the analysis at Tier 2 when a satisfactory estimation of the investigated risks is achieved. Nevertheless, site-specific aspects of particular interest, for which the risk uncertainty could not be reduced in Tier 2, can be analyzed in the third level (i.e. Tier 3), which represents the deepest degree of investigation. However, this effort has to be justified by adequate cost-savings for the site remedial actions.

The developed framework identifies two different phases in the site-specific ERA process. The first phase concerns a comparison of tests (i.e. measurement endpoints) within each Triad LOE and the selection of a suitable set of tests to be applied to the case study in order to collect appropriate information on the ecosystem impairment according to the Triad Lines of Evidence. The results of the measurement endpoints selected in the first phase are then integrated and evaluated in the second phase in order to obtain both quantitative and qualitative evaluations of soil quality.

For each phase, specific Multi Criteria Decision Analysis (MCDA)-based tools were developed and implemented as described in the following section.

10.3 Developed Methodology and Implemented Software System

In order to accomplish the two main functionalities identified by the framework, the DSS-ERAMANIA was structured in two modules: "Comparative Tables" (i.e. Module 1) and "Integrated Ecological Risk Indexes" (i.e. Module 2), as presented in Fig. 10.1.

In the following paragraphs, the methodological approach and the software implementation of each module of the DSS-ERAMANIA are presented in detail.

10.3.1 Module 1: Comparative Tests Tables

The DSS-ERAMANIA Module 1 aims at comparing the different tests belonging to each Triad Line of Evidence to guide the expert and the decision makers in the choice of a suitable set of tests to be applied to the case study at each investigation tier.

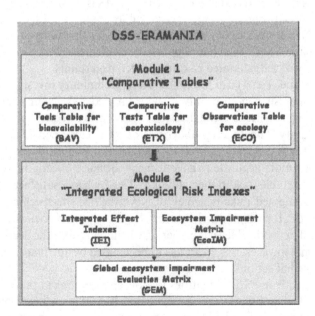

Fig. 10.1 Structure of the DSS-ERAMANIA with indication of main modules and related tools

For this purpose, three specific tables were developed to compare the measurement endpoints belonging to the three LOE: bioavailability assessment tools, ecological observations and ecotoxicological tests. These Comparative Tables were respectively named: Comparative Tools Table for bioavailability (i.e. BAV Table), Comparative Observations Table for ecology (i.e. ECO Table) and Comparative Tests Table for ecotoxicology (i.e. ETX Table) (Fig. 10.1).

They can first of all be regarded as databases for the collection of information related to the specific measurement endpoints. In addition, the tables support experts in: excluding those measurement endpoints which cannot be applied to the case study; comparing the tests belonging to each Triad LOE in order to define their suitability to be applied at the first, second or third tier of investigation; selecting a suitable set of measurement endpoints for the case study.

The multiple measurement endpoints (i.e., bioavailability tools, ecotoxicological tests and ecological observations) included in the Comparative Tables are compared by means of an MCDA-based procedure where multiple criteria are used to support the tests comparison performed by different experts. Three categories of criteria were identified: discriminant, descriptive and comparative. Discriminant criteria were included to make a preliminary selection of tests/observations/tools that really concern the case study. Descriptive criteria provide additional information that can support the experts in the evaluation of the comparative criteria. Comparative criteria were established to compare tests/observations/tools in order to evaluate their suitability to be applied at the previously defined tiers of investigation. The proposed comparative criteria were further classified into different groups: "qualitative" criteria, which include information provided by experts according to their knowledge of tests/observations/tools and their specific expertise; "symbolic", "boolean" and "numerical" criteria, which include "fixed" (i.e. objective) information associated with the test protocols or related documents. Moreover, while boolean and symbolic criteria are characterized by a rating of two or more linguistic attributes respectively (e.g. the linguistic attributes "Chronic" or "Acute" for the criteria "exposure time"), the numerical criteria are described by numerical values. The list of all criteria used in Module 1 of the DSS-ERAMANIA is reported and described in Critto et al., 2007 and Semenzin et al., 2007.

The developed procedure is composed of the following five steps:

- Step 1: Assignation of weights to comparative criteria;
- Step 2: Assignation of numerical equivalents to boolean and symbolic criteria;
- Step 3: Assignation of numerical values to qualitative criteria;
- Step 4: System's normalization of numerical criteria;
- Step 5: System's aggregation and output

described in detail by Critto et al. (2007) and Semenzin et al. (2007).

In the DSS-ERAMANIA Module 1 both numerical and non-numerical criteria need to be considered and, using an ad hoc data entry system all those are converted into a common numerical scale. Moreover, two categories of actors are required, the so called System Expert (SE) and Data Expert (DE). The SEs assign the weights to the comparative criteria for each tier of investigation, and evaluate the relative importance of each proposed attribute of boolean and symbolic criteria compared to the other ones. The DEs have to insert the numerical judgment for the qualitative criteria for each tool (i.e. tests, observations and tools). In both cases, the direct assignment method is applied and the final score is obtained by the subsequent application of the weighted averaging operator.

Three outputs are then available: score, ranking and concordance. For each tier, a score, resulting from the mathematical combination of expert judgments by means of the Simple Additive Weights (SAW) method (Vincke, 1992), is assigned to each test. Then, for each tier and each LOE (i.e. chemistry/bioavailability, ecology, ecotoxicology), a ranking of the measurement endpoints is provided, according to the calculated scores. Finally, an estimation of the judgments' concordance among DEs and among SEs is performed by means of specific indices (i.e. Mean expert concordance and Global mean concordance). On the basis of this information, the Decision Makers (DMs), that should represent the responsible people for the risk management of the contaminated site, can proceed with the selection of a suitable set of tests to be applied to the contaminated site of concern.

10.3.1.1 Software System: DSS-ERAMANIA Module 1

The developed comparison procedure offers a structured solution to one of the most difficult and arbitrary processes in the site-specific ERA application. The implemented set of ad-hoc criteria and Multi Criteria Decision Analysis (MCDA) integrates the evaluation provided by different experts (i.e., Data Experts and System Experts, respectively). The procedure is very flexible allowing easy modification and enlargement of the number of experts, site specific characteristics, and an increasing number of additional or alternative tests/observations/tools according to the users' demands. Obviously, as expected in any site-specific ERA, the obtained results depend on the experts' subjectivity and expertise. For this reason, it is crucial that the procedure makes the entire process transparent and traceable by handling the evaluations of multiple experts and estimating their concordance degree by means of concordance indexes. Moreover, the concordance evaluation is useful in promoting discussion and reaching a consensus among the experts on the assigned values.

To support the application of the proposed procedure, ETX, ECO and BAV Tables were developed and implemented in the "DSS-ERAMANIA Module 1: Comparative Tables" software, in Visual Basic (IDEAS, 2004) (see main interface in Fig. 10.2). Using a matrix format, the comparative tables collect the input information related to the different tests potentially useful for a Triad approach. For this reason they can be considered as databases of tests/observations/tools potentially applicable for site-specific ERA of contaminated sites.

Specifically, the comparative tables were structured with tests/observations/tools reported in different rows and the selected criteria reported in different columns, as described in detail by Critto et al. (2007) and Semenzin et al. (2007).

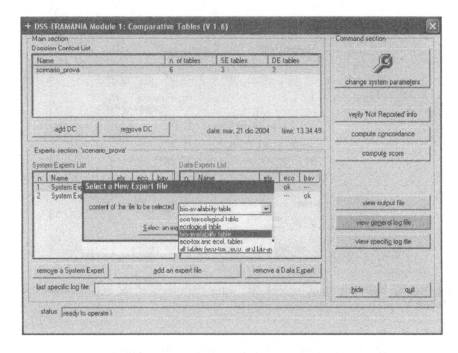

Fig. 10.2 Software interface of the DSS-ERAMANIA Module 1: Comparative tables

For a considerable number of existing ecotoxicological tests, ecological observations and bioavailability tools, the Tables are already populated with information related to the discriminant, descriptive and comparative criteria, the evaluations of which are already provided by specific documents. On the contrary, the Tables' columns referring to the qualitative criteria are empty, in order to be filled by the Data Experts in the third step of the comparative step procedure.

10.3.2 Module 2: Integrated Ecological Risk Indexes

The DSS-ERAMANIA Module 2 provides quantitative and qualitative tools that allow the assessment of terrestrial ecosystem impairment (i.e. the impairment occurring on biodiversity and functional diversity of the terrestrial ecosystem) by integrating the complementary information obtained by the application of the measurement endpoints selected in Module 1. As shown in Fig. 10.1, it includes the following tools: Integrated Effect Index (IEI), Ecosystem Impairment Matrix (EcoIM) and Global Ecosystem Impairment Evaluation Matrix (GEM).

The developed methodology is composed of 9 steps structured as shown in Fig. 10.3 and described in detail by Semenzin et al. (2008a).

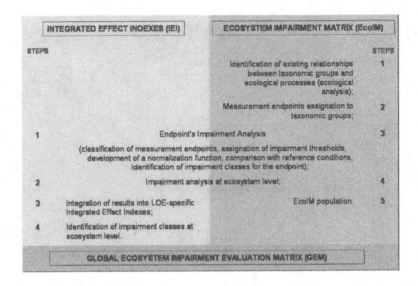

Fig. 10.3 Stepwise procedure implemented in the three tools of the DSS-ERAMANIA Module 2

The core of the procedure implemented in Module 2 is the so-called Ecosystem's Impairment Analysis (step 1 and 3 for IEI and EcoIM respectively, see Fig. 10.3), where the measurement endpoint results, obtained from both contaminated and reference samples, are evaluated in terms of impairment for the tested endpoints and normalized on a 0–1 scale, according to test-specific thresholds (i.e. Negligible Impairment threshold and Relevant Impairment threshold) defined by the involved experts. This way, for each LOE, the scaled test results can be aggregated into a single index (i.e. IEI). The adopted MCDA aggregation operator is a weighted average, where the weights are related to criteria that take into account the relevance of the information provided by each measurement endpoint to the assessment endpoint (i.e. the terrestrial ecosystem). The calculated IEIs quantify the impairment of each contaminated sample by integrating the complementary information obtained by the measurement endpoints belonging to each Triad LOE, compared to the reference conditions. The comprehensive understanding of the quantitative estimation provided by the calculated IEIs is supported by EcoIM. In fact, using the same scaled results obtained from the Ecosystem's Impairment Analysis, EcoIM provides a qualitative assessment of the impairment at the ecosystem level evaluating both biodiversity and functional diversity of the terrestrial ecosystem and highlighting the ecosystem factors stressed by soil contamination. In order to protect both biodiversity and functional diversity of the terrestrial ecosystem, the main elements, which better characterize these components, were identified. Moreover, according to the international literature, the main taxonomic groups occurring in the terrestrial ecosystem and supporting the aforementioned elements were identified and grouped according to the macrogroups: Microorganisms, Plants, Invertebrates and Vertebrates. The ecosystem impairment is then visualised and evaluated by assigning the measurement endpoints' results to the existing relationships between taxonomic groups and performed functions. MCDA methods and soil ecology have been used to develop this tool for supporting risk management. Finally, the outputs of both IEI and EcoIM tools, obtained for the analyzed contaminated and reference sites, can be reported and summarized by the experts in the GEM, allowing a comprehensive evaluation of the impairment occurring on the terrestrial ecosystem of concern.

10.3.2.1 Software System: DSS-ERAMANIA Module 2

To support the application of the proposed procedure, the "DSS-ERAMANIA Module 2: Integrated Ecological Risk Indexes" software was implemented [in Excel; IDEAS, 2004]. Specifically, three tools were developed: IEI, EcoIM and GEM.

As for IEI, the main interface, shown in Figs. 10.4, 10.5 and 10.6 and to be applied for each Triad LOE, supports the collection of all needed data (including those required by the Endpoint's Impairment Analysis), the calculation of each Integrated Effect Index (IEI) and the visualization of the overall results (see Semenzin et al., 2008a for details). As far EcoIM, in order to facilitate the qualitative evaluation of the ecosystem's impairment, it was structured as a matrix with three entrances (see Figs. 10.7, 10.8 and 10.9: (1) the main ecological functions and processes of the terrestrial ecosystem on the top X axis, (2) the taxonomic groups supporting ecological functions and processes on the left Y axis and (3) the applied measurement endpoints, selected by applying Module 1, on the right Y axis (see Semenzin et al., 2008b for details). Finally, GEM was structured in three modules (i.e. EcoIM summary module, IEI summary module and Global Impairment Evaluation Module, see Fig. 10.10) in order to support the expert in summarising the broad and detailed qualitative information obtained by EcoIM, coupling it with the calculated Integrated Effect Indexes (IEIs) and communicating and discussing the results with the decision makers (see Semenzin et al., 2008b for details).

10.4 Case Study Application

The developed DSS-ERAMANIA was preliminary applied to the Acna di Cengio contaminated site for its validation.

The former Acna plant is located in Cengio, a 10.000 inhabitants municipality of the Savona province (Italy). Extending over ca. 550,000 m², it is located in a hilly land named Langhe, in the North West of Italy, and specifically in Valbormida, a valley of the Bormida river, which is a confluent of the Tanaro river and a tributary of the Po river. From 1882 to 1999 its production

Table 10.1 Results of DSS-ERAMANIA Module 1 application to the Acna site for the ecotoxicological LOE: score and ranking for each tier of analysis

Ecotoxicology	Tier 1	Score	Tier 2	Score	Tier 3	Score
1	Root elongation in *Cucumis sativus, Lepidium sativum, Sorghum saccharatum*	0.857	Reproduction of *Lumbricus rubellus*	0.845	Reproduction of *Lumbricus rubellus*	0.839
2	Microtox test elutriate	0.853	Reproduction of *Enchytraeus crypticus*	0.838	Reproduction of *Eisenia species*	0.835
3	Seed germination in *Cucumis sativus, Lepidium sativum, Sorghum saccharatum*	0.844	Reproduction of *Eisenia species*	0.833	Reproduction of *Enchytraeus crypticus*	0.826
4	Chronic toxicity with *Pseudokirchneriella subcapitata*	0.821	Reproduction (hatching juveniles) *Eisenia species*	0.806	Reproduction (hatching juveniles) *Eisenia species*	0.814
5	Mitotic anomalies in *Pisum sativum, Allium porrum L., Zea mais root*	0.811	Growth of *Eisenia species*	0.805	Growth of *Eisenia species*	0.801
6	Nuclear DNA content in *Pisum sativum, Allium porrum L., Zea mais root*	0.811	Mortality of *Lumbricus rubellus*	0.789	Mortality of *Lumbricus rubellus*	0.770
7	Meristematic activity in *Pisum sativum, Allium porrum L., Zea mais root*	0.802	Survival of *Enchytraeus crypticus*	0.782	Survival of *Eisenia species*	0.767
8	Reproduction of *Eisenia species*	0.800	Survival of *Eisenia species*	0.777	Survival of *Enchytraeus crypticus*	0.758
9	Reproduction of *Lumbricus rubellus*	0.793	Root elongation in *Cucumis sativus, Lepidium sativum, Sorghum saccharatum*	0.759	Reproduction of *Folsomia candida*	0.741
10	Growth of *Eisenia species*	0.779	Reproduction of *Folsomia candida*	0.744	Growth of *Folsomia candida*	0.707

Table 10.1 (continued)

Ecotoxicology	Tier 1	Score	Tier 2	Score	Tier 3	Score
11	Survival of *Eisenia species*	0.758	Seed germination in *Cucumis sativus, Lepidium sativum, Sorghum saccharatum*	0.736	Root elongation in *Cucumis sativus, Lepidium sativum, Sorghum saccharatum*	0.691
12	Mortality of *Lumbricus rubellus*	0.751	Growth of *Folsomia candida*	0.716	Survival of *Folsomia candida*	0.679
13	Reproduction of *Enchytraeus crypticus*	0.741	Mitotic anomalies in *Pisum sativum, Allium porrum L., Zea mais* root	0.703	Seed germination in *Cucumis sativus, Lepidium sativum, Sorghum saccharatum*	0.657
14	Microtox test soild phase	0.740	Nuclear DNA content in *Pisum sativum, Allium porrum L., Zea mais*root	0.703	DNA damage of *Lumbricus rubellus*	0.648
15	Reproduction (hatching juveniles) *Eisenia species*	0.740	Leaves growth of *Lactuca sativa*	0.696	Micronuclei frequency of *Lumbricus rubellus*	0.648
16	Seedling germination of *Lactuca sativa*	0.734	Survival of *Folsomia candida*	0.694	Mitotic anomalies in *Pisum sativum, Allium porrum L., Zea mais* root	0.627
17	Leaves growth of *Lactuca sativa*	0.710	Meristematic activity in *Pisum sativum, Allium porrum L., Zea mais* root	0.689	Nuclear DNA content in *Pisum sativum, Allium porrum L., Zea mais* root	0.627
18	Survival of *Enchytraeus crypticus*	0.699	Leaves growth of *Raphanus sativus*	0.656	Leaves growth of *Lactuca sativa*	0.612
19	DNA damage of *Dictyostelium discoideum*	0.693	Tuber growth of *Raphanus sativus*	0.656	Meristematic activity in *Pisum sativum, Allium porrum L., Zea mais* root	0.610
20	Cellular replication rate of *Dictyostelium discoideum*	0.665	Chronic toxicity with *Pseudokirchneriella subcapitata*	0.615	Leaves growth of *Raphanus sativus*	0.595

Table 10.1 (continued)

Ecotoxicology

	Tier 1	Score	Tier 2	Score	Tier 3	Score
21	Leaves growth of *Raphanus sativus*	0.655	Seedling germination of *Lactuca sativa*	0.614	Tuber growth of *Raphanus sativus*	0.595
22	Tuber growth of *Raphanus sativus*	0.655	Aggregation capacity of *Dictyostelium discoideum*	0.609	Aggregation capacity of *Dictyostelium discoideum*	0.568
23	Reproduction of *Folsomia candida*	0.654	DNA damage of *Lumbricus rubellus*	0.576	Lysosomal membranes stability of *Lumbricus rubellus*	0.551
24	Cellular vitality rate of *Dictyostelium discoideum*	0.644	Micronuclei frequency of *Lumbricus rubellus*	0.576	Metallothionein concentration of *Lumbricus rubellus*	0.546
25	Lysosomal membranes stability of *Dictyostelium discoideum*	0.637	Growth inhibition of *Heterocypris incongruens*	0.571	Lysosome/cytoplasm of *Lumbricus rubellus*	0.535
26	Endocytosis rate of *Dictyostelium discoideum*	0.637	Cellular replication rate of *Dictyostelium discoideum*	0.554	Neutral lipid lysosomal concentration of *Lumbricus rubellus*	0.535
27	Growth of *Folsomia candida*	0.633	DNA damage of *Dictyostelium discoideum*	0.550	Lipofuscin lysosomal concentration of *Lumbricus rubellus*	0.535
28	Aggregation capacity of *Dictyostelium discoideum*	0.631	Microtox test elutriate	0.545	Ca2+ ATPase activity of *Lumbricus rubellus*	0.535
29	Growth inhibition of *Heterocypris incongruens*	0.621	Lysosomal membranes stability of *Lumbricus rubellus*	0.520	Growth inhibition of *Heterocypris incongruens*	0.513
30	Survival of *Folsomia candida*	0.618	Microtox test soild phase	0.520	Cellular replication rate of *Dictyostelium discoideum*	0.501
31	Mortality rate of *Heterocypris incongruens*	0.598	Metallothionein concentration of *Lumbricus rubellus*	0.492	Chronic toxicity with *Pseudokirchneriella subcapitata*	0.500
32	PAM (Pulse Amplitude Modulation) of *Selenastrum capricornutum*	0.537	Lysosome/cytoplasm of *Lumbricus rubellus*	0.485	DNA damage of *Dictyostelium discoideum*	0.493

Table 10.1 (continued)

Ecotoxicology	Tier 1	Score	Tier 2	Score	Tier 3	Score
33	DNA damage of *Lumbricus rubellus*	0.486	Neutral lipid lysosomal concentration of *Lumbricus rubellus*	0.485	Seedling germination of *Lactuca sativa*	0.491
34	Micronuclei frequency of *Lumbricus rubellus*	0.486	Lipofuscin lysosomal concentration of *Lumbricus rubellus*	0.485	Microtox test soild phase	0.421
35	Lysosomal membranes stability of *Lumbricus rubellus*	0.467	Ca2+ ATPase activity of *Lumbricus rubellus*	0.485	Cellular vitality rate of *Dictyostelium discoideum*	0.409
36	Metallothionein concentration of *Lumbricus rubellus*	0.424	Cellular vitality rate of *Dictyostelium discoideum*	0.483	Lysosomal membranes stability of *Dictyostelium discoideum*	0.393
37	Lysosome/cytoplasm of *Lumbricus rubellus*	0.419	Mortality rate of *Heterocypris incongruens*	0.477	Endocytosis rate of *Dictyostelium discoideum*	0.393
38	Neutral lipid lysosomal concentration of *Lumbricus rubellus*	0.419	Lysosomal membranes stability of *Dictyostelium discoideum*	0.471	Mortality rate of *Heterocypris incongruens*	0.389
39	Lipofuscin lysosomal concentration of *Lumbricus rubellus*	0.419	Endocytosis rate of *Dictyostelium discoideum*	0.471	Microtox test elutriate	0.374
40	Ca2+ ATPase activity of *Lumbricus rubellus*	0.419	PAM (Pulse Amplitude Modulation) of *Selenastrum capricornutum*	0.336	PAM (Pulse Amplitude Modulation) of *Selenastrum capricornutum*	0.194

Table 10.2 Results of DSS-ERAMANIA Module 1 application to the Acna site for the ecological LOE: score and ranking for each tier of analysis

Ecology					
Tier 1	Score	Tier 2	Score	Tier 3	Score
1　Bacteria community structure (DNA)	0.854	Bacteria community structure (DNA)	0.785	Bacteria community structure (DNA)	0.761
2　Bacteria leucine incorporation (leu)	0.816	Bacteria leucine incorporation (leu)	0.761	Bacteria pollution Induced Community Tolerance (PICT)	0.745
3　Bacteria thymidine incorporation (Tdr)	0.816	Bacteria thymidine incorporation (Tdr)	0.761	Bacteria metabolic community profiles (CLPP)	0.728
4　BSQ Index (invertebrates)	0.794	BSQ Index (invertebrates)	0.758	Vegetation survey	0.725
5　Enchytraeidae community analysis	0.737	Bacteria pollution Induced Community Tolerance (PICT)	0.757	Bacteria biomass	0.706
6　Earthworm community analysis	0.735	Bacteria biomass	0.733	BSQ Index (invertebrates)	0.702
7　Bacteria C-mineralization rates	0.726	Vegetation survey	0.713	Bacteria leucine incorporation (leu)	0.696
8　Bacteria N-mineralization rates	0.726	Bacteria C-mineralization rates	0.705	Bacteria thymidine incorporation (Tdr)	0.696
9　Bacteria pollution Induced Community Tolerance (PICT)	0.726	Bacteria N-mineralization rates	0.705	Bacteria C-mineralization rates	0.656
10　Enchytraeidae Number and Biomass	0.717	Bacteria metabolic community profiles (CLPP)	0.685	Bacteria N-mineralization rates	0.656
11　Earthworm Number and Biomass	0.716	Enchytraeidae community analysis	0.642	Enchytraeidae community analysis	0.560
12　Vegetation survey	0.702	Earthworm community analysis	0.641	Earthworm community analysis	0.560
13　Bait Lamina (earthworm)	0.682	Enchytraeidae Number and Biomass	0.617	Enchytraeidae Number and Biomass	0.532
14　Fungi biomass	0.673	Earthworm Number and Biomass	0.616	Earthworm Number and Biomass	0.532
15　Bacteria biomass	0.671	Fungi biomass	0.597	Bait Lamina (earthworm)	0.525
16　Fungi hyphal lenght	0.659	Fungi hyphal lenght	0.578	Survey/monitoring of micro- and macro-arthropods	0.519

Table 10.2   (continued)

Ecology					
17 Nematodes community analysis (MATURITY INDEX)	0.632	Bait Lamina (earthworm)	0.573	Fungi biomass	0.499
18 Bacteria metabolic community profiles (CLPP)	0.589	Nematodes community analysis (MATURITY INDEX)	0.570	Nematodes community analysis (MATURITY INDEX)	0.493
19 Mites/Springtails ratio	0.551	Survey/monitoring of micro- and macro-arthropods	0.497	Fungi hyphal lenght	0.477
20 Oribatid/non oribatida ratio	0.530	Mites/Springtails ratio	0.475	Mites/Springtails ratio	0.457
21 Survey/monitoring of micro- and macro-arthropods	0.502	Oribatid/non oribatida ratio	0.443	Oribatid/non oribatida ratio	0.413

changed from dynamite and tri-nitrotoluene to lighting gas, chemical products (e.g. nitric acid, phenol, sulphuric acid), dyes and pigments and finally to β-naphthol and phtalocyanine, causing relevant environmental problems. In December 1998, it was identified as one of the 14 contaminated sites of national interest to be reclaimed. Since 2001, remedial actions have been carried out at the Acna site. As a result, in the internal areas, the surface layer of the con-taminated soil was largely removed and reclaimed, or buried in a controlled dumping ground.

The preliminary application of Module 1 to the Acna site, carried out by the Authors acting as one Data Expert (DE) and one System Expert (SE), allowed the ranking of 40 ecotoxicological tests and 21 ecological observations (see IDEAS, 2004 for details) and 14 bioavailability assessment tools (see Semenzin et al., 2007 for details) for the three tiers of the proposed framework. The obtained results are reported in Table 10.1, 10.2 and 10.3 respectively.

For the application to the Acna site of the complete set of measurement endpoints selected by Module 1 the reader should refer to Semenzin et al., 2008a and 2008b. In this chapter only an example of Module 2 application conducted on a subsample of measurement endpoints and sampling stations is presented and discussed. Specifically, the ecotoxicological tests, ecological observations and bioavailability assessment tools reported in Table 10.4 were selected for a first Triad tier of application. Within the ecotoxicological tests (see Table 10.1), all the plant tests obtaining a high score were selected, moreover some suitable tests on microorganisms and invertebrates were selected in order to cover the different taxonomic groups occurring in the system, while the biomarkers on

Table 10.3 Results of DSS-ERAMANIA Module 1 application to the Acna site for the chemical/bioavailability LOE: score and ranking for each tier of analysis

Bioavailability

	Tier 1	Score	Tier 2	Score	Tier 3	Score
1	Extractions by moderately polar organic solvent	0.785	Tissue accumulation in invertebrates (laboratory test)	0.811	Tissue accumulation in invertebrates (laboratory test)	0.888
2	Regression model: OC; %sand; %clay; CEC	0.758	Extractions by moderately polar organic solvent	0.797	Field survey: whole organism bioaccumulation	0.868
3	Organic carbon content (f_{oc}; f_{om})	0.734	Semipermeable Membrane Devices (SPMD)	0.768	Pore water measurements: soil solution sampling and conventional analysis	0.838
4	Soil texture (%sand; %clay; %loam)	0.729	Pore water measurements: soil solution sampling and conventional analysis	0.767	Semipermeable Membrane Devices (SPMD)	0.797
5	CEC (Cation Exchange Capacity)	0.729	Polymeric resins such as XAD-2, XAD-4 and Tenax TA (a 2,6-diphenyl-p-phenylene oxide based polymer)	0.742	Extractions by moderately polar organic solvent	0.774
6	Regression models for plants	0.723	Regression model: OC; %sand; %clay; CEC	0.730	Polymeric resins such as XAD-2, XAD-4 and Tenax TA (a 2,6-diphenyl-p-phenylene oxide based polymer)	0.758
7	Semipermeable Membrane Devices (SPMD)	0.710	Field survey: whole organism bioaccumulation	0.727	Geochemical speciation model "Windermere Humic Aqueous Model" (WHAM)	0.747
8	Pore water measurements: soil solution sampling and conventional analysis	0.702	Geochemical speciation model "Windermere Humic Aqueous Model" (WHAM)	0.698	Regression model: OC; %sand; %clay; CEC	0.719
9	Polymeric resins such as XAD-2, XAD-4 and Tenax TA (a 2,6-	0.690	Regression models for plants	0.693	Regression models for plants	0.660

Table 10.3 (continued)

Bioavailability

	Tier 1	Score	Tier 2	Score	Tier 3	Score
	diphenyl-p-phenylene oxide based polymer)					
10	Tissue accumulation in invertebrates (laboratory test)	0.690	X-ray diffraction (XRD) with associated energy dispersive spectroscopy	0.681	X-ray diffraction (XRD) with associated energy dispersive spectroscopy	0.619
11	X-ray diffraction (XRD) with associated energy dispersive spectroscopy	0.690	Organic carbon content (f_{oc}; f_{om})	0.626	Organic carbon content (f_{oc}; f_{om})	0.513
12	Geochemical speciation model "Windermere Humic Aqueous Model" (WHAM)	0.656	Soil texture (%sand; %clay; %loam)	0.620	Elemental Analysis (EA)	0.504
13	Elemental Analysis (EA)	0.646	CEC (Cation Exchange Capacity)	0.620	Soil texture (%sand; %clay; %loam)	0.504
14	Field survey: whole organism bioaccumulation	0.569	Elemental Analysis (EA)	0.568	CEC (Cation Exchange Capacity)	0.504

Table 10.4 Measurement endpoints applied to the two selected sampling stations: A (reference sample) and C (contaminated sample)

Ecotoxicology	Ecology	Bioavailability
Root elongation and seeds germination of *Cucumis sativus, Lepidium sativum, Sorghum saccharatum* (UNICHIM 1651, 2003)	BSQ index (invertebrates) (Parisi, 2001; Angelini et al., 2002; Gardi et al., 2002)	Chemical extraction (by $CaCl_2$ 1 M) (Petruzzelli et al., 1993; Aten and Gupta, 1996)
Microtox test (*Vibrio fischeri*) on both elutriate and solid phase (ANPA, 2004)	Mites/Springtails ratio (Gorny and Grum, 1993)	
Growth inhibition of *Pseudokirchneriella subcapitata* (ANPA, 2004)	Oribatida/non Oribatida ratio (Gorny and Grum, 1993)	
Biomarkers on roots of pea (*Pisum sativum*), leek (*Allium porrum L.*), maize (*Zea mais*) (Berta et al., 1990; Hooker et al., 1998; Cesaro et al., 2005; Fusconi et al., 2006; Lingua et al., 2001a, 2001b)	Survey/monitoring of micro- and macro-arthropods: indexes of abundance, taxonomic richness and biodiversity (Shannon, Margalef, Pielou) (Magurran, 1988)	
Mortality and reproduction of Lumbricidae (*Eisenia andrei*) (OECD, 1984)		
Biomarkers, mortality and aggregation capacity on amoeba (*Dictyostelium discoideum*) (Dondero et al., 2006; UNIPMN, 2004)		
Growth and mortality of *Heterocypris incongruens* (solid phase) (ANPA, 2004)		

Lumbricidae and the tests on *Folsomia candida* were left out because they were determined to be more suitable for a second or a third Triad tier. Within the ecological observations (Table 10.2), the Biological Soil Quality (BSQ) index and the other observations on micro- and macroarthropods were selected because of the higher score obtained for a first Triad tier compared to tiers 2 and 3 and because they were so easily applicable to the site of concern compared to the other observations. Finally, within the bioavailability tools (Table 10.3), the method of contaminant extraction obtaining the higher score (i.e. higher rank) for tier 1 was selected. The measurement endpoints listed in Table 10.4 were performed by Piedmont Environmental Protection Agency (ARPAP), Eastern Piedmont University and CNR Institute for Ecosystem study on the reference sampling station A and the contaminated sample C. The reference sample A is located upstream (approximately 2 km away from the Acna site) along the river's border while C is located within the Acna site, and the soil sample was collected in a thin grass strip (2 × 10 m) between the road and the buildings. The samples were collected during the campaign carried out in July

Table 10.5 Results of chemical and physico-chemical analyses on fresh soil samples (A, C) collected during the Acna's sampling campaign in July 2004

Sample Site		A		C	
		Concentrations (mg/kg dw)			
Chemicals		Total	Bioavailable (CaCl2 1 M)	Total	Bioavailable (CaCl2 1 M)
Metals	Antimony	5.7	–	8.5	–
	Arsenic	21	1.10	16	1.00
	Cadmium	1.6	0.06	1.9	0.12
	Cobalt	8.1	–	8.6	–
	Chromium	36	1.0	111	1.0
	Mercury	0.8	0.07	16	1.20
	Nickel	24	16.5	60	17.0
	Lead	21	21.0	501	36.5
	Copper	17	3.0	140	3.0
	Selenium	<1	–	<1	–
	Barium	119	–	293	–
	Zinc	116	–	311	–
	Iron	29,566	–	38,626	–
	Aluminum	24,490	–	25,469	–
	Manganese	402	–	321	–
	Molybdenum	<1	–	4.4	–
PAHs	Benzo(a)anthracene	<0.05	–	0.90	–
	Benzo(a)pyrene	0.01	–	0.20	–
	Benzo(b)fluoranthene	<0.05	–	0.61	–
	Benzo(ghi)perylene	0.01	–	0.30	–
	Benzo(k)fluoranthene	<0.05	–	0.610	–
	Chrysene	0.5	–	<0.5	–
	Dibenzo(a,h)anthracene	<0.01	–	0.07	–
	Indeno(123 cd)pyrene	0.01	–	0.22	–
	Pyrene	<0.5	–	<0.5	–
Phenols	Pentachlorophenol	<0.01	–	<0.01	–
	beta-naphthol	<0.01	–	0.03	–
Halogenated Aromatic Compounds	1,2,4,5-Tetrachlorobenzene	<0.05	–	<0.05	–
	1,2,4-Trichlorobenzene	<0.1	–	<0.1	–

Sample Site		A	C
Physico-Chemical Properties			
Texture	LONGIT	1,436,240	1,434,867
	LATID	4,913,311	4,915,517
	Sand (%dw)	75	73
	Silt (% dw)	13	14
	Clay (% dw)	12	14
	pH	7.9	7.6
	TOC (% dw)	1.4	1.8

ENDPOINT'S IMPAIRMENT ANALYSIS												INTEGRATED EFFECT INDEXES CALCULATION						
						C3 Response scale							A		C			
Taxonomic macrogroup (microrganisms, plants, invertebrates, vertebrates)	Test endpoint (DNA, cellular, organism, population, community)	Test name	U.M.	C1 Result transformation according to lab standard (S) or field reference (R)	C2 Response trend (increasing, decreasing)	a	b	Classification group (1-7)	Th1	Th2	Endpoint sensitivity (1,100)	Measurement endpoint (test) results in the reference site	Impairment evaluation for the endpoint [0,1] Aend,i	Measurement endpoint (test) results in site C	Impairment evaluation for the endpoint [0,1] Aend,i	Impairment evaluation for [illegible] [0,1] Aays,i	note
MICRORGANISMS	ORGANISM	Vibrio fischeri elutriate	I% (Luminescence inhibition)	R	increasing	0	100	5	20	80	10	0	0,00	-4,40	0,00	0,00	
MICRORGANISMS	ORGANISM	Vibrio fischeri solid phase (Inhibition referred to a control of ultrapure sand)	I% (Luminescence inhibition)	R	increasing	0	100	5	20	80	10	0	0,00	40,20	0,34	0,03	
MICRORGANISMS	DNA	DNA damage (D.discoideum)	pixel number	R	increasing	1	∞	6	1,2	4	80	1	0,00	1,97	0,27	0,22	
MICRORGANISMS	CELLULAR	Lysosomal membrane stability (D.discoideum)	density	R	decreasing	0	100	7	80	20	10	100	0,00	49,69	0,51	0,05	
MICRORGANISMS	CELLULAR	Endocytosis rate (D.discoideum)	density	R	decreasing	0	100	7	80	20	10	100	0,00	59,98	0,33	0,03	
MICRORGANISMS	CELLULAR	Replication rate (D.discoideum)	percentage	R	decreasing	0	100	7	80	20	25	100	0,00	64,92	0,25	0,06	
MICRORGANISMS	CELLULAR	Mortality (D.discoideum)	percentage	no	increasing	0	100	1	20	80	25	20,17	0,00	30,65	0,18	0,04	Th1 replaced by reference
MICRORGANISMS	ORGANISM	Aggregation capacity (D.discoideum)	number	R	decreasing	0	100	7	80	20	25	100	0,00	22,52	0,96	0,24	
PLANTS	ORGANISM	Pseudokirchneriella subcapitata elutriate	I% (Growth inhibition)	R	increasing	0	100	5	20	80	70	0	0,00	7,00	0,00	0,00	
PLANTS	ORGANISM	Lepidium sativum solid phase	Ig% (Germination inhibition)	R	increasing	0	100	5	20	80	30	0	0,00	0,79	0,00	0,00	
PLANTS	ORGANISM	Lepidium sativum solid phase	Ia% (Elongation inhibition)	R	increasing	0	100	5	20	80	50	0	0,00	-3,30	0,00	0,00	
PLANTS	ORGANISM	Cucumis sativus solid phase	Ig% (Germination inhibition)	R	increasing	0	100	5	20	80	30	0	0,00	-5,26	0,00	0,00	
PLANTS	ORGANISM	Cucumis sativus solid phase	Ia% (Elongation inhibition)	R	increasing	0	100	5	20	80	50	0	0,00	-2,20	0,00	0,00	
PLANTS	ORGANISM	Sorghum saccharatum root solid phase	Ig% (Germination inhibition)	R	increasing	0	100	5	20	80	30	0	0,00	5,13	0,00	0,00	
PLANTS	ORGANISM	Sorghum saccharatum root solid phase	Ia% (Elongation inhibition)	R	increasing	0	100	5	20	80	50	0	0,00	14,10	0,00	0,00	
PLANTS	ORGANISM	Sorghum saccharatum shoot solid phase	Ig% (Germination inhibition)	R	increasing	0	100	5	20	80	30	0	0,00	5,13	0,00	0,00	
PLANTS	ORGANISM	Sorghum saccharatum shoot solid phase	Ia% (Elongation inhibition)	R	increasing	0	100	5	20	80	50	0	0,00	12,80	0,00	0,00	
PLANTS	DNA	Nuclear DNA content (maize)	percentage	S	decreasing	0,2	0,8	4	0,53	0,35	80	0,57	0,00	0,38	0,83	0,67	
PLANTS	CELLULAR	Meristematic activity (maize)	percentage	S	decreasing	0	12	4	6,0	1,5	70	0,57	0,00	0,87	1,00	0,70	
PLANTS	CELLULAR	Mitotic anomalies (maize)	percentage	S	increasing	0	6,5	3	0,35	3,64	80	0,00	0,00	0,41	0,02	0,02	
PLANTS	DNA	Nuclear DNA content (pea)	percentage	S	decreasing	0,5	1,5	4	1,3	0,9	80	1,34	0,00	0,72	1,00	0,80	
PLANTS	CELLULAR	Meristematic activity (pea)	percentage	S	decreasing	0	14	4	5,2	1,3	70	5,56	0,00	0,93	1,00	0,70	
PLANTS	CELLULAR	Mitotic anomalies (pea)	percentage	S	increasing	0	6,5	3	0,48	4,8	80	0,00	0,00	0,88	0,09	0,07	
PLANTS	DNA	Nuclear DNA content (leek)	percentage	S	decreasing	3,5	10	4	7,81	5,2	80	8,67	0,00	4,39	1,00	0,50	
PLANTS	CELLULAR	Meristematic activity (leek)	percentage	S	decreasing	0	14	4	7,66	1,91	70	9,57	0,00	2,38	0,92	0,64	
PLANTS	CELLULAR	Mitotic anomalies (leek)	percentage	S	increasing	0	6,5	3	0,32	3,22	80	0,10	0,00	0,33	0,00	0,00	
INVERTEBRATES	ORGANISM	Heterocypris incongruens solid phase	M% (Mortality)	no	increasing	0	100	1	20	80	30	18,00	0,00	40,00	0,33	0,10	
INVERTEBRATES	ORGANISM	Heterocypris incongruens solid phase	I% (Growth inhibition)	R	increasing	0	100	5	20	80	50	0	0,00	48,00	0,47	0,23	
INVERTEBRATES	ORGANISM	Mortality (E.andrei)	percentage	no	increasing	0	100	1	20	80	30	30,00	0,00	55,00	0,38	0,15	Th1 replaced by reference
INVERTEBRATES	ORGANISM	Reproduction (E.andrei)	percentage	R	decreasing	0	100	7	80	20	70	100	0,00	82,61	0,00	0,00	

Integrated Effect Indexes: 0,19 IEI_etx_C

Standard deviation: 0,28 SD_etx_C

Fig. 10.4 Software interface of the "DSS-ERAMANIA Module 2: Integrated Ecological Risk Indexes" applied to the Acna di Cengio contaminated sampling station C, reporting the results of Endpoint's Impairment Analysis for ecotoxicological tests and calculation of the Integrated Effect Index for ecotoxicological LOE (IEI_etx) with the corresponding standard deviation. Legend for colours in Table 10.6. U.M. = Unit of Measurement; Th1 = Negligible impairment threshold; Th2 = Relevant impairment threshold

Taxonomic macrogroup (microrganisms, plants, invertebrates, vertebrates)	Test endpoint (DNA, cellular, organism, population, community)	Test name	U.M.	C1 Result transformation according to lab standard (S) or field reference (R)	C2 Response trend (increasing, decreasing)	C3 Response scale a	C3 Response scale b	Classification group (1-7)	Th1	Th2	Sensitivity of test for ecosystem effects [1,100]	A — Measurement endpoint (test) results in the reference site	A — Impairment evaluation for the endpoint Aend,i [0,1]	C — Measurement endpoint (test) results in site C	C — Impairment evaluation for the endpoint Aend,i [0,1]	C — Impairment evaluation for ecosystem Asys,i [0,1]	note
INVERTEBRATES	community	Shannon Biodiversity index	adimensional	R	decreasing	100	0	7	0.7	0.3	20	1	0.00	0.77	0.00	0.00	
INVERTEBRATES	community	Margalef Biodiversity index	adimensional	R	decreasing	100	0	7	0.7	0.3	20	1	0.00	1.28	0.60	0.00	
INVERTEBRATES	community	Pielou Biodiversity index	adimensional	R	decreasing	100	0	7	0.7	0.3	20	1	0.00	1.15	0.00	0.00	
INVERTEBRATES	community	Abundance	number of organisms/volume	R	decreasing	100	0	7	0.7	0.3	20	1	0.00	0.19	1.00	0.20	
INVERTEBRATES	community	Taxonomic richness	number of taxonomic units	R	decreasing	100	0	7	0.7	0.3	20	1	0.00	0.44	0.64	0.13	
INVERTEBRATES	community	Biological soil quality index (QBS)	adimensional	R	decreasing	100	0	7	0.7	0.3	70	1	0.00	0.39	0.78	0.55	
INVERTEBRATES	community	Mites/Springtails ratio	adimensional	R	decreasing	100	0	7	0.7	0.3	20	1	0.00	0.71	0.00	0.00	
INVERTEBRATES	community	Oribatida/Non oribatida ratio	adimensional	R	decreasing	100	0	7	0.7	0.3	10	1	0.00	1.07	0.00	0.00	

Integrated Effect Index: IEI_eco C = 0.11

Standard deviation: SD_eco_C = 0.19

Fig. 10.5 Software interface of the "DSS-ERAMANIA Module 2: Integrated Ecological Risk Indexes" applied to the Acna di Cengio contaminated sampling station C, reporting the results of Endpoint's Impairment Analysis for ecological observations and calculation of the Integrated Effect Index for ecological LOE (IEI_eco) with the corresponding standard deviation. Legend for colours in Table 10.6. U.M. = Unit of Measurement; Th1 = Negligible impairment threshold; Th2 = Relevant impairment threshold

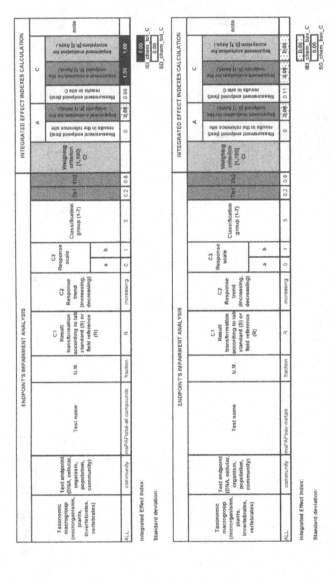

Fig. 10.6 Software interface of the "DSS-ERAMANIA Module 2: Integrated Ecological Risk Indexes" applied to the Acna di Cengio contaminated sampling station C, reporting the results of Endpoint's Impairment Analysis for msPAF (multisubstances Potentially Affected Fraction) and calculation of the Integrated Effect Index for chemical LOE from both total contaminant concentrations (IEI_chem_tot) and bioavailable fraction of metals (IEI_chem_bav), with the corresponding standard deviations. Legend for colours in Table 10.6. U.M. = Unit of Measurement; Th1 = Negligible impairment threshold; Th2 = Relevant impairment threshold

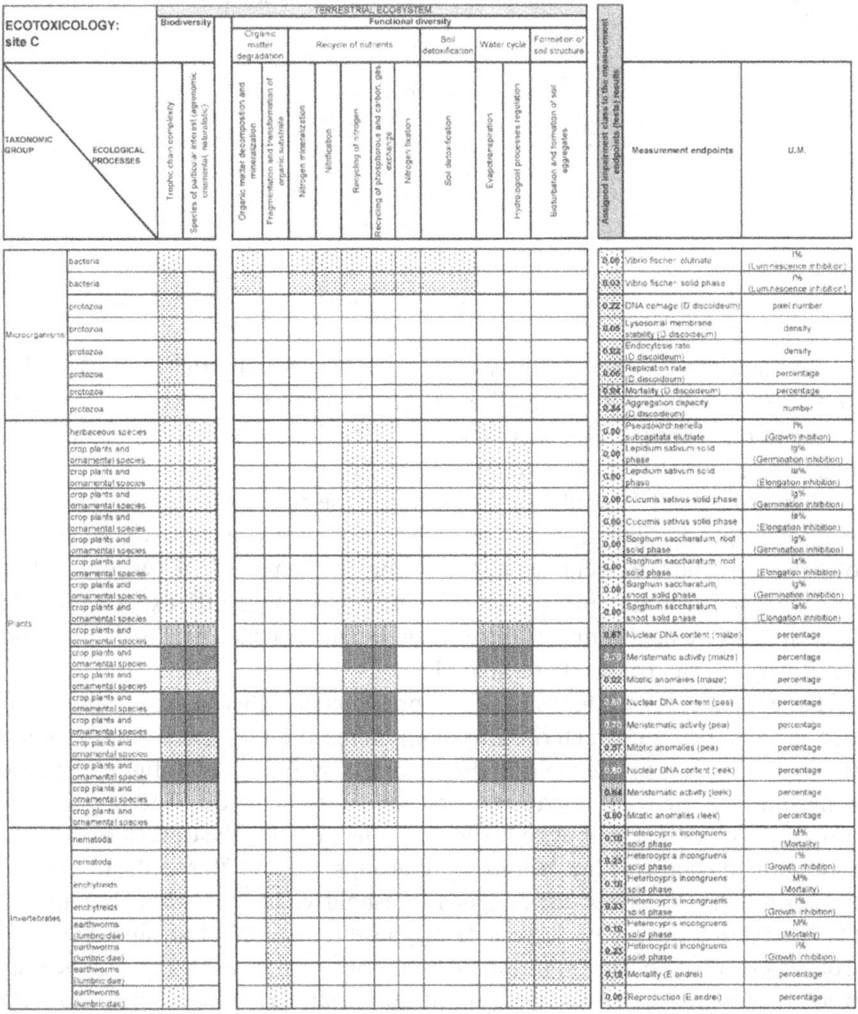

Fig. 10.7 Software interface of the Ecosystem Impairment Matrix (EcoIM) after application to the Acna di Cengio contaminated sampling station C (sample A being utilized as normalization reference). The results refer to the ecotoxicological LOE; legend for colours in Table 10.6. U.M. = Unit of Measurement

2004. The list of chemical and physico-chemical analyses performed on the two sampling sites is presented in Table 10.5.

The results of the measurement endpoints application were processed according to Module 2 in order to obtain both a quantitative and qualitative evaluation of soil quality in sample C (compared to the reference A). First, for the ecotoxicological, ecological and chemical LOE, the Endpoints' Impairment Analysis was applied and the Integrated Effect Indexes (IEI) was calculated as reported in Figs. 10.4, 10.5 and 10.6 respectively. Then, the Ecosystem

Table 10.6 Impairment classes used for each impairment at ecosystem level (A$_{i,sys}$) and each Integrated Effect Index (IEI), with indication of range of values, linguistic evaluation and colour

Ecosystem's Impairment classes (A$_{i,sys}$-IEI)	Linguistic evaluation (symbol)	Colour
0.0	Negligible (N)	
0.0 < A$_{i,sys}$-IEI ≤ 0.3	Intermediate I (II)	
0.3 < A$_{i,sys}$-IEI ≤ 0.7	Intermediate II (II)	
0.7 < A$_{i,sys}$-IEI < 1.0	Intermediate III (III)	
1.0	Relevant (R)	

Impairment Matrix (EcoIM) was applied for the three LOEs (Figs. 10.7, 10.8 and 10.9) and finally the overall evaluation was formulated by the Authors acting as one expert and it was reported in the Global ecosystem impairment Evaluation Matrix (GEM) as shown in Fig. 10.10.

Regarding the IEI calculation, according to the proposed stepwise procedure (Semenzin et al., 2008a), first the applied ecotoxicological tests, ecological observations and the calculated msPAFs (for the chemical LOE) were classified into the seven groups of measurement endpoints. Secondly, the experts assigned to each measurement endpoint the Negligible Impairment Threshold (Th1) and the Relevant Impairment Threshold (Th2).

The ecotoxicological tests (Fig. 10.4) presented a higher variety of classification groups, being characterized by both increasing and decreasing response trends (according to the measured endpoint), by finite and infinite response scales and by different equations for result normalization. The impairment threshold values assigned by the experts are the same for the tests characterized by a finite [0,100] scale (i.e. belonging to groups 1, 5 and 7); specifically, to the increasing response tests, the assigned threshold values were 20 (Th1) and 80 (Th2), inverted in case of decreasing response tests to 80 (Th1) and 20 (Th2). Conversely, the impairment threshold values assigned by the experts to the tests belonging to groups 3, 4 and 6 are specifically related to the measured endpoint and to the response scale obtained by normalizing the test result to the laboratory standard (groups 3 and 4) or to the field reference (group 6).

All the applied ecological observations (Fig. 10.5) belong to the group of measurement endpoints 7 (i.e. the impairment is highlighted by normalizing the result to field reference), the response trend is decreasing and the test response is finite [100,0]. For these observations, in fact, it is necessary to characterize the site under natural conditions (reference) in order to appropriately evaluate the magnitude of impairment from contaminated samples. For these observations, the experts evaluated as negligible (Th1) a result equal to 70% compared to the field reference (i.e. calculated by the equation [site*100/reference]) and as relevant (Th2) a result equal to 30%.

As far as the chemical LOE is concerned (Fig. 10.6), the two msPAFs, calculated from the total contaminant (i.e. metals and organics) concentrations

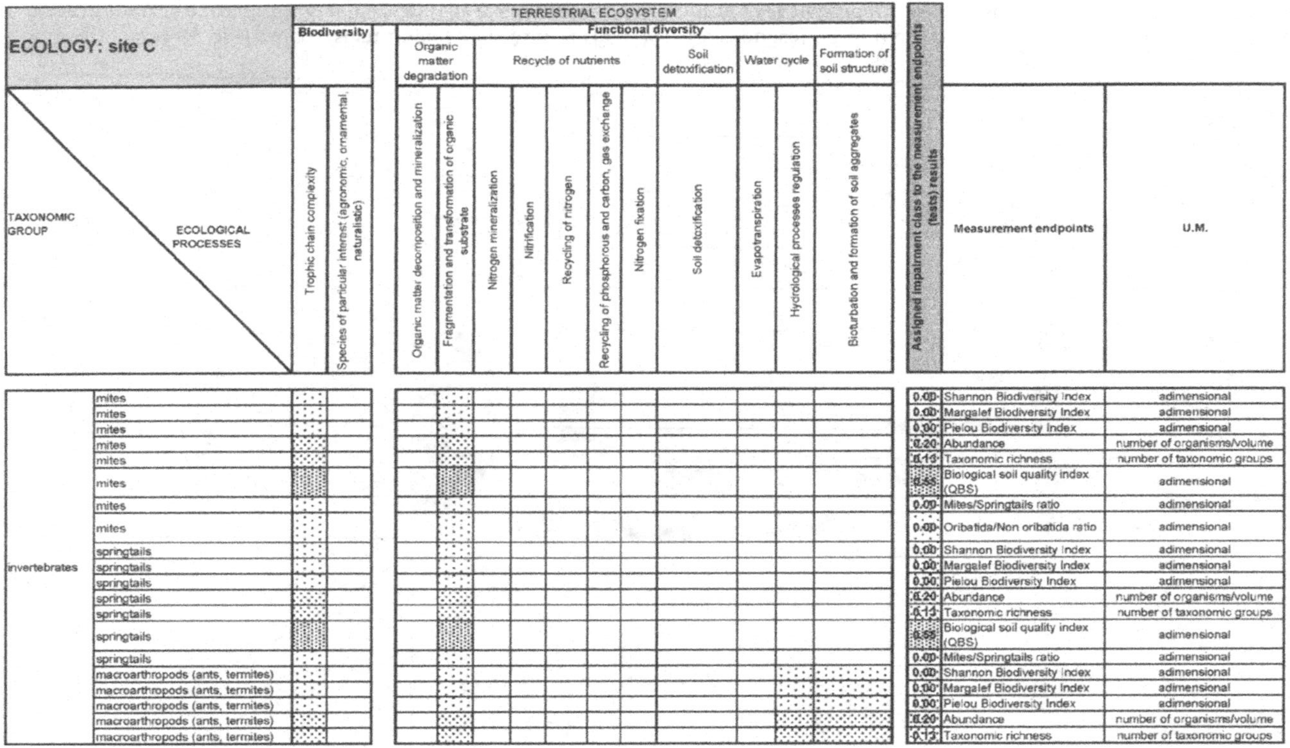

Fig. 10.8 Software interface of the Ecosystem Impairment Matrix (EcoIM) after application to the Acna di Cengio contaminated sampling station C (sample A being utilized as normalization reference). The results refer to the ecological LOE; legend for colours in Table 10.6. U.M. = Unit of Measurement

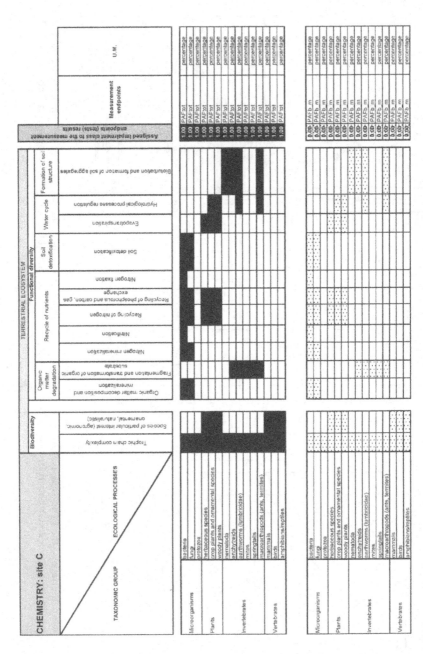

Fig. 10.9 Software interface of the Ecosystem Impairment Matrix (EcoIM) after application to the Acna di Cengio contaminated sampling station C (sample A being utilized as normalization reference). The results refer to the chemical LOE; legend for colours in Table 10.6. U.M. = Unit of Measurement

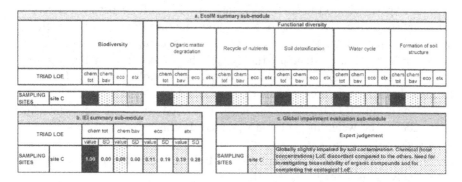

Fig. 10.10 Software interface of the Global ecosystem impairment Evaluation Matrix (GEM) after application to the Acna di Cengio contaminated sampling station C, reporting the results as obtained by Figs. 10.4, 10.5, and 10.6 (**a**) EcoIM summary sub-module, by the Integrated Effect Index (IEI) tool (**b**) IEI summary sub-module (Figs. 10.7, 10.8 and 10.9), followed by the overall impairment evaluation (**c**) Global impairment evaluation sub-module; legend for colours in Table 10.6. LOE = Line of evidence; SD = Standard Deviation

in soil (msPAFtotal-all compounds) and from the bioavailable concentrations of metals (msPAFbav-metals) respectively, belong to group 5 because they need to be normalized to a field reference, and are characterized by an increasing response trend and by a finite response scale. The impairment threshold values assigned by the experts were 0.2 and 0.8 for Th1 and Th2 respectively.

Continuing with the IEI procedure of calculation (see Semenzin et al., 2008a, Semenzin et al., 2008b for details), according to the assigned impairment threshold values, for each measurement endpoint, a specific normalization function was built to estimate the impairment for the endpoint. Moreover, the reference result acceptability was evaluated and the obtained results were coloured according to the five impairment classes reported in Table 10.6. The results of these three operations are also reported in Figs. 10.4, 10.5 and 10.6, for ecotoxicological, ecological and chemical LOE respectively.

Consequently, each result was processed in order to evaluate the impairment at ecosystem level (A_{sys}) highlighted by the Triad LOEs.

Specifically, for the ecotoxicological LOE, for each measurement endpoint, the impairment for the endpoint (A_{end}) obtained by the Endpoint's Impairment Analysis was weighed by means of the values assigned by the expert (i.e. the authors acting as Data Expert, DE) to the criterion "Endpoint sensitivity". These values are reported in Fig. 10.4 and show that the DE considered more sensitive the genetic endpoints (e.g. the DNA damage in amoeba *D. discoideum* and the analysis of mitotic anomalies in plants) than the reproduction and survival endpoints. The obtained A_{sys} for each ecotoxicological test are also reported in Fig. 10.4.

For the ecological LOE, for each measurement endpoint, the impairment for the endpoint (A_{end}) obtained by the Endpoint's Impairment Analysis was weighed by means of the values assigned by the expert (i.e. the authors acting as Data Expert, DE) to the criterion "Sensitivity of test for ecosystem effects". These values are reported in Fig. 10.5 and show that the DE assigned the highest numerical value to the observation concerning more complex evaluation of the invertebrate community structure (i.e. BSQ), which was considered very informative in terms of risk for the whole ecosystem. The obtained A_{sys} for each ecological observation are also reported in Fig. 10.5.

For the chemical LOE, according to the proposed procedure (see Semenzin et al., 2008a), no weighing criterion was used. In this case, for each measurement endpoint, the impairment at ecosystem level (A_{sys}) was considered equal to the impairment for the endpoint (A_{end}) obtained by the Endpoint's Impairment Analysis (as reported in Fig. 10.6).

Finally, the impairments at ecosystem level highlighted by the different measurement endpoints were mathematically aggregated, within each LOE, in order to obtain four Integrated Effect Indexes (IEI_{eco}, IEI_{etx}, IEI_{chem_tot} and IEI_{chem_bav}). The indexes results were coloured according to the Ecosystem's Impairment classes reported in Table 10.6. The obtained results and their correspondent standard deviations are also reported in Figs. 10.4 (IEI_{etx}), 10.5 (IEI_{eco}), and 10.6 (IEI_{chem_tot} and IEI_{chem_bav}).

Observing the obtained indexes, according to both ecotoxicological and ecological LOE, sample C was shown to be slightly impaired (i.e. IEI_{etx} and IEI_{eco} belonging to the first intermediate class; Table 10.6), while the two chemical indexes (IEI_{chem_tot} and IEI_{chem_bav}) highlighted very different situations. According to IEI_{chem_tot} site C was shown to be affected by a relevant impairment (black colour in Table 10.6) but this evidence was not confirmed by IEI_{chem_bav}, whose result was equal to zero, corresponding to a Negligible Ecosystem's Impairment class (see Table 10.6 for colour).

Therefore, summarizing the quantitative estimation of soil quality (i.e. IEI calculation), sample C of Acna was shown to be slightly impaired compared to the reference conditions of A. The impairment highlighted by the IEI_{chem_tot}, which was not confirmed by ecological and ecotoxicological evidences, can be explained by the low contaminants' bioavailability (experimentally checked for metals) in soil, as highlighted by the calculated IEI_{chem_bav}.

Regarding the EcoIM application, according to the proposed stepwise procedure, each measurement endpoint was assigned to one or more of the taxonomic groups reported on the left Y axis of the matrix. The results of this process are reported in Figs. 10.7, 10.8 and 10.9 for the LOE ecotoxicology, ecology and chemistry respectively. For example, in the ecological EcoIM the Shannon Index value was assigned to all the taxonomic groups included in the "invertebrates" macrogroup which are used for the Shannon Index calculation: mites, springtails and macroarthropods. Finally, EcoIM was filled in by reporting on the column "Assigned impairment class to the measurement endpoints (tests) results," the results of the measurement endpoints (i.e. ecotoxicological tests, ecological observations and msPAFs), which were processed according to the "Endpoint's Impairment Analysis" and the "Impairment analysis at ecosystem level" (A_{sys}), and coloured according to the corresponding Ecosystem's

Impairment class (Table 10.6); then the software system automatically reported the same colour in all the cells of the row where a "taxonomic group-ecological process" relationship occurred.

The obtained matrices are reported in Figs. 10.7, 10.8 and 10.9 for the ecotoxicological, ecological and chemical LOE respectively.

As for the ecotoxicological LOE (Fig. 10.7), it can be observed that none of the investigated ecosystem's elements (reported on the columns) was shown to be significantly impaired in sampling station C. Each element (i.e., trophic chain complexity and presence of species of particular interest for biodiversity, and ecological processes for functional diversity) was represented by taxonomic groups slightly impaired, except for plants obtaining, for some tests, a higher impairment. However, the global functionality of all investigated elements was shown to be uncompromised.

Observing the matrix for the ecological LOE (Fig. 10.8), similar comments can be formulated for both biodiversity's aspects and functional diversity's elements. In this case only trophic chain complexity, fragmentation and trans-formation of organic substrate, hydrological processes regulation, bioturbation and formation of soil aggregates were covered by the applied ecological observations and all of them resulted to be only slightly impaired.

Finally, as for the chemical LOE (Fig. 10.9), functional diversity and biodiversity were investigated by considering, separately (similarly to the IEIs calculation), the potential toxicities based on total contaminant concentrations (msPAF_total) and on bioavailable fraction concentrations of metals (msPAF_bav_metals), referred to all the taxonomic groups occurring in the terrestrial ecosystem. In the first case (total concentration) the elements of both biodiversity and functional diversity were shown to be relevantly impaired in sample C. Conversely, by using the bioavailable fractions, all the investigated ecosystem's elements in the sampling station were shown to be unimpaired.

Regarding the GEM application (Fig. 10.10), according to the proposed stepwise procedure, first the *EcoIM summary sub-module* was filled in for the Triad's LOEs and the results of each obtained EcoIM were analyzed and reported in Figs. 10.7, 10.8 and 10.9. Specifically, for the chemical LOE related to total concentration, all the investigated elements were evaluated as highly impaired (relevant impairment class; black colour in Table 10.6), while for the chemical LOE related to bioavailable fraction all the investigated elements were evaluated as not impaired (negligible impairment class; Table 10.6), again reflecting the corresponding calculated msPAFs (see Semenzin et al., 2008b for details). For the ecological LOE, all the investigated ecosystem's elements were evaluated as slightly impaired (first intermediate impairment class; Table 10.6), except for the functions "Recycle of nutrients" and "Soil detoxification" for which it was not possible to assign any evaluation (white coloured cells in Fig. 10.10). For the ecotoxicological LOE, biodiversity and the functions "Recycle of nutrients" and "Water cycle" were evaluated moderately impaired (second intermediate impairment class; Table 10.6) because of a higher evidence of impairment in the "Plants" taxonomic macrogroup, while the remaining functions were evaluated as slightly impaired (first intermediate impairment class; Table 10.6).

Secondly, the results of the calculation of Integrated Effect Indexes (IEI_{chem_tot}, IEI_{chem_bav}, IEI_{eco}, and IEI_{etx} ± standard deviation) were reported in the *IEI summary sub-module* and coloured according to the impairment classes (as in Table 10.6).

Finally, after the analysis of the results reported into *EcoIM* and *IEI summary sub-modules*, it was possible to fill in, by means of expert judgment, the *Global impairment evaluation sub-module*. This operation allowed evaluating sampling station C as globally slightly impaired. However, the evaluation was shown to be affected by a relatively high uncertainty, due to the discordant result (relevant impairment) obtained by the chemical (total concentration) LOE compared to the others (negligible and intermediate I impairment). For this reason the experts highlighted the need for investigating bioavailability of organic compounds in addition to the analysis carried out on metals bioavailability. Moreover, the experts suggested completing the ecological LOE with measurement endpoints covering the ecological functions "Recycle of nutrients" and "Soil detoxification".

The implemented system was shown to be a valuable tool for supporting the decision making process and as a result of the ERA-MANIA project it was suggested that its application be included in the environmental monitoring plan of the Acna site in order to evaluate, over the course of time, the efficiency of the ongoing remedial actions.

10.5 Conclusions

The preliminary application of DSS-ERAMANIA to the Acna contaminated site led to a transparent selection of measurement endpoints for each Triad LOE (i.e. Module 1 application), and to a comprehensive (quantitative and qualitative) evaluation of the impairment occurring on the site of concern (i.e. Module 2 application). Moreover, Module 1 turned out to be a useful tool for promoting discussions among experts to reach a common judgment on advantages and drawbacks of existing experimental tests. Module 2 confirmed its capability of bridging the gap between experimental test results on the one side and site-specific risk estimation and evaluation for terrestrial ecosystems on the other side.

However, in order to improve, calibrate and validate the developed and implemented MCDA-based procedure, applications to other relevant contaminated sites and further discussions with experts, stakeholders and decision makers would be desirable.

Acknowledgment This work and the ERA-MANIA project were funded by the Italian Government Commissary for the rehabilitation of the Bormida Valley and were developed by University of Venice, Ca' Foscari, Interdepartmental Centre for Sustainable Development

(IDEAS) in collaboration with Italian National Environmental Protection Agency (APAT), Dutch National Institute for Public Health and Environment (RIVM) and Consortium Venezia Ricerche (CVR).

We specially thank Claudio Carlon, Miranda Mesman and Ton Schouten for their contribution to the ERA-MANIA project. We also thank Prof. Silvio Giove for developing the MCDA procedures of the DSS-ERAMANIA, Stefano Silvoni for the implementation of the software "DSS-ERAMANIA Module 1: Comparative Tables" and Stefania Gottardo for her technical support.

We are grateful to the Steering Committee members (Trudie Crommentujin, Bona Piera Griselli, John Jensen, Giuseppe Marella, Cristina Menta, Nick Pacini, Gianniantonio Petruzzelli, Eugenio Piovano, Milagros Vega and Aldo Viarengo) for the helpful discussions during the testing phase of both Module 1 and 2 developments.

References

Angelini P, Fenoglio S, Isaia M, Jacomini C, Migliorini M, Morisi A. Tecniche di biomonitoraggio del suolo. ISBN 88-7479-003-1. ARPA Piemonte-CEDAP; 2002. p. 106. [In Italian].

ANPA (Agenzia Nazionale per la Protezione dell'Ambiente). Guida tecnica sui metodi di analisi per il suolo e i siti contaminati – Utilizzo di indicatori biologici ed ecotossicologici. RTI CNT_TES 1/2004; 2004. [In Italian].

Aten CF, Gupta SK. On heavy metals in soil; rationalization of extractions by dilute salt solutions, comparison of the extracted concentrations with uptake by ryegrass and lettuce, and the possible influence of pyrophosphate on plant uptake. Sci Total Environ 1996;178(1–3): 45–53.

Berta G, Fusconi A, Trotta A, Scannerini S. Morphogenetic modifications induced by the mycorrhizal fungus Glomus strani E3 on the root system of Allium porrum L.. New Phytol 1990;114: 207–215.

Burton Jr. GA, Chapman PM, Smith EP. Weight-of-evidence approaches for assessing ecosystem impairment. Hum Ecol Risk Assess 2002;8: 1657–1673.

CARACAS. Risk Assessment for Contaminated Sites in Europe. Final Publication. Vol. 1-Scientific Basis. LQM Press, Nottingham, UK; 1998.

Cesaro P, Volante A, Massa N, Bona E, Lingua G, Berta G. Phytotoxicity Test to be Used in Ecological Risk Assessment of Heavy Metal Polluted Soils. XVII International Botanical Congress, Vienna, 17–23 July; 2005.

Chapman PM, McDonald BG, Lawrence GS. Weight of evidence frameworks for sediment quality (and other) assessment. Hum Ecol Risk Assess 2002;8: 1489–1516.

CLARINET. Proceeding of the CLARINET Workshop on Ecological Risk Assessment for Contaminated Sites. April 17–19, 2001, Nunspeet, NL; 2001.

CLARINET. Sustainable Management of Contaminated Land: An Overview. A Report from the Contaminated Land Rehabilitation Network for Environmental Technologies; 2002a. p. 128.

CLARINET. Review of Decision Support Tools for Contaminated Land Management, and Their Use in Europe. A Report from the Contaminated Land Rehabilitation Network for Environmental Technologies; 2002b. p. 180.

Critto A, Torresan S, Semenzin E, Giove S, Mesman M, Schouten AJ, Rutgers M, Marcomini A. Development of a site-specific Ecological Risk Assessment: Part I: A multi-criteria based system for the selection of ecotoxicological tests and ecological observations. Sci Tot Env 2007;379: 16–33.

Dondero F, Jonsson H, Rebelo M, Pesce G, Berti E, Pons G, Viarengo A. Cellular responses to environmental contaminants in amoebic cells of the slime mould Dictyostelium discoideum. Comp Physiol Biochem 2006;143(2): 150–157.

ECOFRAME. Ecological Committee on FIFRA Risk Assessment Methods: ECOFRAM Terrestrial Draft Report; 1999.

Fusconi A, Repetto O, Bona E, Massa N, Gallo C, Dumas-Gaudot E, Berta G. Effects of cadmium on meristem activity and nucleus ploidy in roots of Pisum sativum L. cv Frisson seedlings. Env Exp Bot 2006;58: 253–260.

Gardi C, Tomaselli M, Parisi V, Petraglia A, Santini C. Soil quality indicators and biodiversity in northern Italian permanent grasslands. Eur J Soil Biol 2002;38: 103–110.

Gorny M, Grum L. Methods in Soil Zoology. Elsevier, Amsterdam, The Netherlands; 1993.

Hooker JE, Berta G, Lingua G, Fusconi A, Sgorbati S. Quantification of AMF-induced modifications to root system architecture and longevity. In: Varma A, editor, Mycorrhyzal Manual. Springer–Verlag, Berlin Heidelberg, Germany; 1998. pp. 515–531.

IDEAS. Quarta relazione (da Giugno 2004 a Dicembre 2004) del progetto di ricerca ERA-MANIA. Final report of the ERA-MANIA project. Centre IDEAS. Venice: Ca' Foscari University; 2004. pp. 144–166. [In Italian].

INIA – Instituto Nacional de Investigacion y tecnologia Agraria y alimentaria, Desarrollo de criterios para la caracterizacion de suelos contaminados y para determinar umbrales para la proteccion de los ecosistemas. Madrid, Espana; 2000.

Kiker GA, Bridges TS, Varghese A, Seager TP, Linkov I. Application of multicriteria decision analysis in environmental decision making. Integr Environ Assess Manag 2005;1(2): 95–108.

Lingua G, Fusconi A, Berta G. The nucleus of differentiated root plant cells: modifications induced by arbuscular mycorrhizal fungi. Eur J Histochem 2001a;45: 9–20.

Lingua G, D'Agostino G, Fusconi A, Berta G. Nuclear changes in pathogen-infected tomato roots. Eur J Histochem 2001b;45: 21–30.

Long ER, Chapman PM. A sediment quality triad: measures of sediment contamination, toxicity and infaunal community composition in Puget Sound Mar Poll Bull 1985;16: 405–415.

Magurran AE. Ecological Diversity and Its Measurement. Princeton University Press, Princeton, NY; 1988. p. 179.

OECD guideline for testing of chemicals n. 207. Earthworm, Acute Toxicity Tests; 1984.

Parisi V. La qualità biologica del suolo. Un metodo basato sui microartropodi. Acta naturalia de L'Ateneo Parmense 2001;37: 97–106. [In Italian].

Pereira AG, Quintana SC. From technocratic to participatory decision support systems: Responding to the new governance initiatives. J Geogr Inf Decision Anal 2002;6: 95–107.

Petruzzelli G, Lubrano L, Giovannini G, Lucchesi S, Cervelli S. Distribution and behaviour of heavy metals in soils following different clean up procedures. In: Arendt F et al., editors, Contaminated Soil. Kluwer Academic Publishers, Dordrecht; 1993. pp. 545–546.

Rutgers M, Den Besten P. Approach to legislation in a global context, B. The Netherlands perspective – soils and sediments. In: Thompson KC, Wadhia K. and Loibner AP, editors, Environmental Toxicity Testing. Blackwell Publishing CRC Press, Oxford; 2005. pp. 269–289.

Semenzin E, Critto A, Carlon C, Rutgers M, Marcomini A. Development of a site-specific Ecological Risk Assessment: Part II: A multi-criteria based system for the selection of bioavailability assessment tools. Sci Tot Env 2007;379: 34–45.

Semenzin E, Critto A, Rutgers M, Marcomini A. Integration of bioavailability, ecology and ecotoxicology by three lines of evidence into ecological risk indexes for contaminated soil assessment. Sci Tot Env 2008a;389: 71–86.

Semenzin E, Critto A, Rutgers M, Marcomini A. Ecosystem impairment evaluation on biodiversity and functional diversity for contaminated soil assessment. Submitted to Integr Environ Assess Manag 2008b.

Siller D, Blodgett C, Cziganyik N, Omeroglu G. Factors involved in urban regeneration: remediation and redevelopment of contaminated urban sites – a comparison of France and the Netherlands. In: Proceeding of CABERNET 2005: The International Conference on

Managing Urban Land. Compiled by Oliver L, Millar K, Grimski D, Ferber U, Nathanail
CP. Land Quality Press, Nottingham; 2004. ISBN 0-9547474-1-0.

Suter II GW, Efroymson RA, Sample BE, Jones DS. Ecological Risk Assessment for Contaminated Sites. Lewis, Boca Raton, FL; 2000.

UK-EA. Ecological Risk Assessment: A Public Consultation on a Framework and Methods
for Assessing Harm to Ecosystems from Contaminants in Soil; 2003. www.environment-
agency.gov.uk.

UNICHIM 1651. Determinazione dell'inibizione della germinazione e allungamento radicale
in Cucumis sativus L. (Cetriolo), Lepidium sativum L. (Crescione), Sorghum saccharatum
Moench (Sorgo) – saggio di tossicità cronica breve; 2003. [In Italian].

UNIPMN (Università del Piemonte Orientale). Progetto per la realizzazione di un centro di
eccellenza riguardante monitoraggio e bonifica dei siti inquinati. Rapporto finale realiz-
zato nell'ambito delle attvità di ricerca finanziate dal Commissario delegato per la bonifica
dell'ACNA di Cengio, Alessandria, Italy; 2004. [In Italian].

US EPA – U.S. Environmental Protection Agency, Guidelines for ecological risk assessment.
EPA/630/R-95/002F. Risk Assessment Forum, Washington, DC; 1998.

Vincke P. Multicriteria Decision-Aid. Wiley & Sons, Chickester, UK; 1992.

VROM. Soil Policy Letter. Circular BWL/2003 096 250, Ministry of Housing, Spatial Planning and the Environment, The Hague, The Netherlands; 2003. Download: international.
vrom.nl/docs/internationaal/7178menarev_1.pdf.

Chapter 11
SADA: Ecological Risk Based Decision Support System for Selective Remediation

S. Thomas Purucker, Robert N. Stewart and Chris J.E. Welsh

Abstract Spatial Analysis and Decision Assistance (SADA) is freeware that implements terrestrial + criteria were applied to determine a spatially explicit remedial design that reduced shrew exposures to protective levels.

11.1 General Introduction

Ecological risk assessments are increasingly factored into remediation decisions as assessments becomes more formalized and more recognized as integral in completing site cleanups adequately. However, significant challenges remain in the practical implementation of such assessments (Bradbury et al. 2004, Dearfield et al. 2005, DeMott et al. 2005). Among these challenges is a need for greater availability of spatially explicit risk characterization capabilities. This is due to the spatial variability inherent in ecological exposures (Suter 1993, Hope 2000, Landis and McLaughlin 2000), site-specific considerations about proximity to ecologically significant resources (Anderson et al. 2004), and the fact that commonly used, non-spatial, generic models are less capable of realistically predicting environmental stressor levels and risks than georeferenced models (Hope 2001, Wind 2004). Given the increasing variety and number of assessments, there is also a need for more streamlined, efficient implementation (Bradbury et al. 2004). Decision support systems that combine ecological risk assessment capabilities with Geographical Information System (GIS) features provide enhanced ability to perform spatially explicit characterizations and offer opportunities to streamline the implementation of such assessments.

The Spatial Analysis and Decision Assistance (SADA) software provides ecological risk assessment capabilities that can be implemented in its integral

S.T. Purucker (✉)
U.S. Environmental Protection Agency, 960 College Station Road, Athens, GA 30605, USA
e-mail: purucker.tom@epa.gov

A. Marcomini et al. (eds.), *Decision Support Systems for Risk-Based Management of Contaminated Sites*, DOI 10.1007/978-0-387-09722-0_11,
© Springer Science+Business Media, LLC 2009

GIS (Purucker et al. 2008, Stewart et al. 2007). The integration of the GIS with estimated ecological risk allows for the spatial presentation of data in the GIS, comparison of model results to an array of ecological benchmarks, exposure modeling in a spatial context, and provides the groundwork for developing movement-based exposure models. These capabilities allow for easier and faster implementation of screening-level benchmark and exposure-hazard quotient (HQ) approaches that are commonly performed for terrestrial ecological risk assessments. While these latter approaches have been criticized for being measures of levels of concern, not actual measures of risk, they are often the only ecological measures used in assessment and, therefore, function as the primary driver of whether ecological-based remediation is necessary at most contaminated sites (Tannenbaum et al. 2003). A shortcoming of most implemented screening methods is that they typically are conducted with spatially homogenous summary statistics. A GIS, in contrast, can overlay maps of screening output by sample location or interpolated values onto maps of ecologically significant resources so that remedial decisions can be conditioned on proximity to important habitat. This approach also yields estimates of the correlations between different stressors and their levels and the evaluation of habitat quality to determine if additional study is needed (Purucker et al. 2007). Since most terrestrial landscapes are characterized by multiple sources and stressors from anthropogenic effects, mapping the available information of known sources and stressors can show significant intersections of multiple stressors in space and time. These spatial approaches can improve the quality of information presented to decision-makers about whether remedial action or additional study is needed without significantly increasing the complexity of the ecological analysis.

This chapter documents the ecological risk assessment functionality of SADA with its relevant GIS and database features. Decision support systems for ecological risk assessment face a conundrum as higher tiers of the ecological risk assessment process are accessed. To an extent, ecological risk assessments are standardized within a given regulatory setting. Significant similarities exist among the different regulatory approaches for lower tiers of the assessment process. Software implementations of decision support follow the branches of specified decision trees. The decision tree is reasonably well-known and predictable for the lower tiers of assessment, and can be used to design decision support software. All that is necessary is to maintain adaptability for the user. That is why allowing site-specific modifications to input parameters, accessing multiple sources of benchmarks, and using a plurality of statistical approaches are all necessary for robust use of ecological risk software.

For the higher tiers of the risk assessment process, it is a different story altogether. Ecological systems are well-known for their variability, and higher tiers of assessment need flexibility to be implemented robustly. Software decision support systems experience an opposite pressure. Consensus in approach and a reasonably finite decision tree process are necessary to achieve reproducibility. As a result, decision support systems lose generality if they attempt to tailor their assessment capabilities to the higher tiers. The focus in this text and in development of the SADA software, therefore, has been on bringing increased flexibility to screening tools in a spatial environment in order to maximize the information that can be gathered at that stage. The use of a decision support tool with these latter features cannot be viewed as a replacement for higher tiered assessment when those are necessary. Many aspects of higher tiered ecological risk assessment require flexibility that simply cannot be achieved in a hard-wired decision support software implementation.

11.2 Description of Framework and Functionalities

SADA is developed at The Institute for Environmental Modeling at the University of Tennessee with funding from the US Department of Energy, the US Environmental Protection Agency, and the US Nuclear Regulatory Commission. SADA can be used within various regulatory contexts to perform the lower tiers of ecological risk assessment (e.g., USEPA 1998). Estimates vary widely, but the total number of contaminated sites in the United States has been estimated as high as 600,000 (USOTA 1985) and there are estimates of similar magnitude in Europe (e.g., Prokop et al. 2000). The size distribution of contaminated soil sites is unknown, but we assume that smaller contaminated sites comprise the majority of existing sites. The scale of analysis for SADA applications is usually the smaller contaminated sites, since stationarity (statistical properties do not vary over space) assumptions inherent in its interpolation methods typically do not hold for larger sites; however, it is possible to break up a larger site into smaller areas within the GIS. At many contaminated sites, ecological risk assessments are not performed because of limited analysis and cleanup budgets, their perceived unimportance relative to human health risk, a persisting idea that protecting human health is also sufficient to protect ecological receptors, and the lack of efficient ecological risk assessment "mass production" software availability. However, tools to assess ecological risk efficiently are becoming more available, including SADA, which has had over 20,000 downloads since its initial release.

SADA is freeware that integrates data analysis – emphasizing spatial analysis capabilities – with risk assessment methods. The software characterizes environmental contamination in multiple media to support decision-making within the context of ecological risk assessment. SADA operates as an environmental decision support system (Purucker et al. 2008) with a user interface to facilitate sequential parameterization and implementation of environmental assessment models (Stewart et al. 2007). An overview of the main SADA modules relevant to ecological risk assessment is presented in Fig. 11.1.

GIS functionality is provided via a mapping window that displays sample results and modeled exposure and risk output relative to the site features that are displayed by GIS layers imported from AutoCAD DXF or ESRI shape files. Imported data sets can be queried by contaminant, media, and date for display

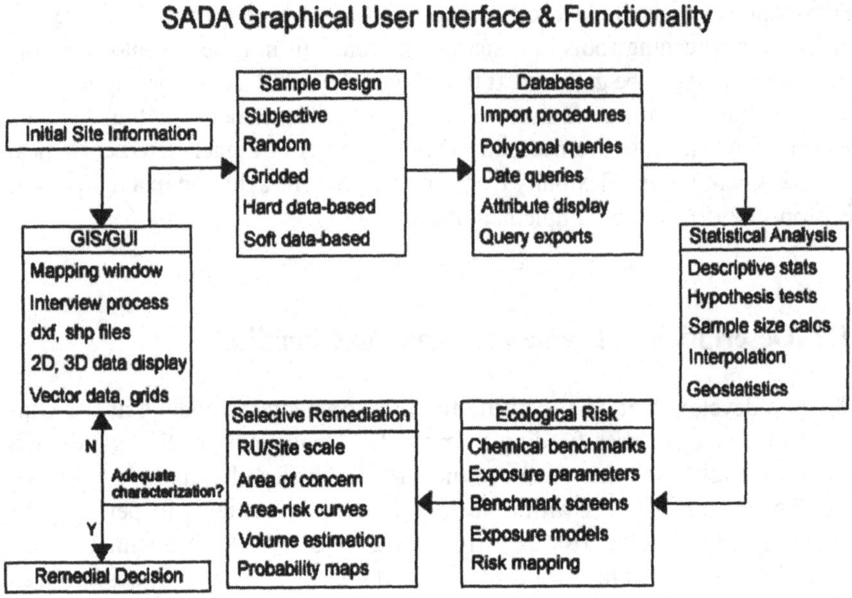

SADA Graphical User Interface & Functionality

Fig. 11.1 The figure displays primary components of the graphical user interface and the functionality of the SADA software. Displayed functionality is limited to SADA components that are germane to the terrestrial ecological risk assessment and the remedial decision-making process (Purucker et al. 2008)

and assessment. Also, output from other science models that simulate release, fate, and transport of hazardous contaminants can be imported to assess exposure and risk.

Sampling designs are available, based on commonly accepted approaches, or determining the number of samples needed and for spatially locating the samples to ensure that collected data sufficiently support the decisions to be evaluated (USEPA 2002). Univariate statistics for evaluating contamination distributions and non-parametric hypothesis tests for drawing conclusions from data relative to environmental decision criteria are also available (USEPA 2006). For sites with contamination that is significantly spatially auto-correlated, the software provides spatial statistics, i.e., variography (Gotway 1991), interpolation methods (Goovaerts 1997), and cross-validation algorithms to compare the fit of alternative interpolation approaches to data (Legendre and Legendre 1998).

11.3 Structure of the System

Ecological risk assessments are used at contaminated sites to determine the need for and extent of remediation. Ecological risk assessment is a tiered process that begins with screening and proceeds to full characterization of the ecological risks necessary to support remediation decisions at contaminated sites. If ecological problems are present, providing support for the design and evaluation of remedial alternatives often entails producing contaminant-specific concentration goals for identifying site-specific contaminants of concern (Screening Values – SVs) and estimating feasible clean-up levels (Preliminary Remediation Goals – PRGs). Use of these latter concentration values allows quick determination of whether additional site assessment is needed, and the contaminants list to be assessed to be pared down to manageable size. PRGs allow for the calculation of site- and receptor-specific cleanup goals. There are many sources of SVs and PRGs because numerous regulatory agencies are creating them for application in their domains. A decision support tool with access to various SVs and the ability to calculate PRGs is very useful for site assessment, since typical implementation may require access to multiple data sources. The SADA software supports many steps in a typical ecological risk assessment. The screening and ecological risk modules are integrated into the environmental information system and allow users to compare summary statistics of analytical data for site sub-areas to a variety of SVs (Fig. 11.2), and calculate doses and risks for a number of terrestrial receptors (Fig. 11.3).

An early step in ecological risk assessment is hazard identification, a process that consists of comparing (or screening) environmental measurements to compilations of SVs. Site contaminants that exceed these SVs are kept for further examination. If no exceedances of relevant SVs are observed, this fact can be used to justify a no further action determination. SADA contains one of the most complete, publicly available compilations of SV sources, and its screening module allows users to perform site comparisons against a large array of SVs in a hierarchical manner. SVs are available for 50+ benchmark sources for sediment, surface water, tissue residue, and soil; values are presented as functions of environmental variables (e.g., pH, hardness, organic carbon content) where appropriate. Screen results, ratios, and the SVs themselves can be viewed in tabular or histogram form. Terrestrial SVs included with the software are the Dutch Intervention and Target (Crommentuijn et al. 2000a,b, Swartjes 1999), USEPA Eco ESLs (USEPA 2003), USEPA Region 4 (USEPA 2001a), USEPA Region 5 ESLs (USEPA 1999), ORNL invertebrates, microbes (Efroymson et al. 1997b), and plants (Efroymson et al. 1997a). Barron and Wharton (2005) provide a comprehensive overview of many of these SV sources and a discussion of their derivation.

The ecological exposure and risk models follow USEPA (1992, 1997, 1998, 2001b) guidance, and can be customized to fit site-specific exposure conditions for generating risk results and calculating PRGs. Exposure and risk assessment calculations are based on an initial conceptual model for the site that includes the contaminated media, routes of contaminant transport, representative ecological receptors, and pathways of exposure for these receptors. Calculations are limited to those contaminants that exceeded relevant SVs in the screening process. The next step is determining the bioavailability of these chemicals to exposed receptors; this can be done through physiological or food chain models that model the bioaccumulation and biomagnification of the contaminants in

(a)

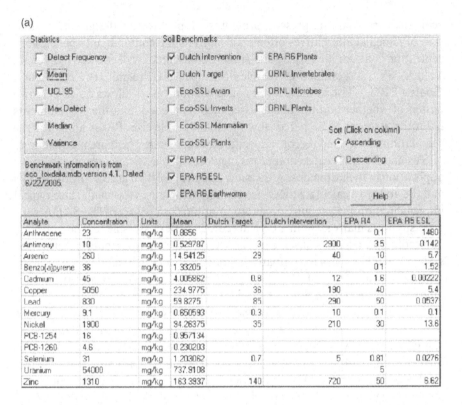

Analyte	Concentration	Units	Mean	Dutch Target	Dutch Intervention	EPA R4	EPA R5 ESL
Anthracene	23	mg/kg	0.8656			0.1	1480
Antimony	10	mg/kg	0.529787	3	2900	3.5	0.142
Arsenic	260	mg/kg	14.54125	29	40	10	5.7
Benzo[a]pyrene	38	mg/kg	1.33205			0.1	1.52
Cadmium	45	mg/kg	4.005862	0.8	12	1.8	0.00222
Copper	5050	mg/kg	234.9775	36	190	40	5.4
Lead	830	mg/kg	59.8275	85	290	50	0.0537
Mercury	9.1	mg/kg	0.650593	0.3	10	0.1	0.1
Nickel	1900	mg/kg	94.26375	35	210	30	13.6
PCB-1254	16	mg/kg	0.95/1.34				
PCB-1260	4.6	mg/kg	0.230203				
Selenium	31	mg/kg	1.203062	0.7	5	0.81	0.0276
Uranium	54000	mg/kg	737.9108			5	
Zinc	1310	mg/kg	163.3937	140	720	50	8.82

(b)

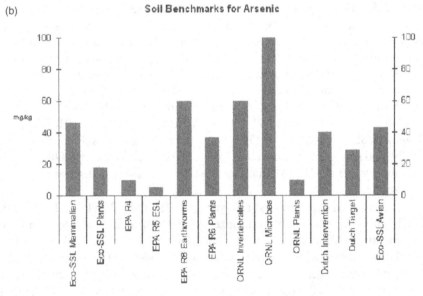

Fig. 11.2 Ecological Screening Values present in the SADA decision support tool. (**a**) Tabular view of screening concentration (maximum detected value), summary analytical data statistics (e.g., mean) for identified contaminants of concern at the K-770 site, and a selectable list of SVs. (**b**) Histogram of SVs for all available sources for a given contaminant (Arsenic)

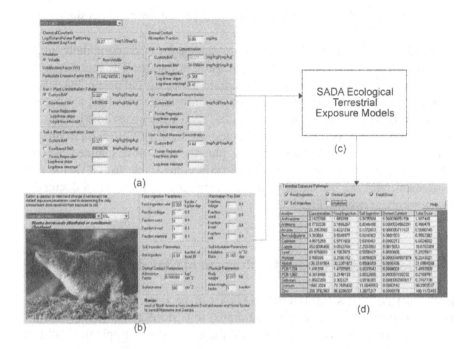

Fig. 11.3 Terrestrial exposure modeling dose results for the short-tailed shrew. Contaminant-specific inputs for physical constants and bioconcentration model selection and parameters (**a**) are combined with species exposure inputs (**b**) to parameterize the dose exposure models (**c**). Tabular model output (**d**) for all contaminants of concern at the K-770 site is generated in formats that can be exported to popular word processors and spreadsheet programs

different trophic levels of the food chain. The magnitude of exposures to the individual receptors can then be calculated using exposure models for the relevant exposure pathways. In the next step, a toxicity reference value (TRV) from a dose-response model is needed. This may call for inter-species extrapolation of the contaminant effects from a laboratory study to a species of interest at the contaminated site. Results of the exposure modeling are then compared to decision criterion (the TRV); the exposure models can then be modified for different exposure scenarios to determine the feasibility of available alternative remedial actions.

The SADA software provides terrestrial dose exposure models to estimate the daily doses of contaminants at a site. Modeling dose to wildlife receptors requires numerous chemical-specific and species-specific exposure parameters. The ecological dose modeling capabilities are located in the ecological risk module, and default exposure parameters are distributed with the software for over 20 terrestrial species (Table 11.1) representing a variety of wildlife receptors, including characteristic herbivores, insectivores, and carnivores. These species can be parameterized individually for males, females, and juveniles, as well as USEPA Soil Screening Level (SSL) defaults (USEPA 2003) that combine male and female parameters. Default values for males and females are based on available literature sources (e.g., USEPA 1993, 2003), but can be user modified to reflect site-specific conditions.

Routines are available in the exposure models for calculating daily intake rates for the selected receptors at each location in the site. The routines in the ecological risk module access the contaminant matching and data management functions, and model parameters and toxicity information are fully parameterized from USEPA guidance when available. SADA calculates dose (mg intake per kg body weight per day) from food ingestion, soil ingestion, dermal contact, and inhalation for terrestrial exposures, plus the total dose summed over all pathways selected. These results are directly comparable to TRVs for risk assessment, and are presented in tabular form commonly used to document risk assessment results (Fig. 11.3d).

11.4 Decision Aspects and Involvement of Stakeholders

Ecological risk assessments using SADA can be implemented under many existing regulatory constructs that are similar to USEPA guidance for contaminated sites. In a regulatory environment, there are often multiple stakeholders with input relevant to the selection of a remedial alternative. To a certain extent, environmental decision-making continues to be decentralized as the participation of environmental agencies, organizations, and individuals increases. Publicly available freeware has an inherent advantage over proprietary software in

Table 11.1 Terrestrial species with default exposure parameters

Common name	Scientific name	Receptor group
American kestrel	Falco sparverius	Avian carnivore
American robin	Turdus migratorius	Avian insectivore
American woodcock	Scolopax minor	Avian ground insectivore
Black-tailed jackrabbit	Lepus californicus	Mammalian herbivore
Burrowing owl	Speotyto cunicularia	Avian carnivore
Deer mouse	Peromyscus maniculatus	Mammalian omnivore
Eastern cottontail	Sylvilagus floridanus	Mammalian herbivore
Great Basin pocket mouse	Perognathus parvus	Mammalian granivore
Kit fox	Vulpes macrotis	Mammalian carnivore
Little brown bat	Myotis lucifugus	Mammalian insectivore
Long-tailed weasel	Mustela frenata	Mammalian carnivore
Meadow vole	Microtus pennsylvanicus	Mammalian herbivore
Mexican free-tailed bat	Tadarida brasiliensis	Mammalian insectivore
Mourning dove	Zenaida macroura	Avian granivore
Northern bobwhite	Colinus virginianus	Avian granivore
Prairie vole	Microtus ochrogaster	Mammalian herbivore
Red fox	Vulpes vulpes	Mammalian carnivore
Red-tailed hawk	Buteo jamaicensis	Avian carnivore
Short-tailed shrew	Blarina brevicauda	Mammalian insectivore

situations with multiple decision-makers because it enhances access to project-related environmental data and subsequent analyses. Sharing SADA freeware files containing data and assessment results allows individuals involved in the decision-making process to visualize results on their own time and examine the sensitivity of analysis results to changes in model parameterization.

11.5 Case Study Application

The following example application illustrates the capabilities of a decision support system with respect to terrestrial ecological risk assessment and selective remediation. The K-770 Scrap Metal Yard is less than 8 ha in size and is located on the western end of the Powerhouse peninsula at East Tennessee Technology Park (ETTP) in Roane County, Tennessee. The Yard operated during the 1940s as an oil storage area, and has operated since the 1960s as a scrap facility, although it is currently inactive. Tens of thousands of tons of metal were stored in piles at the site, but have since been removed. Most contamination at the site originated from the scrap piles. Considerable sampling has been conducted at this site for a full suite of analyses that includes metals and polychlorinated biphenyls (PCBs).

The short-tailed shrew, a small insectivorous mammal that inhabits most regions of the United States, is one of the receptors of interest selected for the site conceptual model. Extensive, frequent disturbance of soils from construction and remediation activities has produced a plant community that is highly adapted to disturbance. Areas near ETTP and the Clinch River that are not paved or graveled include shrubs and small trees, many of which are non-native and invasive. Short-tailed shrews inhabit a wide variety of habitats and are common in areas with abundant vegetative cover (Miller and Getz 1977). The shrews have high metabolic rates and can eat their approximate body weight in food each day. Short-tailed shrews are primarily insectivores, although they do eat some plant material and small mammals. Soil invertebrates make up a large component of their diet (USEPA 1993). For this assessment, it is assumed that shrews eat 100% soil invertebrates associated with soil at ETTP sites.

For this example application, exposures were calculated for PCBs at the ETTP site. PCBs are a family of man-made chemicals consisting of 209 individual compounds with varying toxicity (ATSDR 1989). Because of their insulating and nonflammable properties, PCBs were widely used in industrial applications such as coolants and lubricants in transformers, capacitors, and electrical equipment prior to 1977 (ATSDR 1989). PCBs are known to bioaccumulate and biomagnify to toxic concentrations in animals (Eisler 1986, ATSDR 1989). PCBs with higher chlorine (Cl) content, such as 1254 or 1260, tend to persist in the environment longer than those with lower Cl content. Chronic exposures are a particular concern. Although relatively insoluble in water, PCBs are generally soluble in nonpolar organic solvents and in biological lipids (USEPA 1980).

Mink appear to be among the most sensitive mammals to PCBs, with dietary levels as low as 0.1 ppm wet weight causing death and reproductive toxicity (Eisler 1986). Ringer et al. (1981) reported an LC50 for chronic exposures of 6.65 ppm PCB-1254 for mink exposed over an 8-month period. Exposure of mink for 6 months to 1 ppm PCB-1254 resulted in no significant difference from controls in the number of offspring or offspring mortality (Wren et al. 1987). Halbrook et al. (1999) investigated toxicity of PCB-contaminated Poplar Creek fish to mink by feeding the mink five different fish diets over a 7-month period. No adverse effects on mink reproduction were observed at fish dietary PCB concentrations of 1 ppm or a dose of 0.12 mg/kg/day, but dietary concentrations of 1.36 ppm (0.23 mg/kg/day dose) resulted in increased liver Ethoxyresorufin O-deethylase (EROD) activity and a trend toward decreases in adult and kit body weights and litter size at birth. Aulerich and Ringer (1977) exposed mink to 1, 5, and 15 ppm PCB-1254 in their diet over a 4.5 month period. The number of offspring born alive was significantly reduced at the 5 and 15 ppm levels, but not at 1 ppm. Because the study exposure was 4.5 months, and included critical life stages (reproduction), Sample et al. (1996) considered the 1 and 5 ppm doses (0.14 and 0.69 mg/kg/d) to be the chronic No Observed Adverse Effect level (NOAEL) and Low Observed Adverse Effect Level (LOAEL), respectively. We used these TRVs for the shrews in our SADA application example herein.

For our ETTP example application, PCB-1254 soil concentration interpolations were performed at the site using the nearest neighbor, natural neighbor, inverse distance (with three different powers), and ordinary kriging (with fitted exponential, spherical, and gaussian theoretical variograms) methods and the available analytical data. Cross-validation results for the nine methods are presented in Table 11:2, sorted by mean squared error. Ordinary kriging results had a larger correlation range and a tendency to underestimate the high sample concentration locations in the south; they were the only three interpolants with negative errors. Inverse distance performed relatively well on the mean squared error, but tended to overestimate the influence of the high concentration sample locations due to the smaller effective correlation range. Natural neighbor and nearest neighbor, which did not have defined static correlation relationships over the site, performed poorly relative to the inverse distance and ordinary kriging method implementations. Total PCB-1254 mass estimates at the site are also presented in Table 11.2, based on assumptions of 1500 kg/m^3 soil density and a 1 m contaminated zone thickness. Although the cross-validation errors underestimated the concentrations at the high concentration sample locations due to smoothing, the ordinary kriging results still yielded some of the highest estimates of total PCB-1254 mass at the site, since the effects of these high concentration samples influenced a large portion of the site. Total mass estimates of PCB-1254 at the site ranged from 58 to 71 g.

Figure 11.4 presents the modeled shrew spatial exposure dose estimate results at the site. Mapping exposure in this way can also be used for the short-tailed shrew as a point of departure for producing individual-based movement models

Table 11. 2 Cross-validation and PCB-1254 mass estimates for PCB-1254 interpolation

Interpolation method	Error	Absolute error	Mean squared error	PCB-1254 mass (g)
Ordinary Kriging (Gaussian)	−7.70E-03	1.09	5.6	70.82
Inverse Distance (power = 1)	1.48E-02	1.14	6.11	71.25
Ordinary Kriging (Spherical)	−1.68E-03	1.16	6.37	67.60
Ordinary Kriging (Exponential)	−1.67E-03	1.17	6.44	67.28
Inverse Distance (power = 2)	2.60E-02	1.22	7.06	68.89
Inverse Distance (power = 3)	3.06E-02	1.3	8.12	66.63
Natural Neighbor	6.00E-02	1.41	9.5	62.98
Nearest Neighbor	3.18E-03	1.48	14.1	57.93

that simulate realistic movement patterns for individuals in an effort to model the distribution of the exposed population dose (Chow et al. 2005).

Discernible spatial auto-correlation can enhance methods for identifying priority zones for remediation at contaminated sites. The selective remediation of contaminated sites, the process of remediating specific sub-areas of a contaminated site to achieve quantitative cleanup goals and specified error tolerances for the larger area (Blacker and Goodman 1994, Brakewood and Grasso 2000), requires estimating the spatial distribution of contamination and exposures. Selective remediation requires the use of interpolation estimates and associated uncertainty to specify the minimal (optimal) amount and locations

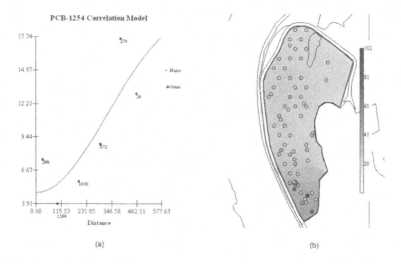

Fig. 11.4 Semivariogram and ordinary kriging interpolation for PCB-1254 concentrations at the ETTP site. The semivariogram (**a**) shows a Gaussian theoretical model fitted to the experimental variography using the SADA software. The ordinary kriging results based on this fitted variogram are then shown as spatially modeled daily dose (mg/kg/day) to PCB-1254 for short-tailed shrew inhabiting different portions of the site in (**b**) using the spatial continuity model. Legend on right shows the scale, light areas have lower PCB-1254 dose levels and darker areas have higher dose levels, up to 102 mg/kg/day

of soil in need of remediation to achieve a local and/or site-wide objective. Therefore, the application of spatial statistical tools to environmental risk assessment is a potentially fruitful area for improving decision-making processes that consider the spatial distribution of contamination at a site.

Selective remediation design proceeds by discretizing the site into a grid and identifying cleanup areas by remediating individual grid blocks in inverse order of magnitude (worst to least) until the cleanup objective is met. For implementation of a contaminant not-to-exceed concentration (NTEC) objective, the interpolation estimate at every grid node is compared to the cleanup criterion,

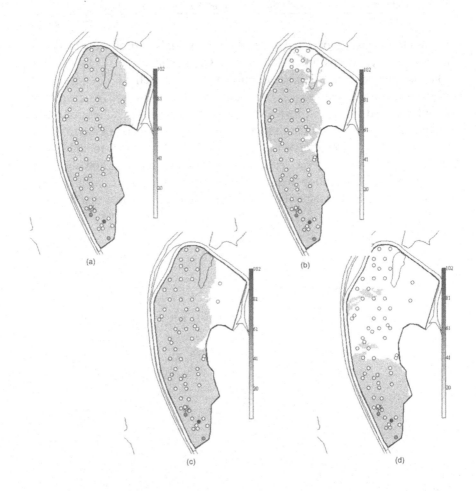

Fig. 11.5 Visual depictions of areas of concern (*gray areas*) for PCB-1254 that exceed an NTEC or site average for LOAEL and NOAEL. Legend shows calculated dose levels at locations where concentration data were collected, up to a maximum of 102 mg/kg/day. (**a**) NTEC for NOAEL of 0.14 mg/kg/day dose yields a cleanup area of 64,000 m³. (**b**) Site average for NOAEL of 0.14 mg/kg/day yields a cleanup area of concern of 52,000 m³. (**c**) NTEC representative of an LOAEL for a dose of 0.69 mg/kg/day gives an area of 59,000 m³. (**d**) Site average for LOAEL yields an area of concern of 25,000 m³

and remediation is conducted at each grid node where the estimate exceeds the criterion. The NTEC for ecological risk applications is typically the LOAEL or other less conservative effects estimate. This same approach can then be used in conjunction with a more conservative effects estimate, such as the NOAEL, to implement remediation at a site so that a representative statistic for the entire site is below this effects level. The grid blocks are again remediated in inverse order of magnitude (worst to least) until the site-wide objective is met. The statistic is recalculated as subsequent grid blocks are cleaned and compared to the effects level until the decision criterion is reached.

In SADA, implementation of selective remediation is based on production of area of concern maps. These maps are functions of contaminant interpolation maps, previously discussed, in conjunction with an ecological decision threshold value for an ecological PRG, benchmark, or calculated receptor dose level. These area of concern maps can be produced based on an NTEC, site-wide statistic, or combination of both, and result in a spatially-explicit remediation design with an associated area or volume of required soil treatment specified.

Figure 11.5 presents our calculated areas of concern based on the two previously discussed PCB-1254 daily dose cleanup criteria, 0.14 (NOAEL) and 0.69 (LOAEL) mg/kg/d. The 0.14 mg/kg/day value is quite low compared to our modeled shrew doses at the site, and results in most of the site being declared an area of concern.

11.6 Future Development

SADA is in active development, and a number of improvements and new features are planned for the next version. For ecological exposure and risk assessment, planned additions include habitat-dependent receptor movement models and the incorporation of contaminant octanol-air partition (Koa) coefficients into the ecological database and exposure methods.

Use of receptor movement models allows exposure analyses to build upon the contamination interpolation capabilities already present in the software. This addition will simulate individual receptors moving freely on a gridded landscape containing interpolated contaminant concentrations or output from spatial fate and transport models. The modeler/assessor controls the number of individual receptors, movement rules, and parameterization of the terrestrial exposure models. Individual receptors can move randomly or move based on the spatial distribution of habitat quality. Previous work (Purucker et al. 2007) has demonstrated that significant positive or negative spatial correlations between contaminant concentrations and habitat quality can affect the estimated exposed population dose distribution. Population ranges can be defined with GIS spatial delineation tools present in the software; home range sizes can also be defined for individuals within the population. The software tracks the exposure of individuals and produces a population exposure distribution that can be compared to toxicity reference values for determining the necessity for and efficacy of alternative remedial actions.

Another area for future development is improving the exposure methods relative to the biomagnification of persistent organic pollutants in terrestrial food webs. Traditionally, the determination and modeling of contaminants that accumulate have been directly or indirectly based on the octanol-water partition coefficient (Kow), with high Kow values indicating significant potential for accumulation. This relationship is based on aquatic laboratory tests demonstrating that lipid-water partitioning is a principal factor in bioconcentration (e.g., Mackay 1982). However, recent studies (Kelly et al. 2007) have indicated that contaminants with moderate Kow values and high Koa values can significantly biomagnify in terrestrial receptors due to the low rate of elimination in air-breathing receptors. Modifications to the ecological exposure methods will include the addition of Koa values in the ecological database distributed with SADA and the ability to specify Koa-based regressions for food chain bioaccumulation calculations for organic contaminants.

11.7 Conclusions

SADA provides data management capabilities, descriptive statistics for characterizing contaminated areas, screening benchmarks, terrestrial exposure methods, and toxicity reference values sufficient to produce tabular output and hypotheses tests to document the majority of commonly used ecological risk assessment endpoints. In addition, the SADA GIS provides a platform for additional exposure/risk assessment methods that account for spatial dependence, including spatially relevant descriptive statistics, moving window spatial statistics, modified hypotheses tests, correlation modeling, and interpolation methods. These tools provide methods that can minimize remedial action decision errors, provide spatial designs for more efficient and cost effective remedial design under selective remediation conditions, and provide a rationale and context for additional sampling efforts at contaminated sites. The integral combination of spatial assessment approaches with screening and exposure assessment capabilities has the advantage of improving the quality of ecological risk-based decisions without overly complicating the assessment.

Utilizing the spatial correlation present in the analytical contaminant data and applying the cleanup decision criteria based on that spatial distribution of the contamination can result in a flexible map of selective remediation areas rather than binary "clean" or "not clean" decisions. This approach can reduce decision errors due to forced assumptions concerning data independence and the parametric distribution of the data. Ultimately, however, implementing remediation based on estimated ecological risk is impacted by feasibility issues. Implementing selective remediation based on ecological risk assessment may help to limit costs, thus making remediation more likely when needed. More generally, situating ecological risk assessment capabilities within a GIS system capable of incorporating models of other, non-chemical, stressors allows for integrated environmental management of multiple stressors. Integrating stressor-specific assessment modules within specific GIS systems is becoming feasible as models are developed to assist with spatially explicit resource management (Wang et al. 2006) and to address problems posed by ecosystem service approaches to addressing environmental issues (Carpenter et al. 2006).

11.8 Availability

SADA version 4.1 is freeware. A full-featured installation file for Microsoft Windows operating systems is available for free download at http://www.tiem. utk.edu/~sada/download.shtml.

Acknowledgments Comments from Lou Gross, Tom Hallam, Suzanne Lenhart, Gary McCracken, Fran Rauschenberg, and Robert Swank improved the manuscript. This chapter has been reviewed in accordance with the U.S. Environmental Protection Agency's peer and administrative review policies and approved for publication. Mention of trade names or commercial products does not constitute endorsement or recommendation for use.

References

Agency for Toxic Substances and Disease Registry (ATSDR), 1989. Toxicological profile for selected PCBs (Aroclor 1260, 1254, 1248, 1242, 1232, 1221 and 1016), Atlanta, Georgia, ATDSR/TP-88/21.

Anderson, M.C., Thompson, B., Boykin, K., 2004. Spatial risk assessment across large landscapes with varied land use: lessons from a conservation assessment of military lands. Risk Analysis, 24(5):1231–1242.

Aulerich, R.J., Ringer, R.K., 1977. Current status of PCB toxicity to mink, and effect on their reproduction. Archives of Environmental Contamination and Toxicology, 6:279–292.

Barron, M.G., Wharton, S.R., 2005. Survey of methodologies for developing media screening values for ecological risk assessment. Integrated Environmental Assessment and Management, 1(4):320–332.

Blacker, S.B., Goodman, D., 1994. Case study: application at a Superfund cleanup. Environmental Science and Technology, 28(11):471A–477A.

Bradbury, S.P., Feijtel, T.C.J., van Leeuwen, C.J., 2004. Meeting the needs of ecological risk assessment in a regulatory context. Environmental Science and Technology, 38:463A–470A.

Brakewood, L.H., Grasso, D., 2000. Floating spatial domain averaging in surface soil remediation. Environmental Science and Technology, 34:3837–3842.

Carpenter, S.R., DeFries, R., Dietz, T., Mooney, H.A., Polasky, S., Reid, W.V., Scholes, R.J., 2006. Millennium ecosystem assessment: research needs. Science, 314:257–258.

Chow, T.E., Gaines, K.F., Hodgson, M.E., Wilson, M.D., 2005. Habitat and exposure modeling for ecological risk assessment: A case study for the raccoon on the Savannah River Site. Ecological Modelling, 189:151–167.

Crommentuijn, T., Sijm, D., de Bruijn, J., van Leeuwen, K., de Plassche, E., 2000a. Maximum permissible and negligible concentrations for some organic substances and pesticides. Journal of Environmental Management, 58:297–312.

Crommentuijn, T., Sijm, D., de Bruijn, J., van Leeuwen, K., de Plassche, E., 2000b. Maximum permissible and negligible concentrations for metals and metalloids in the Netherlands, taking into account background concentrations. Journal of Environmental Management, 60:121–143.

Dearfield, K.L., Bender, E.S., Kravitz, M., Wentsel, R., Slimak, M.W., Farland, W.H., Gilman, P., 2005. Ecological risk assessment issues identified during the U.S. Environmental Protection Agency's examination of risk assessment practices. Integrated Environmental Assessment and Management, 1(1):73–76.

DeMott, R.P., Balaraman, A., Sorensen, M.T., 2005. The future direction of ecological risk assessment in the United States: Reflecting on the U.S. Environmental Protection Agency's "Examination of Risk Assessment Practices and Principles." Integrated Environmental Assessment and Management, 1(1):77–82.

Efroymson, R.A., Will, M.E., Suter, G.W., Wooten, A.C., 1997a. Toxicological Benchmarks for Screening Contaminants of Potential Concern for Effects on Terrestrial Plants: 1997 Revision. Oak Ridge National Laboratory, Oak Ridge, TN. ES/ER/TM-85/R3.

Efroymson, R.A., Will, M.E., Suter, G.W., 1997b. Toxicological Benchmarks for Contaminants of Potential Concern for Effects on Soil and Litter Invertebrates and Heterotrophic Processes: 1997 Revision. Oak Ridge National Laboratory, Oak Ridge, TN. ES/ER/TM-126/R2.

Eisler, R., 1986. Polychlorinated Biphenyl Hazards to Fish, Wildlife, and Invertebrates: A Synoptic Review. U.S. Fish and Wildlife Service, Patuxent Wildlife Research Center, Laurel, MD. Biological Report 85 (1.7).

Goovaerts, P., 1997. Geostatistics for Natural Resource Evaluation. Oxford University Press, New York.

Gotway, C., 1991. Fitting semi-variogram models by weighted least squares. Computers and Geosciences, 17(1):171–172.

Halbrook, R.S., Aulerich, R.J., Bursian, S.J., Lewis, L., 1999. Ecological risk assessment in a large river-reservoir: 8. Experimental study of the effects of polychlorinated biphenyls on reproductive success in mink. Environmental Toxicology and Chemistry, 18:649–654.

Hope, B.K., 2000. Generating probabilistic spatially-explicit individual and population exposure estimates for ecological risk assessments. Risk Analysis, 20:573–590.

Hope, B.K., 2001. A case study comparing static and spatially explicit ecological exposure analysis models. Risk Analysis, 21(6):1001–1010.

Kelly, B.C., Ikonomou, M.G., Blair, J.D., Morin, A.E., Gobas, F.A.P.C., 2007. Food web-specific biomagnifications of persistent organic pollutants. Science, 317:236–239.

Landis, W., McLaughlin, J., 2000. Design criteria and derivation of indicators for ecological position, direction, and risk. Environmental Toxicology and Chemistry, 19(4):1059–1065.

Legendre, P., Legendre, L., 1998. Numerical Ecology. Elsevier, Amsterdam.

Mackay, D., 1982. Correlation of bioconcentration factors. Environmental Science and Technology, 16:274–278.

Miller, H., Getz, L.L., 1977. Factors influencing local distribution and species diversity of forest small mammals in New England. Canadian Journal of Zoology, 55:806–814.

Prokop, G., Schamann, M., Edelgaard, I., 2000. Management of contaminated sites in Western Europe. Europe Environment Agency, Topic Report No. 13.

Purucker, S.T, Welsh, C.J.E., Stewart, R.N., Starzec, P., 2007. Use of habitat-contamination spatial correlation to determine when to perform a spatially explicit ecological risk assessment. Ecological Modelling, 204(1–2):180–192.

Purucker, S.T., Stewart, R.N., Wulff, J., 2008. A spatial decision support system for efficient environmental assessment and remediation. In: Madden, M. & Allen, E. (eds.), Landscape Analysis Using Spatial Tools. Springer-Verlag.

Ringer, R.K., Aulerich, R.J., Blevins, M.R., 1981. Biologic and toxic effects of PCBs and PBBs on mink and ferret- a review. In: Khan, M.A.Q. & Stomton, R.M.H. (eds.), Toxicology of Halogenated Hydrocarbons: 329–343. Pergamon, New York, USA.

Sample, B.E., Opresko, D.M., Suter, G.W., 1996. Toxicological Benchmarks for Wildlife: 1996 Revision. Oak Ridge National Laboratory, ES/ER/TM-86/R3, Oak Ridge National Laboratory, Oak Ridge, TN.

Stewart, R.N., Purucker, S.T., Powers, G.E., 2007. SADA: A Freeware Decision Support Tool Integrating GIS, Sample Design, Spatial Modeling, and Risk Assessment. Proceedings of the International Symposium on Environmental Software Systems, Prague, Czech Republic.

Suter, G.W. II, 1993. Ecological Risk Assessment. Lewis Publishers, Boca Raton, FL.

Suter, G.W. II, Vermeire, T., Munns, W.R. Jr., Sekizawa, J., 2003. Framework for the integration of heath and ecological risk. Human and Ecological Risk Assessment, 9(1):281–301.

Swartjes, F.A., 1999. Risk-based assessment of soil and groundwater quality in the Netherlands: standards and remediation urgency. Risk Analysis, 19(6):1235–1249.

Tannenbaum, L.V., Johnson, M.S., Bazar, M., 2003. Application of the hazard quotient method in remedial decisions: A comparison of human and ecological risk assessments. Human and Ecological Risk Assessment, 9(1):387–401.

U.S. Environmental Protection Agency (USEPA), 1980. Ambient Water Quality Criteria for Polychlorinated Biphenyls. Office of Water Regulations and Standards. Office of Research and Development. Carcinogen Assessment Group. Environmental Research Laboratories. EPA/440/5-80-068.

U.S. Environmental Protection Agency (USEPA), 1992. Framework for Ecological Risk Assessment. Risk Assessment Forum, Washington DC. EPA/630/R-92/001.

U.S. Environmental Protection Agency (USEPA), 1993. Wildlife Exposure Factors Handbook. Office of Research and Development, U.S. Washington, DC, EPA/600/R-93/187a.

U.S. Environmental Protection Agency (USEPA), 1997. Ecological risk assessment guidance for superfund: process for designing and conducting ecological risk assessments – interim final. EPA 540-R-97-006, OSWER 9285.7-25.

U.S. Environmental Protection Agency (USEPA), 1998. Guidelines for Ecological Risk Assessment. EPA 630/R-95/002F.

U.S. Environmental Protection Agency (USEPA), 1999. Ecological screening levels for RCRA Appendix IX hazardous constituents. Washington DC, Region V. Work draft, August 22, 2003 update.

U.S. Environmental Protection Agency (USEPA), 2001a. Supplemental Guidance to RAGS: Region 4 Bulletins, Ecological Risk Assessment. Originally published: EPA Region IV. 1995. Ecological Risk Assessment Bulletin No. 2: Ecological Screening Values. U.S. Environmental Protection Agency Region 4, Waste Management Division, Atlanta, GA.

U.S. Environmental Protection Agency (USEPA), 2001b. The role of screening-level risk assessments and refining contaminants of concern in baseline ecological risk assessments. ECO Update. Washington DC, Office of Solid Waste and Emergency Response. EPA 540/F-01/014.

U.S. Environmental Protection Agency (USEPA), 2002. Guidance on Choosing a Sampling Design for Environmental Data Collection. Office of Environmental Information, Washington, DC. EPA QA/G-5S.

U.S. Environmental Protection Agency (USEPA), 2003. Guidance for Developing Ecological Soil Screening Levels. Office of Solid Waste and Emergency Response, U.S. Environmental Protection Agency, Washington, DC. OSWER Directive 9285.7-55. November 2003.

U.S. Environmental Protection Agency (USEPA), 2006. Data Quality Assessment: A Reviewer's Guide. Office of Environmental Information, Washington, DC. EPA QA/G-9R.

U.S. Office of Technology Assessment (USOTA), 1985. Superfund Strategy. Washington, U.S. Congress, Office of Technology Assessment, DC. OTA-ITE-252.

Wang, D., Buchanan, N., Berry, M.W., Carr, E., Comiskey, J.E., Gross, L.J., Shaw, S.-L., 2006. A GIS-enabled distribution simulation framework for high-performance ecosystem modeling. Proceedings of the ESRI International User Conference, August 11–15, 2006.

Wind, T., 2004. Prognosis of environmental concentrations by geo-referenced and generic models: a comparison of GREAT-ER and EUSES exposure simulations for some consumer product ingredients in the Itter. Chemosphere, 54:1145–1153.

Wren, C.D., Hunter, D.B., Leatherland, J.F., Stoakes, P.F., 1987. The effects of polychlorinated biphenyls and methylmercury, singularly and in combination on mink. II: reproduction and kit development. Archives of Environmental Contamination and Toxicology, 16:449–454.

Chapter 12
Decision Evaluation for Complex Risk Network Systems (DECERNS) Software Tool

Terry Sullivan, Boris Yatsalo, Alexandre Grebenkov and Igor Linkov

Abstract Environmental management requires decision makers to integrate heterogeneous technical information with values and judgment. Decision Evaluation in Complex Risk Network Systems (DECERNS) is a computer system with the objective of providing a methodology, computer models and software tools that facilitate decision-making in the field of sustainable land use planning, sediment management, and related areas. DECERNS will integrate risk assessment and decision analysis tools from multiple disciplines (e.g., Geographic Information System (GIS) tools for mapping and data analysis, ecological and human health risk models, economic analysis tools for evaluating costs, and tools for incorporating social choices) into a single user-friendly software package. This chapter provides an overview of the approach used in DECERNS and a case study. Future plans on development and testing of the software tool are also discussed.

12.1 Introduction

Decision making for environmental protection and land-use planning is complex and involves multiple interconnected and often competing objectives. Many countries face the need for land management tools that allow balancing local ecological, economic and social processes while assuring environmental security and sustainability. There is a need for an integrated site management approach supporting economic development and a parallel goal of natural resources conservation, restoration of habitats in surrounding ecosystems and biodiversity, as well as reduction of present and future pollution. In order to evaluate available management alternatives, the decision makers could benefit from a framework that would lead them through a systematic process of priorities elicitation, strategic planning and comprehensive assessments, explicitly integrating human

T. Sullivan (✉)
Brookhaven National Laboratory, Upton, NY, USA
e-mail: tsullivan@bnl.gov

A. Marcomini et al. (eds.), *Decision Support Systems for Risk-Based Management of Contaminated Sites*, DOI 10.1007/978-0-387-09722-0_12,
© Springer Science+Business Media, LLC 2009

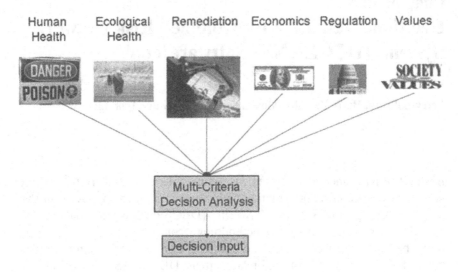

Fig. 12.1 Example of DECERNS Framework for sustainable land use

health risks, ecological risks, and socio-economic measures. Multi-Criteria Decision Analysis (MCDA) tools have begun to see use in environmental applications (Malczewski, 1999, Linkov et al. 2004a, 2006).

Decision Evaluation in Complex Risk Network Systems (DECERNS) is a Spatial Decision Support System (SDSS) with the objective of providing a methodology, computer models and software tools that would facilitate decision-making in the field of environmental management.

DECERNS SDSS has the ability to integrate different models and measures (Fig. 12.1) as well as decision-maker values into a common framework using MCDA and GIS tools. The MCDA tools are incorporated into a systematic and modular framework that allows experts and stakeholders to integrate objective data, model assessments and subjective judgments and then to examine alternatives to rank them, select the "most appropriate", or determine the "best/optimal" or, correctly speaking, a trade-off alternative for a specific (spatial) multi-criteria problem. Sensitivity and uncertainty analysis will be incorporated within the framework to address the robustness of the decision to changing the importance and value of each decision criteria.

This chapter provides an overview of the approach used in DECERNS followed by a brief overview of methods and tools and an application case study. Future plans on DECERNS development and testing are also discussed.

12.2 DECERNS Framework

Decision support can be defined as the assistance for, and substantiation and corroboration of, an act or result of deciding (Bardos et al., 2001); typically this decision will be a determination of an *optimal* or *trade-off* approach which leads naturally to the use of MCDA tools. Decision support integrates specific information about a site and general information such as legislation, guidelines and technical know-how, to produce decision-making knowledge in a way that is transparent, consistent and reproducible. In land use decisions, depending on the size and prominence of the site/territory, the principal stakeholders can include any of the following: land owners or problem holders; regulatory and planning authorities; site users, workers, visitors; financial community (banks, lenders, insurers); site neighbors (tenants or visitors); environmental organizations; consultants, contractors and technology vendors; and, possibly, researchers.

Several different methodological approaches can be employed to assist environmental decision-making. Advanced SDSSs for environmental protection and landuse management include: (a) GIS tools for input/output data/information analysis, processing and presentation, (b) specific methods for decision support (MCDA tools), and (c) models for risk assessment and technological evaluations.

The GIS tools will include all of the basic functions typical for these systems, including the ability to handle vector and raster maps, basic statistical analysis of the data (min, max, mean, average, standard deviation), multiple map layers, and the ability to sort and query the data (e.g., provide all locations with a specific land use such as farmland). These will be discussed in more detail in Section 2.1.

The decision support methods will include several multi-criteria decision analysis tools, cost/benefit analysis (CBA) and cost/effectiveness analysis (CEA) tools. The methods implemented in DECERNS also permit group or individual decision making. The decision support methods are discussed in detail in Section 2.2. The process models will include some basic models of human health risk, ecological risk, and remediation costs. Problem-specific models may be added as required to address unique situations or improve upon the base models. The process models are discussed in Section 2.3. Section 2.4 discusses software implementation.

DECERNS is developed in two versions: as a distributed/web-based SDSS, and as a stand-alone (desk top) application. Both use program libraries that contain information for evaluating contaminated land problems such as databases of risk factors for contaminants and costs for different remedial options. The user accesses DECERNS through a graphical user interface which connects to an application programming interface to access the three main components in DECERNS: GIS tools, Decision Support Tools and Process Models. These three components each have data requirements including site-specific problem dependent information and model input. Figure 12.2 presents the framework and linkages between the different aspects discussed above for DECERNS.

The common graphical user interface provides an intuitive access to features implemented within DECERNS. The Application Programming Interface integrates the three main components (GIS, Decision Support Tools, and Models)

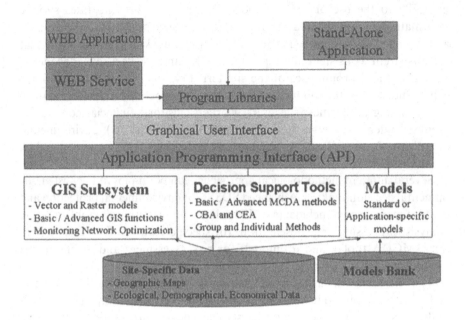

Fig. 12.2 DECERNS SDSS architecture

onto a single platform. All MCDA and GIS functions, as well as models implemented in DECERNS, can be activated from this interface.

12.2.1 GIS Tools

The GIS subsystem is one of the key components of SDSS. It is designed to have
all of the basic GIS features including handling vector and raster maps, realization of scalar operations and distance functions, overlay modeling, data query,
and creating buffer zones (with the subsequent possibility of union, intersec-
tion, or exclusion between different buffer zones). This permits handling of
multiple-criteria siting problems. For example, if a business wants to select a
location that is within 10 km of a hospital, at least 50 m off the road, in an area
that is already zoned for industrial use, buffer layers for each of these three
criteria could be generated and combined to provide a map with all of the
acceptable locations. The GIS system also provides easy to use methods to re-
classify data into different groups. Advanced GIS functions include basic statistical measures of data (mean, median, standard deviation, min, max, range,
skewness, kurtosis, and density distribution functions/histograms); geostatistical
analysis (inverse distance weighting, kriging), and multi-variate analysis (correlation, covariance, and variograms). Figure 12.3 shows an example of the
DECERNS GIS subsystem with multiple layers, and statistical evaluation of
the data.

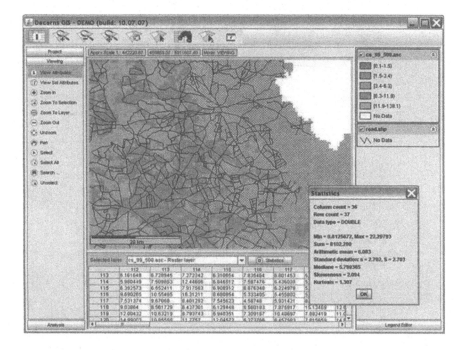

Fig. 12.3 DECERNS SDSS: Main screen of GIS subsystem

12.2.2 Decision Support Tools

The core module of the DECERNS SDSS is the decision evaluation subsystem. It is based on implementation of decision analysis tools, including different MCDA methods and Cost-Benefit Analysis (CBA).

MCDA methods and techniques are at the center of this process and offer the possibility of comparing alternatives taking into account different quantitative and qualitative criteria. The common purpose of MCDA methods is to evaluate and choose among alternatives based on multiple criteria using systematic analysis that overcomes the limitations of unstructured individual or group decision-making (Belton and Steward, 2002, von Winterfeldt and Edwards, 1986). The aim of MCDA in a broad sense is to facilitate decision makers' learning about and understanding of the problem, about their own, other parties' and organizational preferences, values and objectives and, through exploring these in the context of a structured decision analysis framework, to guide them in identifying a preferred course of action.

Figure 12.4 presents a decision process flow chart for typical MCDA problems and used in applying DECERNS. The process includes input from stakeholders, decision makers, and scientists/experts. The process begins with the problem definition. The second step is the development of problem specific management alternatives and criteria specification by the problem holders, stakeholders, and

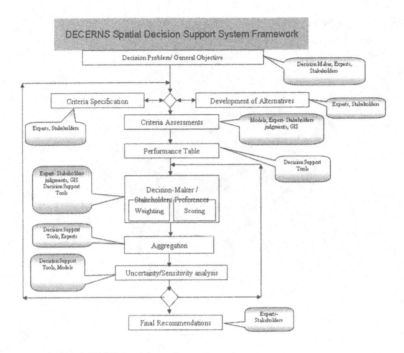

Fig. 12.4 DECERNS SDSS: Decision process flow chart

technical experts. With these, assessments of the different alternatives against the different criteria are conducted using models, GIS tools, and expert/stakeholder judgments. A performance table can be generated from the results of the criteria assessments. The next step involves determining the preferences and weighting of each criterion by the stakeholder community. The MCDA tools take this information, aggregate the results, and perform sensitivity/uncertainty analysis. The resulting ranking of alternatives is reviewed by the stakeholders and recommendations are made for the decision makers. The process can be repeated adaptively to refine any of the steps.

When solving a specific multicriteria problem, the DECERNS users will have the opportunity to choose an appropriate MCDA method or, if possible, compare several methods with an analysis of uncertainties associated with the chosen MCDA approaches (Yatsalo et al., 2007). The following decision analysis tools are either implemented or under consideration.

12.2.2.1 Cost Benefit Analysis (CBA) and Cost Effectiveness Analysis (CEA)

One approach to Multi-Criteria Decision Analysis is to convert all decision metrics into cost. Cost Benefit Analysis (CBA) or Cost Effectiveness Analysis (CEA) aim at assessment of all the costs and benefits of alternative options, using monetary valuations of all the main criteria for each alternative. When solving problems on environmental protection or remediation of contaminated territories, many criteria are not expressed a priori in monetary values. In this case conversion factors which transform advantages (benefits) and disadvantages (damages/losses/risks) into monetary values are used.

CEA is used to compare different options to achieve the same objective and then to choose the option that achieves the objective at the lowest cost or with minimal cost associated with a unit of positive effect/benefit (Grebenkov and Yakushau, 2007). It should be pointed out that CEA cannot be used, in general, for analysis of justification of alternatives, but may be considered as a secondary (auxiliary) method; it is often used if the full CBA cannot be correctly realized (e.g., when not all advantages/disadvantages can be effectively presented in monetary terms). CBA can be an effective method when conversion factors are determined and formalized at a departmental or national level (e.g., as for radiation protection).

Use of CBA/CEA in DECERNS is implemented and can be applied for investigation of problems on land management within remediation/revitalization of radioactively contaminated sites/territories (Yatsalo et al., 1997b, Yatsalo and Bardos, 2002).

Methods for economic valuation of environmental measures (value of biodiversity including genetic biodiversity, value of living, value of nature production, etc.) are also applied. These methods utilize biotic and abiotic (non-living) indices and their monetary equivalents.

CBA methods for ecological impacts, just as for social costs, moneterizes the cost of ecosystem services and biological resources (Phillips, 1998). These two components, which human society pays for sustainability of its habitat and living environment, have a long pay-back period and therefore can not compete (in the near-term on an economic basis) with the near-term immediate benefits of industrial production. Therefore, this approach requires a special discounting rate for all cost and benefits of environmental protection measures.

12.2.2.2 Multi Attribute Decision Methods (MADM)

Within multi-attribute decision methods there are a finite number of (explicitly given) alternatives and a set of elaborated criteria.

MCDA techniques are generally categorized into

- *value function*-based methods, and
- *outranking* methods.

Approaches that use *value functions* form the so-called MAVT methods (multi-attribute *value* theory). MAUT methods (multi-attribute *utility* theory) are also often used. While the methods are not always seen as fundamentally different (von Winterfeldt and Edwards, 1986), they are differentiated on the basis of certainty. A value function describes a person's preference regarding different levels of an attribute under *certainty*, whereas utility theory extends the method to use probabilities and expectations to deal with uncertainty (Belton and Steward, 2002, von Winterfeldt and Edwards, 1986, Figueira et al., 2005).

Outranking Relation Theory involves forming an ordered relation of a given set of alternatives. *Outranking methods* are based on a pairwise comparison of alternatives for each criterion under consideration with subsequent integration of obtained preferences according to a chosen algorithm. The PROMETHEE method developed by Brans and Vincke (Brans and Vincke, 1985) is an outranking technique implemented in DECERNS.

The *Analytic Hierarchy Process (AHP)*, developed by T. Saaty (Saaty, 1980) is implemented in DECERNS. Within AHP, a systematic pairwise comparison of alternatives with respect to each criterion is used based on a special ratio scale: for a given criterion, alternative i is preferred to alternative j with the strength of preference given by $a_{ij} = s$, $1 \leq s \leq 9$, correspondingly, $a_{ji} = 1/s$. Then, the same procedure is implemented for $n(n-1)/2$ pairwise comparisons in the same scale for n criteria.

Advanced methods under consideration for implementation into DECERNS include: MAUT (Multi-Attribute Utility Theory), SMAA (Stochastic Multiobjective Acceptability Analysis) and TOPSIS (Technique for Order Preference by Similarity to Ideal Solution), and realization of Fuzzy set approaches.

12.2.2.3 Multi-Objective Decision Methods (MODM)

MODM and MADM problems are often referred to as continuous and discrete decision problems, respectively. Within MODM, in contrast to MADM, we have an infinite or sufficiently great number of *implicitly given* alternatives. In MODM each alternative is defined implicitly in terms of decision variables and evaluated through the use of objective functions. In this case an attribute is a descriptive variable, and an objective is a more abstract variable (function) with a specification of the relative desirability of the levels of that variable/objective. Within MODM the set of alternatives is defined in terms of causal relationships and constraints on the decision variables (Malczewski, 1999). Some MODM approaches will be realized in DECERNS SDSS for solving problems on monitoring networks optimization, and for location-allocation problems (Malczewski, 1999).

12.2.2.4 Group Decision Making Methods

Recently, interest has been growing in the methods and systems which realize *group decision support* approaches. The ability to use group decision support is facilitated with the development of Web-DSSs. Group decision making is a process of aggregation of individual preferences of different experts into one group preference. Within a group MCDA method, each expert can have a weight, which corresponds to the competence of the expert. Sometimes each expert is given only one weight, but the expert may be given a separate weight for each criterion, showing his/her competence according to this criterion. Group MCDA methods may be divided into elementary/voting and complex methods (which use individual MCDA methods to calculate individual preferences), and methods used under certainty and uncertainty. DECERNS SDSS will implement group decision support approaches.

12.2.3 Environmental Models

The third major component of DECERNS is the library of models that can be used for evaluation of criteria and measures necessary for decision making. The models can be accessed through the SDSS interface. The library will include the following models:

i. human health risk assessment for individuals and population groups incorporating spatial distribution of contaminants and temporal concentration changes;
ii. spatially-explicit ecological risk assessment with explicit incorporation of ecological receptor migration and habitats quality; and
iii. fate and transport models for radionuclides and chemicals in the environment.

The human health risk models will be simple screening models that cover the major exposure pathways for humans. A library of default values for toxicity benchmarks (for chemicals) and dose conversion factors (for radionuclides) as well as exposure parameters is included in DECERNS. The user will be able to change any parameter based on site-specific data.

In addition to spatially varying ecological assessments for stationary receptors (plants), DECERNS will include a model to simulate the migration of animals (receptors) through contaminated regions (Grebenkov et al., 2002, Linkov, 2004b, 2004c). The animals interact with each other, interact with the total environment and adapt to changes in the environment. If an organism can learn, it will be able to modify its behavior based on environmental feedback and potentially increase its survival probability. The main idea of the animal behavior models is based on the assumption that animal behavior is a stochastic chain of behavior states. Probabilities exist to describe the animal's preferences in moving from one behavior (foraging, eating, reproduction, rest, etc.) to another, and the distribution for time spent in each behavior. A simple descriptive model for such systems is a Markov chain, in which animal's behavior at time $t + 1$ depends only on its behavior in previous moment of time t. In addition, it can be assumed that for all steps of time the behavior of animals has a rational character. Rational behavior can be defined as having the ability to make correct decisions with high probabilities among multiple choices, determined, for example, by the quality of habitat (Grebenkov et al., 2006). With these types of models, questions pertaining to the length of time an animal will spend in different regions (with different contamination levels) can be calculated and the exposure can be calculated based on the time integrated contamination levels that the animal experienced.

We are working on inclusion of more sophisticated fate and transport models (e.g., two-phase flow for oil spills), and advanced animal migration models for predicting exposure and risks. DECERNS provides tools for adding user-defined models to address site-specific issues, or new classes of problems. DECERNS will allow generation and visualization of spatially-explicit information.

12.2.4 Software Implementation

Integration of the different software components into a single unit is a challenge. Oftentimes it will be the most efficient to take existing models and implement them into DECERNS SDSS. Existing software may be written in a variety of languages (e.g. C++, Visual Basic, FORTRAN, etc.). These programs are not automatically compatible with web-based programming languages such as JAVA. As part of the DECERNS project, software tools to translate programs from their native language to C++ have been developed. Bridging of software tools using different programming languages is conducted using JAVA-Net tools.

We are currently developing a web based DECERNS version. Its key feature will be the use of GIS functions and decision analysis tools with a client/server computing system. The client requests data and analysis tools from server. The server either performs the analysis and sends the results back to the client through the network or sends the data and analysis tools to the client for processing. Standard HTTP protocol is used to establish the connection between application server and the client. This allows the use of a web-browser to access DECERNS SDSS. TCP/IP protocol is used to establish communication between the information tier (database server) and the application tier (J2EE compliant application server).

12.3 DECERNS Application Case Study

The land contamination by radioactive Cesium (Cs-137) as a result of the Chernobyl accident has led to external and internal exposures of the local population in the Bryansk Region of Russia (Karaoglou, 1996, Yatsalo et al., 1997a, Yatsalo and Bardos, 2002, Yatsalo et al., 2004, Yatsalo, 2007). Monitoring data and modeling indicate that more then 50% of the rural population in the five indicated districts in the Bryansk region live in settlements with a mean dose above 1 mSv/y (international regulatory benchmark) (Yatsalo et al., 2004, Yatsalo, 2007). This case study illustrates DECERNS application for selecting remedial and abatement policies based on radionuclide distribution data and value judgment on the efficiency of different management alternatives.

A library of vector electronic maps of the region has been developed based on land use information for each district and farm using agricultural and

Fig. 12.5 Surface Cs-137
contamination in Bryansk
region of Russia

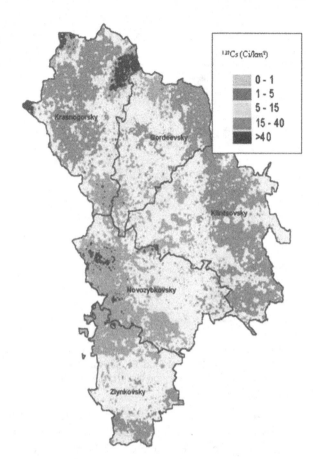

radiation monitoring data along with demographic and economic data. These
maps can be easily integrated into DECERNS; for example, Fig. 12.5 displays
the Cs-137 contamination for the five most contaminated districts in this
region. The inhomogeneous nature of the contamination and the large size
of the Bryansk Region (over 40,000 km^2) make radiation protection decisions
difficult.

The following analyses were performed:

- estimation of agricultural produce contamination (for the given years) for
  each location;
- estimation of wild mushroom and berries contamination for all the farms
  and farming scenarios;
- assessment of internal and external doses to the general population and
  critical (i.e. most exposed) population groups for all the settlements within
  contaminated districts at different time following the Chernobyl accident.
  Contribution of different pathways to internal and total dose was also
  analyzed;

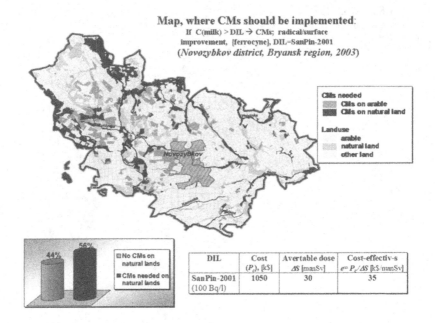

Fig. 12.6 Map of Novozybkovsky district (Bryansk region, Russia) where protective measures on agricultural lands should be implemented and corresponding cost-effectiveness analysis at the derived intervention level (DIL) of 100 Bq/l

- risk-based ranking of settlements, farms and agricultural land;
- analysis of fields and settlements where concentrations exceed legislative limits and countermeasures (CMs) should and can be implemented in accordance with the existing standards, requirements and restrictions;
- assessments of the results of the countermeasure implementation in terms of residual produce contamination and doses over time;
- assessments of countermeasure efficiency in space and time.

Figure 12.6 shows the areas where countermeasures on natural lands (pastures and hayfields) and arable lands should be implemented in the Novozybkovsky district, Bryansk region, in order to reduce exposure from agricultural produce (milk, crops, potato) to permissible levels. The analysis showed that about 55% of the natural lands required implementation of countermeasures. Cost-effectiveness analysis was calculated by determining the cost of implementation and remediation divided by the reduction in the dose to the population due to the counter measures. Avertable dose is the measure for the benefit in applying the counter measures and is defined as the baseline dose minus the dose after implementation of the counter measures. The cost effectiveness was defined as the total cost divided by the total dose reduction and the results are displayed in the bottom of Fig. 12.6.

Figure 12.7 illustrates a multi-attribute value theory (MAVT) implementation example within DECERNS. The goal is to select the best alternative for

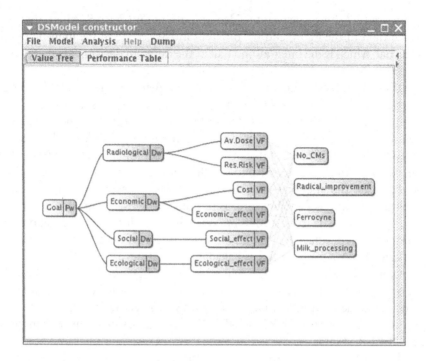

Fig. 12.7 DECERNS decision tree

semi-natural ecosystem (e.g., pasture and farms) treatment. The following four alternatives were found to be feasible for the region (Karaoglou, 1996, Yatsalo et al., 1997b):

a. No Counter measures (No CM) applied (no action alternative),
b. Radical improvement (any combination of several measures to treat the soil including ploughing to reduce contamination in the root zone, adding fertilizer, lime, reseeding or other soil treatments to reduce the uptake of Cs),
c. Adding ferrocine in cow feed (to decrease transfer of Cs from feed to milk/meat),
d. Processing the milk produced from contaminated pastures to make cheese and butter (removing Cs).

These four objectives were evaluated for the following six criteria:

- Averted Dose (Dose reduction);
- Residual risk;
- Cost;
- Economic Effect (potential benefits of increasing farming and/or the value of agricultural products as a result of CM implementation);
- Social Effect (value of lower dose to the local population);
- Ecological Effect (impact of countermeasures on the ecology).

Figure 12.7 shows the resulting decision tree in DECERNS. Alternatives are to the right of the figure and the different attributes are to the immediate left. The goal, which is the integration of all of the different decision attributes, is at the left.

DECERNS uses weights and value functions determined by the stakeholders for each attribute and stores them in a performance table as shown in Fig. 12.8. The top section of this table lists the six criteria and provides general information about them including their weight in the decision process. The second section of the performance table provides the units for each criteria (for example, cost is in roubles), a description of the units (for example, normalized cost per hectare), the minimum and maximum value for each criteria, and information as to whether the objective is to minimize (cost) or maximize (averted dose) the value. The third section provides the value for each criterion on each alternative. The values for some parameters (averted dose, residual risk, and cost) are obtained from subsidiary calculations (Yatsalo et al., 1997b). The values for economic effect, social effect, and ecological effect are preference values and are obtained from the stakeholders. In this example, averted dose and cost were given the highest weights, while social effect was given the lowest weight to the decision (Fig. 12.8 line labelled "weight"). Each of these values can be changed by the DECERNS user.

Figure 12.9 shows calculation results indicating the radical improvement as the "preferred"/trade-off choice (highest overall score) based on the data used in the analysis. Single parameter sensitivity analysis can be performed in DECERNS by varying each of the weights. An example is shown at the bottom

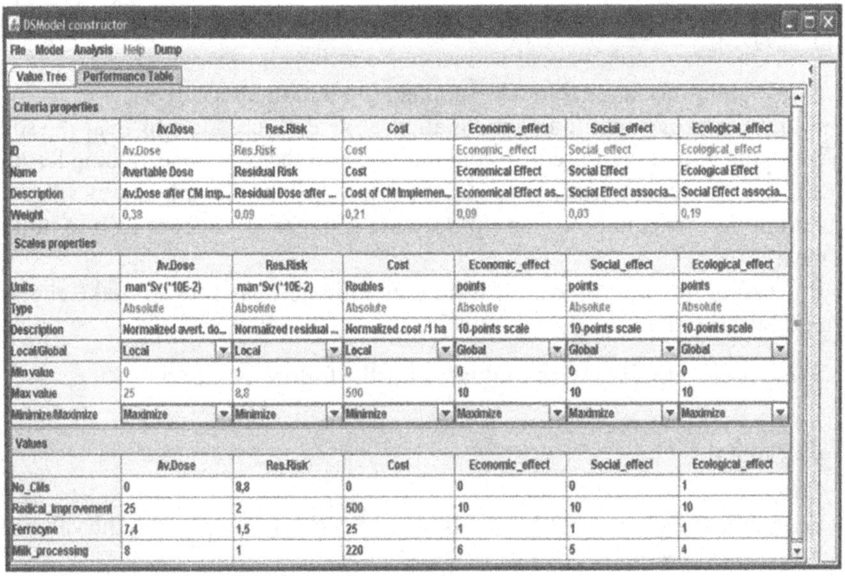

Criteria properties

	Av.Dose	Res.Risk	Cost	Economic_effect	Social_effect	Ecological_effect
ID	Av.Dose	Res.Risk	Cost	Economic_effect	Social_effect	Ecological_effect
Name	Avertable Dose	Residual Risk	Cost	Economical Effect	Social Effect	Ecological Effect
Description	Av.Dose after CM imp...	Residual Dose after ...	Cost of CM Implemen...	Economical Effect as...	Social Effect associa...	Social Effect associa...
Weight	0,38	0,09	0,21	0,09	0,03	0,19

Scales properties

	Av.Dose	Res.Risk	Cost	Economic_effect	Social_effect	Ecological_effect
Units	man*Sv (*10E-2)	man*Sv (*10E-2)	Roubles	points	points	points
Type	Absolute	Absolute	Absolute	Absolute	Absolute	Absolute
Description	Normalized avert. do...	Normalized residual ...	Normalized cost /1 ha	10-points scale	10-points scale	10-points scale
Local/Global	Local	Local	Local	Global	Global	Global
Min value	0	1	0	0	0	0
Max value	25	8,8	500	10	10	10
Minimize/Maximize	Maximize	Minimize	Minimize	Maximize	Maximize	Maximize

Values

	Av.Dose	Res.Risk	Cost	Economic_effect	Social_effect	Ecological_effect
No_CMs	0	8,8	0	0	0	1
Radical_improvement	25	2	500	10	10	10
Ferrocyne	7,4	1,5	25	1	1	1
Milk_processing	8	1	220	6	5	4

Fig. 12.8 DECERNS SDSS: performance table

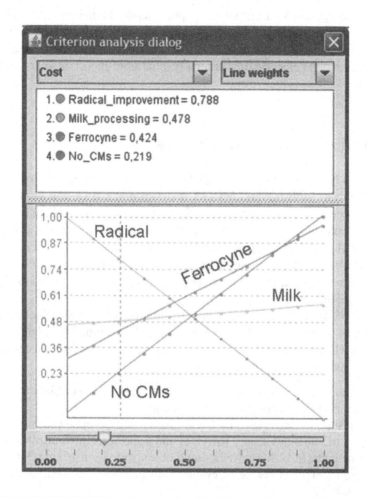

Fig. 12.9 DECERNS SDSS: analysis results with sliding weight for sensitivity analysis

of Fig. 12.9. In accordance with this approach, one weight (weight for the cost criterion in this example) can be changed from 0 to 1, while the other weights are proportionally changed (such that the sum of all weights equals 1); the cost weight of 0.21 which represents the baseline case as determined in the analysis is indicated by the vertical dotted line. If the weight for the cost criterion is between 0 and 0.45, the radical improvement option is considered as the "preferred" one. If the weight for the cost criterion exceeds 0.45, the ferrocyne treatment becomes the "preferred" option. When the weight of the cost criterion exceeds 0.9, the no counter measure option is considered as the "preferred" one. However, it is unlikely that with six criteria, one of them would have over 90% of the weight. In fact such a weighting to cost is in contradiction with justification and optimization principles for radiation protection, which are based on cost-benefit analysis (CBA) implementation. In addition, certain land sections are above the regulatory guidelines and in practice, no counter measures is not a viable option. Nevertheless, the value of the approach is to illustrate that if cost is the dominant decision criteria (>90% of the weight), no countermeasures would be selected. As other decision criteria become more important, other choices would be made.

12.4 Conclusions and Future Directions

The DECERNS SDSS integrates GIS (basic and some extended functions), decision analysis tools (several MCDA methods and CBA), and models for different aspects of land management problems (e.g., ecological and human health risk models, economic analysis tools for evaluating costs, and tools for incorporating social choices) into a single user-friendly software package. DECERNS will be available in either a stand-alone package or in a web based system where it is accessed remotely. Information on the status of DECERNS can be found at (www.DECERNS.com).

DECERNS meets the basic functional requirements for an SDSS tool and is ready for use. It currently contains many advanced functionality and modeling capabilities. However, DECERNS is still under development and will continue to be improved. Major improvements include:

GIS Module – advanced geostatistical methods (kriging), the ability to calculate areas above a contamination threshold.

Decision Support Module – the MCDA tools will be enhanced by adding multi-attribute utility theory (MAUT) and some other multi-attribute decision methods (MADM) as well as multi-objective decision making (MODM) techniques. Techniques being evaluated include PROMETHEE, SMAA, and TOPSIS.

Model Module – the model bank will be extended to include contaminant fate and transport models, animal migration models, additional screening level chemical and radiological risk models, and screening level ecological risk models (Habitat Suitability Index).

DECERNS SDSS will be applied to several additional test cases. This is a key component of the development of DECERNS.

References

Bardos, R.P., Mariotti, C., Marot, F., and Sullivan, T. 2001. Framework for Decision Support used in Contaminated Land Management in Europe and North America, Land Contamination & Reclamation, Vol. 9, No. 1, pp. 149–163.

Belton, V., Steward, T. 2002. Multiple Criteria Decision Analysis: An Integrated Approach. Kluwer Academic Publishers. 372p.

Brans, J.P. and Vincke, P. 1985. A Preference Ranking Organization Method: The PRO-METHEE Method for Multiple Criteria Decision-Making. Management Science, Vol. 31, pp. 647–656.

Economic Values of Protected Areas. 1998. Guidelines for Protected Area Managers. Adrian Phillips, Series Editor. Best Practice Protected Area Guidelines Series No. 2. IUCN – The World Conservation Union.

Figueira, J., Greco, S., and Ehrgott, M. editors. 2005. Multiple Criteria Decision Analysis: State of the Art Surveys. New York (NY): Springer.

Grebenkov, A., Linkov, I., Zibold, G., Andrizhievski, A., and Baitchorov, V. 2002. Approaches to Spatially-Explicit Exposure Modeling and Model Validation / Radioactivity in the Environment, IUR, 1–5 September, Monaco, pp. 461–464.

Grebenkov, A., Loukashevich, A., Linkov, I., and Kapustka, L. 2006. A Habitat Suitability Evaluation Technique and its Application to Environmental Risk Assessment. In: Ecotoxicology, Ecological Risk Assessment and Multiple Stressors. Springer Publisher. – NATO Science Series, Series IV: Earth and Environmental Series,– Vol. 6, pp. 193–201.

Grebenkov, A.J., and Yakushau, A.P. 2007. Ranking of Available Countermeasures Using MCDA Applied to Contaminated Environment. In: Risk Management Tools for Environmental Security, Critical Infrastructure and Sustainability. Springer Publisher. – NATO Science Series, Series IV: Earth and Environmental Series, Vol. 7,pp. 131–135.

Karaoglou, A., Desmet, G., Kelly, G.N., and Menzel, H.G. editors. 1996. The Radiological Consequences of the Chernobyl Accident. Proceedings of the first International Conference. Minsk, Belarus 18–22 March 1996. European Commission. ISBN 92-827-5248-8. ECSC-EC-EAEC, Brussels, Luxembourg, 1192 pp.

Linkov, I., Grebenkov, A., Andrizhievski, A., Loukashevich, A., and Trifonov, A. 2004a. Risk-trace: software for spatially explicit exposure assessment. In: L.A. Kapustka, H. Gilbraith, M. Luxon, and G.R. Biddinger, eds., Landscape Ecology and Wildlife Habitat Evaluation: Critical Information for Ecological Risk Assessment, Land-Use Management Activities, and Biodiversity Enhancement Practices, ASTM STP 1458. ASTM International, West Conshohocken, PA.

Linkov, I., Grebenkov, A., Andrizhievski, A., Loukashevich, A., Trifonov, A., and Kapustka, L. 2004b. Incorporating habitat characterization into Risk-trace: Software for spatially explicit exposure assessment. In: I. Linkov and A. Ramadan, eds., Comparative Risk Assessment and Environmental Decision Making. Kluwer, Amsterdam.

Linkov, I., Varghese, A., Jamil, S., Seager, T.P., Kiker, G., and Bridges, T. 2004c. Multicriteria decision analysis: framework for applications in remedial planning for contaminated sites. In: I. Linkov and A. Ramadan, eds., Comparative Risk Assessment and Environmental Decision Making. Kluwer, Amsterdam.

Linkov, I., Satterstrom, K., Kiker, G., Batchelor, C., Bridges, T. 2006. From Comparative Risk Assessment to Multi-Criteria Decision Analysis and Adaptive Management: Recent Developments and Applications. Environment International, Vol. 32 pp. 1072–1093.

Malczewski, J. 1999. GIS and Multicriteria Decision Analysis. John Wiley & Sons Inc. New York. 392p.

Saaty, T.L. 1980. The Analytic Hierarchy Process. New York: McGraw-Hill. 287p.

von Winterfeldt, D., Edwards, W. 1986. Decision Analysis and Behavioral Research. Cambridge: Cambridge University Press. 604p.

Yatsalo, B., Mirzeabassov, O., Okhrimenko, I., Pichugina, I., Kulagin, B. 1997a. PRANA – Decision Support System for Assessment of Countermeasure Strategy in the Long-Term Period of Liquidation of the Consequences of a Nuclear Accident (Agrosphere). *Radiation Protection Dosimetry* Vol. 73, Nos. 1–4, pp. 291–294.

Yatsalo, B.I., Hedemann Jensen, P., Alexakhin, R.M. 1997b. Methodological Approaches to Analysis of Agricultural Countermeasures on Radioactive Contaminated Areas: Estimation of Effectiveness and Comparison of Different Alternatives. Radiation Protection Dosimetry Vol.74, No. 1/2, pp. 55–61.

Yatsalo, B., Bardos, P. 2002. Decision Support on Risk Based Land Management and Sustainable Rehabilitation of Radioactive contaminated Territories. Radioprotection, Vol. 37, C1, pp. 1087–1092.

Yatsalo, B., Okhrimenko, I., Pichugina, I., Mirzeabassov, O., Didenko, V., Okhrimenko, D., Golikov, V., Bruk, G., Shutov, V. 2004. Decision-Making Support Within Risk Based Land Management and Sustainable Rehabilitation of Radioactive Contaminated Territories. Proc. of the 11 Intern. Congr. IRPA-11, Madrid, Spain, May 23–28, (Programme & Abstract: p. 293; Full text: IRPA-11 CDROM, ISBN: 84-87078-05-2,: Yatsalo B.I., 7b6).

Yatsalo, B.I. 2007. Decision Support System for Risk Based Land Management and Rehabilitation of Radioactively Contaminated Territories: PRANA approach. International Journal of Emergency Management, Vol. 4, No. 3, pp. 504–523.

Yatsalo, B., Kiker, G., Kim, J., Bridges, T., Seager, T., Gardner, K., Satterstrom, K., Linkov, I. 2007. Application of Multi-Criteria Decision Analysis Tools for Management of Contaminated Sediments. Integrated Environmental Assessment and Management, Vol. 3, No. 2, pp. 223–233.

Chapter 13
Decision Support Systems (DSSs) for Contaminated Land Management – Gaps and Challenges

Paola Agostini and Ann Vega

13.1 Introduction

The previous chapters in Section 2 provided a plethora of information regarding decision support systems for risk-based management of contaminated land. Chapter 6 lays the foundation for Section 2 by defining terminology and discussing the broad issues of what is contaminated land in the EU and US; what is a Brownfield; and what is remediation. Chapter 6 also explains the complexity of contaminated land management and emphasizes that it is a multi-dimensional problem. It communicates the need for and the role of DSSs to address risk-based contaminated land management issues. Chapter 7 presents a review of existing DSSs and identifies the phases and process of contaminated site redevelopment including questions that decision makers should ask as they go through the process. Using these questions, existing DSSs are compared to identify which systems address different phases of the process. Subsequent chapters (Chapters 8–12) highlight actively-used DSSs for contaminated land management, with each chapter providing information regarding: framework and functionality; structure of the system; decision aspects and stakeholder involvement; case study application; and ongoing development if applicable. Finally, this chapter (Chapter 13) is intended to summarize identified gaps (i.e., where additional contaminated land management DSSs should be focused) and the challenges faced by anyone attempting to build and implement a DSS. This chapter takes into consideration the contents of all the other chapters of Section 2, and discusses the gaps that may characterize the presented DSSs with respect to contaminated land assessment and management objectives (Section 13.2). Moreover, some relevant challenges still open for these tools, and in general for any decision support system, are presented (Section 13.3).

P. Agostini (✉)
IDEAS (Interdepartmental Centre for Dynamic Interactions between Economy Environment and Society), University of Ca' Foscari, San Giobbe 873, 30121, Venice, Italy
e-mail: paola.agostini@unive.it

A. Marcomini et al. (eds.), *Decision Support Systems for Risk-Based Management of Contaminated Sites*, DOI 10.1007/978-0-387-09722-0_13,
© Springer Science+Business Media, LLC 2009

Discussion of gaps and challenges may help in identifying obstacles for the application of DSSs, as well as in promoting future advancements. In fact, Goosen and colleagues (2007) already discuss several reasons as to why DSSs are not being used by their intended audiences. In their paper they report an overview of some of the pitfalls for decision support development reported in the literature. The pitfalls described include: involving the users in the development of the DSS; user-friendliness/presentation of results; simplicity/over-complexity/transparency; flexibility; and reliability and confidence. That article should be reviewed carefully before embarking on the development of additional DSSs. Some of these pitfalls apply to DSSs for contaminated land management, and in fact are discussed in the following sections.

13.2 Gaps – What is Missing from Existing Tools?

The previous chapters identify existing DSSs for most of the key dimensions of the problem of risk-based management of contaminated land. To review, the five dimensions, as highlighted in Chapter 6, are: liability, risk, technology, socioeconomics, and stakeholder involvement. Of these, the majority of the existing DSSs focus on risk, technology selection, and stakeholder involvement. While liability and socioeconomic issues are addressed in some DSSs, the majority do not incorporate these two dimensions into the overall decision process. In fact, the review did not identify any system with a decision analysis component for liability issues. One possible explanation is that each country, and even states/regions within a country, have different liability rules and regulations. This complicates the development of a DSS intended to be used nationally or internationally. Socioeconomic issues are defined in Chapter 6 as "evaluating impacts and benefits to the land owner . . . and to society at various levels . . . in consideration of a sustainable reuse of the site". Although this issue is an extremely important dimension to the risk-based management of contaminated land, these impacts and benefits are notoriously difficult to quantify (European Commission, 2006). This could explain why existing DSSs are weaker in this dimension than in others.

Existing DSSs are also limited by their spatial analysis capabilities. As reported in Chapter 11 "there is the need for greater availability of spatially explicit risk characterization capabilities". Although some of the presented tools already propose sophisticated solutions for spatially-based assessment and management of contaminated land, making a broad use of GIS spatial features, more efforts should still be committed to improve this important characteristic. In particular, DSSs should be able to perform spatially explicit analyses of interacting processes that operate at different scales. Moreover, many variables used in risk assessment and management practices are still difficult to resolve spatially.

As a general gap that may characterize DSSs, it is worth mentioning the potential obscurity of tools of this kind. The potential for the intended audience to know and widely apply these tools is very often limited by a lack of clear definition and application. It is therefore an extremely important issue for DSS developers that the systems, which in many cases are very good and promising support tools, are presented to potential users timely and openly. As discussed in one of the following challenges, involvement of end-users in DSS development may well respond to this critical limitation.

13.3 Additional Challenges

After having identified some gaps in the tools presented in Section 2, some challenges faced by DSS developers may be equally identified. These challenges are not only specific for DSSs presented in Section 2, but also for any other DSS application. The identified challenges are:

- Balance of complex technical information with user-friendly interfaces and results presentation
- Involving users in development
- Group decision support
- Continued operation and maintenance
- Data accuracy and quality
- Flexibility and adaptability
- Integration of tools

13.3.1 Balancing Complex, Technical Information with User-Friendly Interfaces and Results Presentation

DSSs often have the need to integrate very complex and technical information with non-technical information in order to support multi-dimensional decisions. DSSs often seek information from a variety of different stakeholders (both technical and non-technical), which then must be combined in a decision structure. The challenge is to create a user-friendly interface to elicit detailed, non-technical information that can then be integrated with complex, technical information to produce results and present them in a non-technical way. This process also needs to be transparent to allow users, reviewers, and others to evaluate the inner workings of the system which will enable technical inaccuracies to be corrected and build confidence in its accuracy. For example, while a DSS may incorporate complex fate and transport models in order to identify where on the site there might be a potential risk to humans (and hence, what kind of remediation is needed), the inputs and outputs to these models need to be understood by non-technical users. Technical users must be able to access the inner-workings of the model to evaluate, for example, model parameters, equations, and any measure of uncertainty, in order to have confidence that the model is producing the information needed for the decision to be made.

13.3.2 Involving Users in Development

While it is often challenging and more time consuming to involve lots of people in the development of a DSS, the results are more likely to be rewarding and successful if this occurs. Involving users in the development of a DSS gives them a sense of ownership of the final product. This promotes use of the DSS for the intended purpose. Also, those users, who were involved in development, often become the promoters of the DSS and generate additional users. Finally, the development of on-going relationships with users of the DSS can provide on-going feedback for continued evolution of the DSS, keeping it current and useful.

There are several ways to involve users in the development of a DSS including: workshops, meetings, on-line feedback, surveys, interviews, real-life applications, and "beta" testing (where draft versions of the DSS or parts of the DSS are provided to users for testing and feedback). In all cases, it is important to involve both technical and non-technical experts and understand the needs and wants of all potential users of the DSS. It is important to also check repeatedly with interested parties regarding the usefulness and usability of the developed system and to identify additional needs and wants.

Challenges related to involving users in development include logistics, cost, and time. Ideally, all types of users should be brought together to discuss the DSS development so that all understand preferences and needs of other users and a consensus can begin to form. However, it can be logistically difficult and costly to organize meetings, workshops, etc., in which all users (technical and non-technical) can participate. Using surveys and interviews do not allow users to interact with one another and may therefore require several iterations, which increases the cost of those options and the time they take to implement.

Availability is an additional challenge. Web-based tools can reach anyone with an internet connection, but low band width may decrease functionality for some users or may not allow the DSS to be available to some users at all. DSSs using commercial software may have licensing fees or may need to be purchased, precluding some users from accessing those systems.

While no easy task, it is the opinion of the authors that the benefits of involving DSS users in the development of the DSS (as stated in the first paragraph) far outweigh the challenges. It is strongly recommended that development teams build the logistics, costs and time elements into their development plan to ensure this crucial challenge is not ignored.

13.3.3 Group Decision Support

In many discussions of the Section 2 chapters presenting DSSs, as well as in Section 13.3.2 of this chapter, the need to allow participation of all decision-makers and stakeholders in the process is emphasized. Participation of different people may also mean adoption of group decision making approaches, by which several individual preferences may be aggregated into one group preference. The development of support methodologies for group decision making, coupled with Internet-based communication, opens interesting challenges and opportunities for DSSs developers, in terms of teleconferences and web seminars through the Internet; instant messaging; inclusion of tools for group discussion management and agenda/minutes generation; methodologies for consensus building and aggregation of different values.

13.3.4 Continued Operation and Maintenance

In order to ensure a DSS is usable and useful, continuous operation and maintenance are required to keep it reliable and current. There are many technical challenges facing DSS developers including software and browser updates, power failures, security breaches (both intentional and non-intentional) and bugs in software code. Additionally, changes to regulations, processes, and information can cause a DSS to become quickly outdated if not routinely updated. These challenges must be addressed by the DSS development team and the DSS owners.

13.3.5 Data Accuracy and Quality

The results of DSS use are supposed to support and guide decisions. Therefore, decision makers must be able to trust the accuracy and quality of results. For this reason, the quality of the input data provided by DSS users should be assured. Considering what the computer scientists call the *garbage in – garbage out* syndrome, DSS users should be aware that any data, values and judgments they put in during an application influence the results and ultimately their decisions. DSSs should therefore be constructed to reduce the likelihood of "garbage in" as much as possible. Much of this "garbage in" reduction is an effort required of the users, who should insert data and knowledge of good quality. To this end, the systems may propose guidelines or formats for the correct input and management of data and values. Also the inclusion of uncertainty analyses in some cases may provide a positive management of this weakness.

13.3.6 Flexibility and Adaptability

Flexible tools, that can be adjusted to fit multiple situations and applications, are difficult to develop. For example, as mentioned in Chapter 11, "many aspects of higher tiered ecological risk assessment may require flexibility that simply cannot be achieved in a hard-wired decision support implementation".

The need to structure the elements of a DSS to provide a technologically-based instrument does indeed conflict with the opposite need to allow flexibility in specific analytical steps of the process. Flexibility concerns not only the ability to adapt to different conditions, but also to incorporate dissimilar or changed values by DSS users. Adaptability in diverse contexts may also determine whether these support systems are used by target decision-makers. Often the technological character of these tools, developed to include complex models and advanced research results, may impair the practical adoption for daily decision making.

13.3.7 Integration of Tools

As stated in Chapter 12: "Integration of different software components into a single unit is a challenge ... existing software may be written in a variety of languages". Integration is an important issue for multiple-function DSSs, that address the complexity of contaminated land management by including several models and tools. The technological and procedural aspects of incorporation should be planned and developed carefully. The integration of different software components may avoid fragmentation or duplication of tools as well.

13.4 Conclusions

This Chapter presented some identified gaps in DSSs for contaminated land assessment and management discussed in the whole of Section 2, and presented some of the challenges that for these, and in general for all tools, may be faced by DSS developers. Consideration of these gaps and challenges may enhance potentialities and applicability of Decision Support Systems in this field.

References

European Commission (2006) LIFE in the City: Innovative solutions for Europe's urban environment. Luxembourg, Office for Official Publications of the European Communities, Belgium. p. 64. ec.europa.eu/environment/life/publications/lifepublications/lifefocus/env.htm
Goosen, H., Janssen, R., and Vermaat, J. (2007) Decision support for participatory wetland decision-making. Ecological Engineering 30: 187–199.

Chapter 14
Use of Decision Support Systems to Address Contaminated Coastal Sediments: Experience in the United States

Charles A. Menzie, Pieter Booth, Sheryl A. Law, and Katherine von Stackelberg

Abstract Management of contaminated sediments in coastal and inland waterways poses many complex technical and social issues that must be addressed by multiple stakeholders. These stakeholders are guided by conflicting sets of values, regulatory constraints, and management goals, and have imperfect information on which to base decisions. The environmental quality of coastal and inland waters is usually affected by multiple stressors in addition to the sediment contamination at issue, and navigating these dynamics can be aided when decision makers and stakeholders have a shared vision of goals and a path forward to achieving them. The management process necessarily involves making a variety of decisions of varying complexity, with varying levels of importance to affected stakeholders. That process can benefit from using a systematic decision support system (DSS) to develop management goals, consider alternatives, understand their associated risks, and include the participation of various specialists such as environmental scientists, policy makers, economists, the businesses community, cultural experts, and public interest groups. This chapter summarizes recent developments in quantitative DSSs, both informal and formal, in the context of contaminated sediments in coastal environments, and provides insight into the benefits of incorporating DSS into sediment management.

14.1 Introduction

We consider decision support systems (DSSs) to include a range of tools that convey information to decision makers regarding the merits of various alternatives. This information can include simple approaches, such as matrices of positive, negative, or neutral features of alternatives and more sophisticated approaches, such as multi-criteria decision analysis (MCDA) and multi-attribute utility theory (MAUT). The approaches vary in the extent to which they include

C.A. Menzie (✉)
Exponent, 1800 Diagonal Road, Suite 300, Alexandria, VA, 22314
e-mail: Camenzie@exponent.com

A. Marcomini et al. (eds.), *Decision Support Systems for Risk-Based Management
of Contaminated Sites*, DOI 10.1007/978-0-387-09722-0_14,
© Springer Science+Business Media, LLC 2009

quantitative information, as well as the degree to which they reflect preferences or weighting factors.

This chapter considers how DSSs have been or may be applied to addressing contaminated-sediment issues in coastal environments. We define these environments as to include open coastal areas, as well as estuaries. Some of the largest and most challenging contaminated-sediment sites are located in coastal areas and estuaries. Examples of geographic areas that contain such sites include the Palos Verdes Shelf (California), New Bedford Harbor (Massachusetts), Eagle Harbor (Washington), Passaic River (New Jersey), the Historical Area Restoration Site (HARS) on the continental shelf off New Jersey, and the Hudson River (New York). Contaminated sediments in coastal and inland waterways constitute complex issues that involve multiple stakeholders with differing values and regulatory constraints, and who are making decisions based on limited information. In addition, there may exist a range of management goals for the water body and many other stressors that may influence these goals.

Achieving positive outcomes requires a shared vision of goals and an understanding of how they can be achieved. In coastal areas and estuaries, overall management goals often involve protection of people who use the resources directly or indirectly, commercial and recreational fish species, ecosystem services, particular species or assemblages of biota, and use of these areas for commercial and recreational purposes. Achieving these various management goals often involves negotiating multiple large and small decisions among individuals or regulatory bodies charged with oversight of one or more of the goals. That process can benefit from incorporating a DSS to develop a shared understanding of the suite of management goals, consider alternatives, understand their associated risks, and include participation of various specialists such as ecologists, environmental scientists, policy makers, economists, business persons, cultural experts, and public interest groups.

Making decisions is a complicated process of evaluating tradeoffs characterized by competing stakeholder objectives, regulatory constraints, the environmental context, and management goals. The process is also influenced by the belief systems, personal characteristics, and interpersonal dynamics of the parties involved. Each of these factors is influenced in different ways, depending on the nature of the decision. The use of DSS to manage decision making at contaminated-sediment sites is emerging, and while there are many opportunities for application, there are few demonstrated success stories, especially with respect to the use of more sophisticated DSSs such as MCDA (Kiker et al. 2005; Linkov et al. 2006). The apparently infrequent use of more sophisticated DSSs might reflect the manner in which decision makers prefer to reach decisions, or a lack of experience with these tools, or possibly conformance to regulatory guidance documents and precedent. Because most regulatory programs involving contaminated sediments have been underway for at least a few decades, there may be resistance to trying something different, especially for the first time.

While there are historical precedents and preferences, experience gained over the past few decades has underscored the importance of taking a wider view for decision making at contaminated-sediment sites. The multiple dimensions of managing contaminated sediments are clearly recognized in the U.S. Environmental Protection Agency's *Contaminated Sediment Remediation Guidance for Hazardous Waste Sites* (USEPA 2005), and the challenges of remediating sediments at large sites are discussed in a recent National Research Council report (NRC 2007). These key documents both call for greater use of comparative approaches and analyses for making decisions at contaminated-sediment sites. Implementation of USEPA's (2005) guidance, and accounting for the NRC (2007) findings, can benefit from greater use of DSSs.

The need to consider the multi-dimensional aspects associated with making decisions on environmental issues in coastal areas would seem to invite a more holistic approach to decision making—an approach that does not simply evaluate one aspect of the site in isolation (e.g., chemical contamination) but rather considers options in terms of ecosystem services and benefits (e.g., habitat quality, status of biological populations, regional development efforts, and so on). Although such a holistic approach may seem intractable because it introduces even more elements to consider in the decision-making process, in fact, DSSs are designed to synthesize disparate information about a site within an integrated framework that allows for explicit trade-offs among attributes of importance to stakeholders and decision makers. MCDA represents a specific category of methods within the DSS framework that incorporate formal decision analysis approaches to evaluate management alternatives (e.g., potential decisions) in terms of trade-offs. There are other DSSs that do not necessarily rely on MCDA, including relative risk models (Landis and Weigers 2005), comparative risk assessment (Cura et al. 2004), net environmental benefits analysis (Nicolette and Hutcheson 2004; Efroymson et al. 2003), and other methods, but these methods do not explicitly and quantitatively evaluate the full set of trade-offs that may need to be considered in making sediment management decisions.

We believe that decisions regarding management of contaminated sediments should stem from recognition of the suite of management goals that are inherent in coastal waters and estuaries, because decisions about how to manage contaminated sediments should be affected by other management goals. Some of these goals may be aligned with goals related to sediment management, while other management goals may be neutral (i.e., not influenced by decisions concerning sediments), and still others may be compromised to varying degrees if they are not considered within the decision-making framework for sediments. Therefore, we begin this chapter with a discussion of management goals and associated problem formulation. This is followed by a discussion of regulatory guidance that relates to decision making. We then describe how various DSS tools have been or could be applied to management of contaminated sediments in coastal areas. For that presentation, we group DSS tools into two categories. The first category includes tools that organize and convey information on attributes of alternatives but does not include weighting of attributes or other explicit incorporation of preferences. We refer to this as "DSS Tools without Preferences or Weighting." The second category includes tools that explicitly incorporate preferences and weighting. We refer to this as "DSS Tools with Preferences and Weighting." Within each of these sections, and in our conclusions, we discuss why some tools may be preferred over others.

14.2 Management Goals and Problem Formulation for Coastal Areas

Contaminated-sediment sites in coastal areas and estuaries may be many square kilometers in area and many kilometers long. Environmental conditions and resource use can vary across the site. Therefore, it is useful to divide these environments into subareas that reflect important differences in contaminant levels or sources, environmental conditions, and/or resource use. In such systems, achieving some management goals may require the implementation of a combination of actions in a particular subarea, whereas other goals may be achieved by targeting specific locations within an area. Management goals for coastal areas could include the ability to use natural resources for food or other purposes, provision of public access for recreation and other uses, preservation of natural areas, provision of habitat for fish and wildlife, restoration of impaired natural resources, preservation or enhancement of aesthetic values, reduction in the levels of toxic chemicals that impair the use or sustainability of natural resources, discharge of municipal and industrial wastewaters, use for cooling water and other municipal/industrial purposes, placement of offshore or shoreline facilities for energy generation or other purposes, oil and gas extraction, and sustainability of commercial or recreational navigation. Achieving these goals within a large site can involve making decisions that apply to different spatial scales, and these scales need to be defined to properly orient decisions related to contaminated-sediment management. Various goals for a coastal water body may be constrained by different environmental features, including the presence of contaminants in sediments. For example, some coastal shorelines have physical constraints, such as bulk-heads shorelines, that reduce the availability of intertidal areas for marsh formation. Other areas may have biological constraints, such as the absence of seagrasses or oyster reefs that are needed to support certain types of animal communities. Also, the presence of chemical contamination in sediments can impair the use of the area for fishing and shellfishing and threaten the sustainability of biological communities.

We believe that the presence of contaminated sediments should be viewed within the broader context of a suite of management goals. Within this context, contaminated sediments is one of a number of potential constraints (Table 14.1), and achieving the full suite of goals will require decisions and associated actions that are broader than a narrow focus on sediment contamination. For example, one significant constraint on meeting desired sediment and water quality goals in a river may be ongoing sources. Thus, in charting a path toward meeting the management goals, decision makers are challenged to consider a broad range of

Table 14.1 Examples of potential management goals for a coastal area and the constraints that may impede achieving them

Goals	Constraints				
	Existing shoreline development	Toxic chemicals in surface sediments	Toxic chemicals in deep sediments	Ongoing point and non-point sources	Historical physical modifications of river bottom
Increase Public Access	XXX	X For wadable areas	NA	XX Pathogens and aesthetics	NA
Increase Wildlife Habitat (Tidal Wetlands)	X	XX	NA	X	XXX
Restore Benthic Habitat (primarily as a prey base)	NA	XXX	XX If mixed into surface sediments or re-exposed	XX Sedimentation, organic loading, and input of toxic chemicals	X
Restore Submerged Aquatic Vegetation	NA	X Depends on chemical	NA	XX As these affect sediment loads, turbidity, and nutrient levels	XX As these affect water depth, sediment type, and light penetration
Reduce the Levels of Toxic Chemicals in Fish and Shellfish	NA	XXX	XX If mixed into surface sediments	XX As these are a source of toxic chemicals to the water column and sediments	NA
Navigational Dredging	NA	XX Influences dredging methodology and disposal options	XX Influences dredging methodology and disposal options	NA	NA

X important, *XX* very important, *XXX* extremely important, *NA* not applicable.

potential constraints and to evaluate the sequencing of control actions needed to meet management goals. The iterative and sequential nature of this process is consistent with the USEPA guidance on managing contaminated sediments (USEPA 2005) and supports the idea of using adaptive management strategies.

Problem Formulation can identify how elements of DSS can be incorporated into the analysis. Generally, DSS, involves the following steps:

- Define the management goals and alternatives (problem formulation)
- Develop a conceptual model
- Specify criteria by which alternatives will be judged (decision matrix)
- Screen alternatives to eliminate the clearly inferior and retain the clearly superior
- Evaluate performance of alternatives relative to (weighted) criteria
- Rank alternatives.

An additional element that can be included, external or internal to the DSS, is the elicitation of preferences and value judgments for weighting criteria. As discussed in this chapter, this element can be incorporated explicitly into the DSS. However, some decision makers may elect to keep this aspect separate for the DSS, instead using information organized by the DSS to support their decisions without participating in a formal process to weight their preferences.

Defining the management goals is a key first step that frames the entire decision-making process for a particular site or situation. Typically, the regulatory framework under which the site is being assessed defines management goals. For example, in the case of Superfund (Superfund; 42 U.S.C. s/s 9601 et seq.), a common management goal is to determine the optimum remedial action across a range of alternatives, which may include no action or monitored natural attenuation in addition to more active approaches such as dredging or capping. A number of alternatives will emerge from the management goal. In a natural resource damage assessment (NRDA), the management goal is to compensate for human and ecological service losses resulting from injuries sustained by ecological resources, most often in terms of habitat.

Regulatory frameworks differ in their objectives. For example, the management goals in the first Chesapeake Bay Agreement (1983) aimed to "improve and protect the water quality and living resources of the Chesapeake Bay estuarine systems." However, as more information was gleaned over the years, the list expanded to include, in 1987, 28 specific goals for water quality, aquatic species, population growth, development, public participation, and agency governance. In 2000, the Chesapeake Bay program set five broad, overarching goals (for biological resources, habitats, water quality, land use, and stakeholder participation), with more than 100 specific strategic goals (Gerlak and Heikkila 2006). As new information became available from studies and continuous monitoring, the Chesapeake Bay program was able to revisit their management goals in an iterative process, and as a result, their goals became more specific and were subsequently used as standards by which they can measure the success of their projects.

Defining management goals is an element of problem formulation. This element is characteristically similar to other environmental management frameworks, such as the Problem Formulations in USEPA's ecological risk assessment guidance, the U.S. Army Corps of Engineers' (Corps') risk guidance, and USEPA's data quality objectives (Kiker et al. 2005). However, unlike other DSS methods, MCDA provides for stakeholder agreement about the problems at the beginning of the process. This increases the chances that decisions made at the end of the process will be relevant to the issues identified going in, and proposes possible solutions from the very start.

Conceptual models can be used to facilitate the problem formulation. Conceptual models can provide all stakeholders with an illustrated presentation of the sources (e.g., air, point sources, groundwater), processes occurring at the site (e.g., physical process such as transportation and bioturbation), the exposed media (e.g., water, sediment), intermediary receptors (e.g., benthic invertebrates), final receptors (e.g., predators/scavengers, humans), and assessment endpoints.

Site that are divided into several reaches, areas or zones often require individual conceptual models that show differences in exposure and effects processes. For example, the Chesapeake Bay program initiated a segmentation scheme of the Bay and tributaries, with detailed analysis and discussion of the parameters that differentiate one segment from another (Chesapeake Bay Program 2004). The conceptual model at a contaminated-sediment site should include several depictions of zones that are defined by exposure, risk, and stability. For instance, surficial zones are given the most attention for risk assessment, because this is the zone where most exposure occurs. A second zone, which includes the surface zone, has the potential to be affected by physical process such as erosion or scouring events. A final zone beneath the erosional zone has the least amount of risk but is important for potential future dredging activities. Additional information can be presented in cross sections to show the various zones, such as biological, erosional, contamination, depositional, and potential dredging depths (Kiker et al. 2005). Once conceptual models have been finalized and presented, stakeholders can use the information to help develop management goals.

Each management alternative is evaluated with respect to specific criteria. This provides a consistent basis from which to compare alternatives. Using the Superfund regulatory framework as an example, Fig. 14.1 provides an example of criteria for a contaminated-sediment management decision.

To be effective, sediment management criteria need to be directional (i.e., can distinguish between minimum, maximum, and so on), comprehensive (i.e., no attribute of significance to any stakeholder group is left out), consistent (i.e., no hidden or unexpressed preferences), concise and non-redundant (i.e., use the smallest number of criteria possible and avoid double-counting), and finally, clearly understandable (Yoe 2002). For example, criteria that are relevant to alternatives for the Lower Passaic River would be ecological and human health risk, cost of projects, flood control, and benefits to navigation and recreation

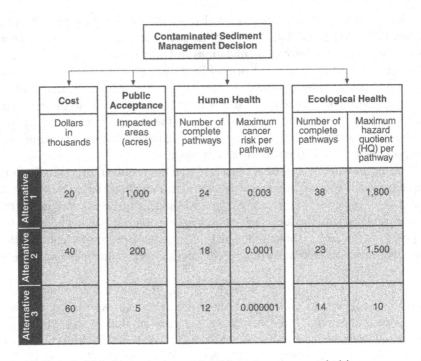

Fig. 14.1 Example of decision criteria used for sediment management decisions

(Kiker et al. 2005). The Chesapeake Bay Program developed criteria, which they called "indicators," that relate all of the programs to each other and provide a coherent way to communicate the conditions at the Bay. Indicators were also used to measure the progress and success of projects. These indicators of project progress were used in each subsequent year to re-evaluate the project itself and make necessary adjustments to work toward management goals (Chesapeake Bay Health and Restoration Assessment presentation 04/17/2007). Guidance for selecting criteria relevant to sediments is presented in (USEPA 2005).

MCDA methods, in particular, transform various criteria, as shown in Fig. 14.1, into a common dimensionless scale representing utility or value. This is accomplished through utility or value functions, which relate the original units to the utility or value scale, which is typically 0–1. Each alternative receives scores for each criterion. The criteria scores can then be weighted (see next section) depending on stakeholder preferences.

If the DSS process includes preferences and weighting, these can be defined by the decision maker or elicited through the use of workshops and surveys with specific stakeholder groups. The weights can then be applied to the normalized (0–1) criteria scores to obtain final scores for each alternative. For example, in the decision matrix shown in Fig. 14.1, meetings in which stakeholders are asked to rank each criterion may reveal that ecological risk and habitat preservation are key criteria against which to evaluate the alternatives. They would then receive a weighting of 1.0. This process allows for stakeholder interests and preferences to be formally included in a decision-making framework, rather than on an ad hoc basis, and provides a mechanism and framework for building consensus across disparate groups. Alternatively, specific criteria can be established relative to stakeholder preferences and then weighted equally. The choice of which approach to take depends on the specific context and method (e.g., multiattribute utility, analytical hierarchy, and so on).

The focus of the alternatives evaluation element in DSS is to determine the type of data that needs to be collected for each criterion and each alternative. Stakeholders should be aware early in this process of the types of data they need and how they will use the information (Bergquist and Bergquist 1999). This step requires technical expertise to provide the information through models, monitoring, risk analysis, cost/benefit analysis, and stakeholder preferences (Linkov et al. 2004). Common tools that are already well established include ecological and human health risk assessments, comparative risk assessments, remedial investigations/feasibility studies, economic assessments, and site characterization. MCDA can be used to integrate the data generated from these studies.

Collecting data in an iterative process will help link the constraints of each alternative to the management goals (Kiker et al. 2005). This element provides enormous potential to engage stakeholders and generate preferences in the planning process. For example, the Chesapeake Bay Bi-State Blue Crab Advisory Committee made decisions on blue crab management by engaging stakeholders to agree to an understanding of the decision process. All stakeholders understood the uncertainties in the technical evaluations and the manner in which these uncertainties and scientific conflicts were to be managed (Beem 2006). All of the steps up to this point can be done using spreadsheets. However, the comparison of alternatives must be done using specialized software (e.g., Expert Choice [www.expertchoice.com], Criterion Decision Plus [www.infoharvest.com], or Decision Lab [www.visualdecision.com]). In some cases, clearly inferior outcomes will emerge, and it is beneficial to identify these early in the analysis so they can be discarded. Alternatively, a clearly dominant best decision may emerge, although in practice, this is unlikely.

Once scoring is complete, and inferior alternatives have been identified and eliminated, the remaining alternatives are ranked according to their overall scores. Different mathematical approaches are used depending on the specific MCDA method (see, for example, Linkov et al. 2004). A variety of graphical techniques allow decision-makers to visualize the results.

The value of MCDA is not necessarily to "pick" or identify the single best decision, but to make explicit particular trade-offs that are possible given management goals, site conditions, and diverse stakeholder preferences. It represents a class of methods that make explicit preferences, criteria, and weights within an objective, integrated framework. MCDA represents one example of different kinds of DSS, but one that holds particular promise for assisting in the decision-making process.

Effective decision-making requires stakeholder involvement at all stages. Once management goals have been established, stakeholders in the decision process are asked to create a list of actions that have the potential to address the problems and potentially achieve management goals. At this early point in the process, technical experts may be needed to introduce concepts and options to the stakeholders. For example, in 1997, the Chesapeake Bay program was directed to identify innovative technologies that could help achieve management goals. Experts may provide a numerous alternatives for stakeholders, some of which are unfamiliar. Alternative actions taken early in the process can provide a guide to data collection and data analysis. For example, restoration in the Lower Passaic River prompted ideas for wetland restoration and replacement of bulk-head areas with tidal flats (Kiker et al. 2005). Restoration of living resources in Chesapeake Bay can include oyster restoration, invasive species removal, and fish passage restoration. A detailed discussion of creating alternative choices can be found in Gregory and Keeney (2002).

Alternatives generated by stakeholders should be based on site-specific conditions at each reach of the study site and should be considered with spatial and temporal scales. As noted in "problem identification," the conditions in different reaches will affect the management goals and present unique constraints. Temporal scales are important because decisions will be based on present conditions, as well as future anticipated conditions. Linkov et al. (2002) compared a spatially and temporally explicit risk assessment on flounder to one that was non-explicit. They found that predicted risk differed by up to one order of magnitude between the two cases. They concluded that defaulting to extremely conservative estimates did not provide adequate information to be used in sediment management decisions.

14.3 Regulatory Guidance Related to Decision Making

Decisions related to the management of contaminated sediments generally arise from three regulatory arenas: contaminated site programs such as Superfund and state hazardous waste management, management of sediments for navigational dredging, and clean water programs (e.g., through the Clean Water Act; 33 U.S.C. ss/1251 et seq.;1977), including the development of Total Maximum Daily Loads (TMDLs) and restoration initiatives. Each of these broad areas requires decisions about whether an intervention is appropriate and, if so, what type of intervention. As a result, various federal and state decision frameworks have been developed in support of these regulatory programs.

Concern over the magnitude of the management problem posed by contaminated sediment has increased steadily over the past 10–15 years. USEPA (1998) reported that approximately 1.2 billion cubic yards of sediment-an amount equal to 10% of the surface sediment underlying the nation's waters-was sufficiently contaminated by pollutants to pose potential ecological and human health risks. Not counting active environmental remediation, annual maintenance dredging by the Corps and other entities includes 3–12 million cubic yards of sediment that is sufficiently contaminated to require special handling and disposal (USEPA 1998).

This concern spurred widespread development and implementation of public policy. At the federal level, contaminated sediment management is governed by six comprehensive acts of Congress, and implementation responsibilities are spread among seven federal agencies (NRC 1997). The two major programmatic drivers for managing contaminated sediment are site cleanup and navigation dredging. Major federal regulations governing contaminated sediment in environmental cleanup actions include the Comprehensive Environmental Response, Compensation and Liability Act of 1980 ([CERCLA] Superfund; 42 U.S.C. s/s 9601 et seq.); the Resource Conservation and Recovery Act (RCRA; 42 U.S.C. s/s 321 et seq.; 1976); and Section 115 of the Clean Water Act (CWA; 33 U.S.C. ss/1251 et seq.; 1977). Other federal regulations governing contaminated sediment management include the Rivers and Harbors Act of 1899 (33 U.S.C. 403); the Marine Protection, Research and Sanctuaries Act (MPRSA, commonly known as the Ocean Dumping Act; 33 U.S.C. 1401); the Coastal Zone Management Act (CZMA; 16 U.S.C. 1451–1456), and the Oil Pollution Act (OPA; 33 U.S.C. 2701 et seq.). Regulations are implemented by numerous federal agencies, including USEPA, USACE, the National Oceanic and Atmospheric Administration (NOAA), and the U.S. Fish and Wildlife Service (USFWS). Many states also have specific programs and rules governing the assessment and management of contaminated sediment (e.g., the aquatic lands cleanup program of the Washington State Department of Ecology, and sediment quality guidelines adopted by many other states). A tangled regulatory web of enabling legislation; mandates of implementing agencies; and regulations and guidance of federal, state, and local jurisdictions has given rise to the development of several programs aimed specifically at harmonizing and streamlining sediment management strategies. Some of these are described briefly below.

14.3.1 U.S. Environmental Protection Agency

The U.S. Environmental Protection Agency first developed a concerted decision-making process for managing contaminated sediment in 1990. This process was refined 8 years later in a strategy document that lays out four goals (USEPA 1998):

- Prevent the volume of contaminated sediment from increasing
- Reduce the volume of existing contaminated sediment
- Ensure that sediment dredging and dredged material disposal are managed in an environmentally sound manner

• Develop scientifically sound sediment management tools for use in pollution prevention, source control, remediation, and dredged material management.

USEPA (1998) describes how the Agency intends to accomplish these goals via assessment, prevention, remediation, dredged material management, research, and outreach, and proposes to use multiple statutes to require remediation by responsible parties, including CERCLA, RCRA, CWA, TSCA, OPA, and the Rivers and Harbors Act. USEPA's Superfund program has been a major driver behind development of contaminated sediment management programs at the Agency. By the end of fiscal year 2005, the Superfund program at EPA had determined remedies at more than 150 sediment sites, including 11 "mega sites" where the cost of sediment remediation was in excess of $50 million (http:// www.epa.gov/ superfund/health/conmedia/sediment/index.htm).

Since the 1998 strategy document and the NRC (2001) report, EPA has developed several programs and guidance documents that specifically address contaminated sediment management, including definition of principles for managing risks from contaminated sediment (Horinko 2002), guidance for conducting sediment assessments and sediment remediation guidance (USEPA 2005), and numerous technical references and manuals related to specific aspects of sediment investigations. [1]

The USEPA (2005) sediment management guidance is designed to support decision making at contaminated-sediment sites. The guidance document brings forward much of USEPA's Superfund guidance regarding involvement of stakeholders and application of the nine USEPA criteria. Consistent with previous guidance, the USEPA sediment management guidance highlights the importance of developing an accurate conceptual site model, which identifies contaminant sources, transport mechanisms, exposure pathways, and receptors at various levels of the food chain; the role of a sediment site in the watershed context, including other potential contaminant sources; key issues within the watershed; and current and reasonably anticipated or desired future uses of the water body and adjacent land. The guidance articulates the existing regulatory framework as appropriate and necessary for remedial decision making. This includes applying the National Contingency Plan (NCP) remedy selection criteria, identifying applicable or relevant and appropriate requirements (ARARs), evaluating effectiveness and permanence, estimating cost, and using institutional controls.

The USEPA (2005) guidance extends beyond existing Superfund guidance with respect to additional sediment-specific factors to be considered when judging alternatives. The guidance notes that, where a remedy is necessary, the best route to overall risk reduction depends on a large number of site-specific considerations, some of which may highly uncertain. The value of DSSs is evident in the USEPA guidance, which states that any decision

---

[1] (see for example http://www.epa.gov/waterscience/cs/pubs.htm).

Table 14.2 Some site characteristics and conditions conducive to remedial approaches for contaminated sediment

MNR	In-situ capping	Dredging/Excavation
Site characteristics		
Anticipated land uses or new structures are not incompatible with natural recovery. Natural recovery processes have a reasonable degree of certainty to continue at rates that will contain, destroy, or reduce the bioavailability or toxicity of contaminants within an acceptable time frame.	Suitable types and quantities of cap material are available. Anticipated infrastructure needs (e.g., piers, pilings, buried cables) are compatible with cap. Water depth is adequate to accommodate cap with anticipated uses (e.g., navigation, flood control). Incidence of cap disrupting human behavior, such as large boat anchoring, is low or controllable.	Suitable disposal sites are available. Suitable area is available for staging and handling of dredged material. Existing shoreline areas and infrastructure (e.g., piers, pilings, buried cables) can accommodate dredging or excavation needs. Navigational dredging is scheduled or planned.

MNR monitored natural recovery.

regarding the specific choice of a remedy for contaminated sediment should be based on careful consideration of the advantages and limitations of each available approach, and a balancing of trade-offs among alternatives. The guidance provides a comparative framework (Highlight 7–2 of the guidance[2]) for gaining insights into the conditions under which one or more of the remedies might be selected. This framework is presented here (Table 14.2) as a basis for evaluating the compatibility and appropriateness of decisions. The guidance underscores the importance of considering the full range of alternatives and the possibility of incorporating remedial methods from multiple alternatives. At sediment contamination sites where we are involved, we are using Table 14.2 as the basis for comparative analysis of alternatives. It can help identify the sets of conditions under which monitored natural recovery, capping, and dredging are more appropriate. While these are the major options that EPA wants to see considered at every site, other remedial methods can also be considered, such as reactive caps and sediment treatment materials.

[2] U.S. EPA (2005) notes that Highlight 7-2 is intended as a general tool for project managers as they look more closely at particular approaches when most of these characteristics are present. Project managers should note that these characteristics are not requirements. It is important to remain flexible when evaluating sediment alternatives and when considering approaches that at first may not appear the most appropriate for a given environment. When an approach is selected for a site that has one or more characteristics or conditions that appear problematic, additional engineering or in-situ caps (ICs) may be available to enhance the remedy.

14.3.2 U.S. Army Corps of Engineers (Corps)

The principal activity of the Corps that involves the management of contaminated sediment is navigational dredging. The Corps' Center for Contaminated Sediments serves as a clearinghouse for technology and expertise concerned with contaminated sediments.[3] As a result of their management of contaminated sediment generated by navigation projects, the Corps has developed numerous guidance documents and engineering manuals related to testing of dredged material and evaluating and designing dredged-material disposal sites for both contaminated and uncontaminated sediment. On a regional basis, the Corps has also developed decision-making tools and programs for managing contaminated sediments, such as the Multiuser Disposal Sites (MUDS) program for Puget Sound (Palermo et al. 2000) and Programs in the Great Lakes (e.g., GLDT 1999).

Yoe (2002) developed a manual intended to provide Corps planners with a framework to aid Civil Works decision making involving several alternatives and multiple evaluation criteria. The manual introduces the concept of trade-offs within the Corps' planning process and presents a decision support framework, simple rules for decision making, methods for weighting and ranking decision criteria, and examples of multicriteria decision-making software. In addition to programmatic decision support systems, the Corps' Institute for Water Resources has been actively involved in developing new tools for decision making in the realm of contaminated sediment management, including application of multicriteria decision analysis.

14.3.3 Private-Sector Initiatives

The private sector, consisting primarily of the regulated community, has been actively involved in initiatives to streamline and improve decision making regarding contaminated sediment. In particular, early in 1998, the Sediment Management Work Group (SMWG) formed as an ad hoc group open to industry and government entities that are responsible for managing contaminated sediment.[4]

The SMWG has been the principal focal point for advancing discussions of decision support systems within the regulated community, and it has undertaken several projects aimed at advancing contaminated sediment decision making, that include environmental regulators. These projects have focused on DSSs that provide methodical approaches for evaluating technical aspects of contaminated-sediment management, primarily from the perspective of determining the optimum cleanup alternative from a set of fixed alternatives—for example, a decision tree outlining the general technical and regulatory processes applied to contaminated-sediment sites (Table 14.3).

---

[3] (http://el.erdc.usace.army.mil/dots/ccs/).

[4] (http://www.smwg.org/).

Table 14.3 Examples of cleanup alternatives and anticipated results

MNR	In-Situ Capping	Dredging/Excavation
Human and ecological environment		
Expected human exposure is low and/or reasonably controlled by IC. Site includes sensitive, unique environments that could be damaged irreversibly by capping or dredging.	Expected human exposure is substantial and not well controlled by ICs. Long-term risk reduction outweighs habitat disruption, and/or habitat improvements are provided by the cap.	Expected human exposure is substantial and not well controlled by ICs. Long-term risk reduction of sediment removal outweighs sediment disturbance and habitat disruption.
Hydrodynamic conditions		
Deposition of sediment is occurring in the areas of contamination. Hydrodynamic conditions (e.g., floods and ice scour) are not likely to compromise natural recovery.	Hydrodynamic conditions (e.g., floods and ice scour) are not likely to compromise the cap, or can be accommodated in the design. Rates of groundwater flow in cap area are low and not likely to create unacceptable contaminant releases.	Water diversion is practical, or current velocity is low and can be minimized to reduce resuspension and downstream transport during dredging.
Sediment characteristics		
Sediment is resistant to resuspension (e.g., cohesive or well armored sediment).	Sediment has sufficient strength to support cap (e.g., has high density and low water content).	Contaminated sediment is underlain by clean sediment (so that over-dredging is feasible). Sediment contains low incidence of debris (e.g., logs, boulders, scrap material) or is amenable to effective debris removal prior to dredging or excavation.
Contaminant characteristics		
Contaminant concentrations in biota and in the biologically active zone of sediment are moving toward risk-based goals. Contaminants readily degrade or transform toward lower toxicity forms. Contaminant concentrations are low or cover diffuse areas. Contaminants have low potential to bioaccumulate.	Contaminants have low rate of flux through cap. Contamination covers contiguous area (e.g., to simplify capping).	Higher contaminant concentrations cover discrete areas. Contaminants are highly correlated with sediment grain size (i.e., to facilitate separation and minimize disposal costs).

MNR monitored natural recovery, *IC* in-situ capping

14.4 DSS Tools Without Preferences or Weighting

Many simple tools that have been used historically for ranking alternatives for sediment remediation. For the most part, these have been aligned with NCP or state criteria. These usually take the form of matrices or graphics with accompanying narratives.

Although there is no official software used to aid decision making for contaminated sediments, many programs have been developed to aid as generic DSS tools. Programs created for remedial investigation/feasibility studies (RI/FSs) include RAAS, MEPAS, and ReOpt. The Remedial Action Assessment System (RAAS) was developed at Pacific Northwest Laboratories to expedite RI/FSs at waste sites at Department of Energy operable units.[5] The Multimedia Environmental Pollutant Assessment System (MEPAS), and Remedial Options (ReOpt), were designed for each phase of the RI/FS process. MEPAS is a human health risk model that provides baseline data within the RAAS methodology. ReOpt provides the database of remedial technologies, contaminant information, and regulations. Both programs work within RAAS to quantify site cleanup strategies, remedial constraints, and assumptions. The software identifies feasible remedial options by linking alternatives to meet the objectives defined by the user.

Other software programs for DSS include those that are based primarily on cost estimating or forecasting. For example, the Modular Oriented Uncertainty SystEm (MOUSE), developed by USEPA for general risk assessments, incorporates Monte Carlo simulation for costs and waste management. The Independent Cost Estimating Contingency Analyzer (ICECAN) was developed for cost contingencies, and typically is used at the end of the remedial design phase. ICECAN was created by the U.S. Department of Energy. Range Estimating (REP) was created by Decision Sciences Corporation for risk analysis and consideration of hidden and contingent costs of environmental projects. Costs associated with remediation are considered in HAZRISK and RACER. HAZRISK, by Independent Project Analysis, Inc., is used by the oil and gas industries, who participated in developing the program with the Corps. The program aims to streamline the evaluation of cost and scheduling risks during the RI/FS and remedial design/remedial action phases of Environmental Restoration projects. The Remedial Action Cost Engineering and Requirements System (RACER)/ENVEST, by Delta Research Corporation, estimates costs for studies, remedial design, and remedial action for restoration projects.[6]

DSS tools that can be applied to sediments, and that extend beyond simple matrices, include Comparative Risk Analysis (CRA) and the Relative Risk Model and Net Environmental Benefits Analyses (NEBA). Each provides information that can be used to compare alternatives.

---

[5] (http://www.osti.gov/ bridge/servlets/purl/6263513-x0b8ug/6263513.PDF).

[6] (http://www.p2pays.org/ref%5C%5C01%5C%5C00 047/index.htm).

14.4.1 Comparative Risk Analysis

Contaminated-sediment site managers have increasingly been adopting new developments in the field of risk assessment to enhance the quality and usability of information for decision making. One of these tools is CRA, a relatively new tool within the larger, but still developing, scientific discipline of environmental risk assessment. There is ongoing debate over the definition of CRA and how it should be applied in the context of environmental management.

CRA focuses strictly on the results of human health or ecological risk assessments as the basis for making decisions. Risks are estimated for a set of alternatives that can then be compared directly. An important feature of CRA is that all risks are considered across the spectrum of spatial and temporal scales. For example, some management alternatives may involve transportation of waste, and potential risks from those kinds of exposures need to be considered as well as risks at the site.

Cura et al. (2004) summarize the origins of CRA and philosophical differences among technicians and policymakers regarding its implementation and use. The use of CRA for environmental decision making has spanned a spectrum from: (a) a purely technical application, where decisions are based solely on a ranking of human health and environmental risks among management alternatives (often termed "hard CRA"), to (b) a more holistic approach wherein CRA results are used in conjunction with the results of a broader spectrum of analyses to inform decision making while incorporating economic, social, and technological criteria. Successful application of hard CRA rests on the premise that good results are achievable in cases where decisions can rely solely on quantitative data and involve only the scientific community as decision makers (Cura et al. 2004). The latter approach recognizes that the results of CRA are fraught with myriad technical and scientific value judgments, and that risk is only one of many factors that must be considered in complex environmental decision making.

USEPA, various USEPA Regions, and several states have used CRA in a policy analysis setting and to prioritize regulatory activities. CRA has been used to prioritize contaminated sites, rank hazards from specific contaminants, evaluate the relative importance of hazards from a variety of stressors, and evaluate remedial options at contaminated sites. USEPA's efforts to evaluate and develop CRA tools began within the Office of Policy Analysis in the mid-1980s (e.g., USEPA 1987) and by the early 1990s, the application of CRA began to mirror aspects of MCDA. For example, USEPA (1993) presents a "weighted scoring approach," with the following steps:

1. Identify criteria for evaluating risks
2. Score each problem for each criterion
3. Assign weights to each criterion
4. Multiply the criteria score by the weights, and sum
5. Rank problems according to total score.

USEPA (2000) offers a framework as a flexible guide to address environmental problems of different size, scope, and location and incorporates risk comparison as the basis for ranking and selecting problems for further consideration. Cura et al. (2004) recognize that the critical step in CRA is risk ranking, because it is at this step that most cross-discipline communication takes place among participants in the decision-making process.

One of the earliest examples of the use of CRA for contaminated-sediment management was in the Puget Sound area, where Tetra Tech (1986) proposed a "multiattribute tradeoff system" as a means for aggregating the results of human health and ecological risk assessments into a common measure for evaluating alternatives for dredged-material disposal. This document suggests applying coefficients to risk indices (e.g., hazard quotients) to reflect the "judgment of the relative importance of the risk variables"; however, it does not provide any guidance on developing risk categories, using criteria for ranking, or applying weighting values (Cura et al. 2004).

Although CRA has been applied to address environmental problems, this has been done primarily to support policy decisions regarding resource management at relatively large scales (e.g., watershed). For example, Ohlson and Serveiss (2007) use CRA in conjunction with core principles of decision analysis in evaluating management alternatives for four watersheds within the Greater Vancouver Water District in British Columbia, Canada.

Notwithstanding the considerable debate in professional and academic circles, this review did not reveal any application of CRA within a larger DSS specifically for contaminated-sediment sites. Driscoll et al. (2002) present a framework for a screening-level human health and ecological risk assessment, to compare risks associated with various management alternatives for contaminated dredged material. The authors state that the framework can be used to help managers identify risk drivers associated with the alternatives, select preferred alternative(s) for highly contaminated dredged material, implement risk reduction measures, and identify whether additional work is needed to better characterize risk.

The framework relies on standard approaches to risk assessment, including development of conceptual site models for each disposal alternative for both ecological and human receptors, identification of significant pathways of exposure, and identification of specific receptors. Receptor-specific risk is expressed in terms of chemical-specific hazard quotients. Risks among alternatives were compared on the basis of six criteria: ratio of area affected to capacity of facility, number of complete ecological exposure pathways, magnitude of ecological hazard quotients, number of complete human exposure pathways, magnitude of maximum cancer risk, and ratio of estimated chemical concentration in fish to risk-based levels. The authors recommend that future comparative risk assessment efforts of this nature incorporate other important aspects of the decision-making process, such as engineering feasibility, cost, stakeholder acceptance, reliability, beneficial uses, and economic impacts.

Kiker et al. (2007) cite the work of Driscoll et al. (2002) by example to propose a generalized framework for integrating CRA and MCDA. The author's proposed framework consists of a process outline for what they define as the core components of any complex decision-making process: people, process, and tools. The outline illustrates points during the process for key input from policy makers, scientists and engineers, and stakeholders (the "people" element), and application of tools (environmental assessment and decision analysis).

14.4.2 Relative Risk Model

The Relative Risk Model (RRM) is a matrix and ranking method that has been applied to a wide variety of environmental issues in coastal areas and other ecosystems. These applications include evaluating declines of Pacific herring (Landis et al. 2004), assessing environmental conditions in the Willamette and McKenzie rivers in Oregon (Luxon and Landis 2005), evaluating contaminant and other stressors in a portion of the Amazon River watershed (Morares et al. 2002) and in the Delaware River estuary (Stahl, personal communication, January 2009), and evaluating oyster restoration options for the Chesapeake Bay (Versar and Exponent 2008). RRM has been suggested as a means for addressing regional issues with multiple stressors (Menzie et al. 2007). RRM uses ordinal ranks to classify the relative importance or magnitude of stressors, effects, and estimates of ecological consequences (Fig. 14.2). The RRM does not predict the actual magnitude of changes or risks, such as the increase or decrease in abundance or the risk of extinction; nevertheless, it allows a comparison of the relative positive and negative changes in stressors or environmental conditions among the various alternatives.

The RRM is being used to identify priorities and to guide environmental decisions in coastal areas. Two recent applications include the assessment of historical impacts on, and current environmental conditions in, the Delaware River Estuary and the comparison of alternatives for restoring oysters in Chesapeake Bay. The assessment for the Delaware River is employing RRM as a regional risk assessment approach to identify the major environmental stressors within the watershed (Stahl, personal communication, January 2008).

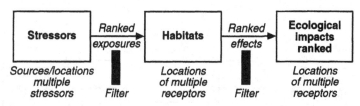

Fig. 14.2 The regional approach for application of the relative risk model (Landis and Weigers 2005)

The approach is focused on identifying and then linking the suite of physical, chemical, and biological stressors to current environmental conditions, and where feasible, placing in perspective those stressors that have had, or may currently have, impacts on the estuary. The ultimate goal of the Delaware Estuary application of RRM is to help scientists and decision makers ascertain what might be done to mitigate or improve conditions in the future—whether those include improvements to water quality, increased habitat restoration, and/or other options. The application for the Delaware relies on existing physical, chemical, and biological data in a typical ecological risk assessment approach. The initial results suggest that physical stressors represent more than 50% of the relative risk posed by all stressors in the estuary (Stahl, personal communication, January 2008.)

RRM was applied within the ecological risk assessment for the proposal to restore oysters in Chesapeake Bay. For this application, the decision makers wanted to know the influence that the different options would have on other ecological components of the Bay. In particular, they requested insight into the relative positive or negative influence that the various options would have on establishing oyster hard bottom communities and the associated influences that oysters would have on plants, invertebrates, and fish and wildlife, as well as on overall water quality.

The RRM, as applied to the proposed action and alternatives, provided a means for integrating the various major influences and portraying the results to decision makers in graphical and tabular form. These matched the form and structure of information requested by the decision makers. The decision makers declined to use more sophisticated analytical processes to help them judge the available information on economics, and sociological and ecological factors.

The RRM analysis for evaluating ecological implications considered four types of stressors or environmental conditions: habitat, food, predation, and water quality. The direction and magnitude of an influence are reflected in a derived RRM value that ranges between –5 and 5 (Fig. 14.3).

With the exception of dissolved oxygen, all modeled or predicted ecological influences followed a scale based on an overall order-of-magnitude range (i.e., a change of 10×), such that a change approximating or greater than an order of magnitude was assigned a value of 5 or –5, depending on the direction of the effect. The RRM score for each ecological receptor and alternative was calculated as the sum of the scores related to habitat, food, and water quality. An

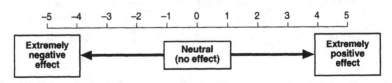

Fig. 14.3 Direction and magnitude of an influence in a derived RRM

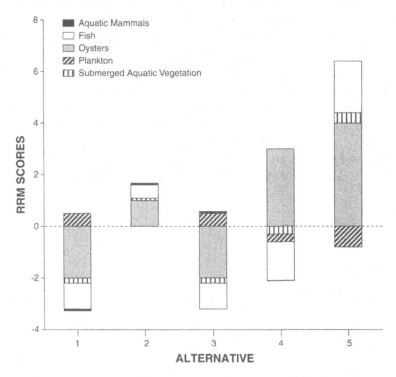

Fig. 14.4 Example of using the Relative Risk Model to evaluate changes in ecological services associated with various alternatives. This figure is presented as a visual example and does not display actual results of the analysis

example of a comparison of alternatives for one of the geographic regions of Chesapeake Bay is given as Fig. 14.4.

14.4.3 Net Environmental Benefits Analyses

NEBA provides an approach for evaluating management alternatives in terms of ecosystem services. The management alternatives define different future levels of contaminants and other kinds of stressors, including physical stressors, that result from modification to an ecosystem through implementation of a remedial alternative (e.g., dredging). While NEBA is usually considered as a process for evaluating alternatives on the basis of ecosystem services, it can also provide an objective way of attributing value to various ecosystem services.

The starting point is to identify those attributes of the site that are the focus of management goals (assessment endpoints). This involves holding a series of conversations with regulators to identify the ecosystem attributes that are considered worth protecting and that represent the focus of management

efforts. This iterative dialogue will establish the conceptual overview of the
NEBA by making explicit those ecosystem attributes that are of primary concern. A conceptual site model (CSM) is then developed that describes the
relationship between assessment endpoints and various stressors. The analysis
proceeds by quantifying specifically how the alternatives under consideration
will influence the assessment endpoints. The results are converted to rankings
and tabulated in a matrix across alternatives, and then summed. The alternatives can then be ranked according to the number of points each receives. This
approach is described in greater detail in Efroymson et al. (2003).

Figure 14.5 provides an example of a graphical representation of the results
of a NEBA. The black and white circles depict human health and ecological

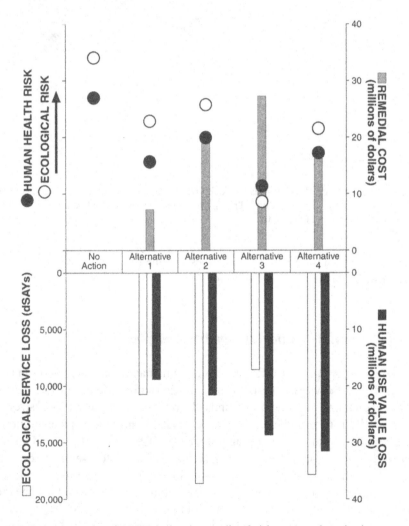

Fig. 14.5 Example results of a NEBA showing predicted risk, cost, and service loss across
remedial alternatives (adapted from Nicolette and Hutcheson 2004)

risks, respectively, with corresponding costs for each alternative depicted by the black bars. Ecological and human services (aesthetic and recreational) are depicted using white and black bars, respectively. Ecological services are estimated using discounted service acre years (dSAYs), although other metrics are possible and should be encouraged, particularly as measurement and valuation of changes in ecosystem services becomes a more established technique. In some NEBAs, clear "cut-offs" across alternatives will be obvious, while others require additional analysis to determine which management alternative maximizes benefits overall. The results are well-suited for inclusion in a formal MCDA, which would allow particular endpoints to be weighted in different ways (for example, stakeholders have expressed that minimizing dSAY losses should be the priority).

Another way to look at net environmental benefits is shown in Fig. 14.6. This figure depicts net environmental services versus cost over time. Under this scenario, it is possible to compare the area under the curve for service benefits and losses (which can be expressed as cost) for each alternative.

A NEBA provides a framework for comparing ecosystem service losses and benefits across management alternatives. For example, NEBA can be used to evaluate the service losses imposed by the partial removal of a forested wetland for remediation of a contaminant. Although contaminant risks decrease, there is also significant habitat loss that, when compared to the cost of remediation, may not be offset by the decrease in risk. The opposite may be true for a site for which initial ecosystem services are very low, and would increase over time following remediation.

14.5 DSS Tools that Incorporate Preferences or Weighting

Multi Criteria Decision Analysis (MCDA) is an important DSS tool because it explicitly incorporates preferences and weighting (Kiker et al. 2005; Linkov et al. 2006; Yoe 2002; Yatsalo et al. 2007). In MCDA, decision makers establish a

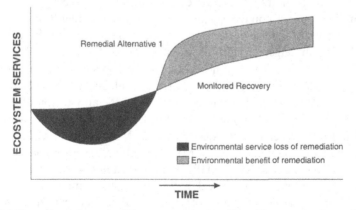

Fig. 14.6 Example of the tradeoff between service benefit and loss across alternatives

decision matrix to evaluate management alternatives in terms of specific criteria. Alternatives are scored according to the criteria, and then are ranked according to the scoring matrix and the decision rule, which is typically expressed in terms of maximizing a particular environmental benefit or service and minimizing cost. MCDA assumes that decision makers are rational, that preferences (of stakeholders and decision makers) do not change appreciably over time (e.g., what is valued now will be valued in the future), and that the decision maker has "perfect" knowledge and will be consistent regarding judgments. The goal of the decision maker, then, is to maximize utility, which is a function of costs and benefits.

The term multi-criteria decision analysis (MCDA) is meant to capture a collection of formal approaches aimed at helping people account explicitly for many conditions as part of a decision-making process (e.g., Belton and Stewart 2002). Mendoza and Martins (2006) identify several characteristics of MCDA that make it particularly helpful for decision making concerning natural resource management. It:

1. Enables explicit accounting of multiple, conflicting criteria
2. Assists in developing a structure for the management problem
3. Provides a conceptual model for focusing discussion
4. Offers a process for making rational, justifiable, and explainable decisions.

These characteristics are well suited for supporting decision making at contaminated sites in general, and contaminated-sediment sites, in particular. In order for policy and decision makers to effectively choose among various project alternatives for contaminated sites, policy and decision makers almost inevitably must weigh tradeoffs among social, political, economic, and environmental interests.

NRC (1997) concluded that formal decision analysis may be particularly valuable for addressing contaminated-sediment issues, because it can accommodate numerous disparate variables (and the uncertainty in estimating values for variables), as compared to typical decision-making tools that result in single outcomes, such as risk assessment, cost/benefit analysis, and feasibility assessment. In principle, MCDA is well suited for grappling with the multidimensional nature of contaminated sediment management projects that require resolution of complex technical, social, and economic considerations.

Recent reviews of the use of decision analysis tools to help resolve contaminated sediment management problems have revealed that formal decision analysis, including MCDA, has not been widely applied to the problem of contaminated sediment management (e.g., Kiker et al. 2005; Linkov et al. 2004, 2005). Stahl et al. (2002) and Stahl (2003) reviewed the application of decision analysis within USEPA. A fairly recent effort by USEPA focused on developing a DSS called Multi-Criteria Integrated Resource Management (MIRA), with MCDA at its core (Stahl et al. 2002). MIRA combines decision analysis using the Analytical Heirarchy Process (AHP), with a data automation interface and a module for developing geostatistical indicators. This combination of tools enables the MIRA approach to incorporate methods and concepts used by ecologists, toxicologists, economists, statisticians, and sociologists to frame the varied aspects of environmental decision-making problems. A key aspect of MIRA is its use of AHP in an exploratory and iterative decision-making process. MIRA uses AHP more as a learning tool that fosters consensus-building by revealing stakeholder preferences and helping individuals understand the positions and priorities of their fellow stakeholders. MIRA's iterative use of AHP also provides opportunity for modifying decision sets and revealing new options that may not have been identified at the outset of the process.

The U.S. Army Corps of Engineers is charged with making decisions concerning the management of sediment that is dredged for navigational purposes. The Corps' environmental decision-making processes are fashioned to conform with seven environmental principles, two of which exemplify the complex decision-making environment within which the Corps must fulfill its mandates (Corps 2004):

1. Strive to achieve environmental sustainability. An environment maintained in a healthy, diverse and sustainable condition is necessary to support life.
2. Build and share an integrated scientific, economic, and social knowledge base that supports a greater understanding of the environment and impacts of our work.

To these ends, the Corps implements a six-step process for developing and evaluating alternative plans based on a complex analysis of the costs and benefits of each alternative, in terms of both dollars and non-monetary measures, such as effect on environmental quality, safety, etc. (Linkov et al. 2004). Numerous mechanistic and deterministic tools are used by the Corps to develop information for evaluating project cost and risk, including traditional tools such as human and ecological risk assessment and cost/benefit analysis. An NRC review of the water resources planning procedures used by the Corps recommended using additional decision analysis tools to evaluate the environmental implications of restoration, flood control, and navigation projects (NRC 1999). Implementation guidance in the Corps' *Environmental Operating Principles* specifically advises the use of multi-criteria tradeoff methods to develop alternatives that are acceptable based on a combination of economic development and ecosystem restoration attributes (Corps 2003).

In addition, recent emphasis on basin-level planning within the Corps has provided impetus for the development and adoption of decision analysis tools to help accommodate the diverse and often competing interests of stakeholders at the regional level. Yoe (2002) presents a multi-criteria decision analysis approach for evaluating alternative plans and compares that approach with the Corps' traditional six-step process. The Corps' Institute of Water Resources has developed the Shared Vision Planning (SVP) methodology as a decision tool that includes evaluation of stakeholder preferences in combination with the simulation software package STELLA^{TM}.[7] In their review of the Corps' use of

---

[7] (http://www.svp.iwr.usace.army.mil/).

decision analysis tools, Linkov et al. (2004) conclude that there is still a need for a systematic framework for implementing these tools within the Corps.

At this writing, USEPA Region 10 and the Seattle District of the Corps are considering the use of MCDA in their dredged-material management program, specifically with regard to bioaccumulative substances. Explicit use of MCDA is one of several proposals being considered for dredged-material management going forward. The goal is to undertake a multi-agency/stakeholder process to identify a long-term solution that protects the health of the Puget Sound ecosystem and incorporates both maritime interests and the need for tribal subsistence. It has been argued that MCDA's role in the development process would facilitate defensible public policy decisions by transparently comparing a number of factors:

- Human and ecological risks
- Administrative feasibility of changing rules
- Equity (who or what would be affected)
- Economic efficiency (economic benefits vs. economic costs)
- Environmental efficiency (environmental impacts vs. environmental benefits)
- Ability to adaptively manage dredged material sites for recovery
- Other regulations and programs.

The U.S. Department of Energy (DOE) has applied and developed several models for assisting with their contaminated-site decision-making process. Because of their mandate, many DOE decision-making tools are designed to specifically address radiologically contaminated sites and sites with mixed chemical and radiological contamination. Linkov et al. (2005) describe the limited use of a multi-attribute model by DOE as part of its Environmental Restoration Priority System (ERPS) developed in the late 1980s. The ERPS decision process relies on assigning system weights to the following primary criteria: health risk, environmental risk, socioeconomic impact, regulatory responsiveness, cost, and uncertainty reduction. ERPS was abandoned by DOE in 1993, generally because of difficulty in implementing a transparent process within the constraints of national security, difficulty harmonizing the decision approach with the decision process for congressional funding for environmental restoration projects, and stakeholder opposition due to a failure to adequately involve the public in the early stages of the project (Jenni et al. 1995).

In 1998, DOE recommended the use of Multi-Attribute Utility Theory (MAUT), because it provides a quantitative and rigorous framework for taking quantitatively dissimilar measures of variables such as cost, risk, benefit, and stakeholder preferences, and developing aggregated measures that can be used for decision making (DOE 1998). MAUT is also preferred for facilitating discussion and mediating conflict.

Several authors have noted the utility of MCDA and other decision analysis tools for stakeholder engagement and consensus building. Stakeholder engagement and consensus building are of particular importance at large contaminated-sediment sites, partly due to the diverse and often competing interests that occur in urban watersheds, where large contaminated-sediment sites are often located. Linkov et al. (2004) note that MCDA can be used as a framework for stakeholders to structure their thoughts and value judgments about the advantages and disadvantages of remedial options at contaminated sites. The authors present several examples where MAUT was used in conjunction with information from stakeholder surveys and workshops to improve the comfort levels of participants relative to the decision-making framework. Linkov et al. (2004) find applications of MAUT and other decision analysis tools for stakeholder engagement and consensus building in the arenas of water resource management, regional forest planning, air quality valuation, mining, wilderness preservation, estuary management, and agriculture. As noted above, Stahl et al. (2002) describe the usefulness of AHP for informing decision makers, and as a tool for building the comfort level of stakeholders during the decision-making process.

14.6 Conclusions

We have examined the application of decision tools for guiding environmental decisions in coastal areas. Our evaluation focused on decisions concerning management of contaminated sediments. However, we also considered a few applications related to setting priorities and guiding the selection of restoration options. We find that decision makers make little use of formalized analytical methods for guiding decisions, even when sophisticated tools are made available to them. We speculate that such rejection of potentially useful decision-making tools may reflect a reluctance to change long-standing procedures, lack of familiarity with the methods and resulting lack of confidence in their results, unwillingness to incorporate disparate individual and group preferences, or possibly a desire to gather additional information before initiating the decision-making process.

There appears to be greater acceptance by decision makers of tools to organize information in a way that aids understanding. These tools include various ways to rank or display information and tradeoffs. These same objectives are embedded within MCDA, but use of that method appears to be unappealing to many decision makers at contaminated-sediment sites. How can MCDA and other more sophisticated decision support systems be made more useful? We offer the following suggestions:

1. Where guidance or regulations are available, structure the analysis around their elements, so that the analysis is familiar. With this approach, the major criteria are already defined. A good example involves combining the CERCLA criteria with the guidelines for managing contaminated sediments. These regulations and guidance documents contain most of the criteria that would be relevant to decision making at contaminated-sediment sites.

2. Use a MCDA structure to organize and track information. The form of this structure should be clear to all participants. It would be beneficial to avoid displaying this structure as part of the decision-making process. The MCDA elements can simply be embedded within the more familiar risk, economic, and other analyses.
3. Once information is organized, simply use MCDA to place the information into various perspectives, which may take the form of scenarios.
4. If the decision maker(s) are comfortable with the process, the group can exercise what-if scenarios or apply weighting factors. The information would already be in a form that would permit these applications.

Environmental decisions are often complex and can benefit from aids to organize and display various types of information and analyze relationships among them. Simpler tools are commonly used for this process. We believe that considerable benefit can be derived by incorporating MCDA within existing decision-making processes.

References

Beem B (2006) Planning to learn: Blue crab policymaking in the Chesapeake Bay. Coast Manage 34(2):167–182

Belton V, Stewart TJ (2002) Multiple criteria decision analysis: An integrated approach. Kluwer Academic Publishers, Boston

Bergquist G, Bergquist C (1999) Post decision assessment. In: Tools to Aid Environmental Decision Making, Dale VH, English MR (eds), Springer, Berlin

Chesapeake Bay Agreement (1983) http://www.chesapeakebay.net/pubs/1983Chesapeake-BayAgreement.pdf

Chesapeake Bay Program (2004) Analytical segmentation scheme. Revisions, decisions and rationales 1983–2003. October 2004. http://www.chesapeakebay.net/pubs/segment scheme.pdf

Chesapeake Bay Health and Restoration Assessment (2007) file:/// C:/Documents%20and %20Settings/slaw/Local%20Settings/Temporary%20Internet%20Files/Content.IE5/ 29PZGZS9/Bay%2520Restoration%2520Effort%2520Indicators%5B1%5D.ppt#388,1, Chesapeake Bay 2006 Health and Restoration Assessment, Part Two: Restoration Efforts, April

Corps (2003) USACE environmental operating principles and implementation guidance. U.S. Army Corps of Engineers, http://www.hq.usace.army.mil/CEPA/7%20Environ%20-Prin%20web%20site/Page1.html

Corps (2004) River basins and coastal systems planning within the U.S. Army Corps of Engineers: methods of analysis and peer review for water resources planning. National Research Council

Cura JJ, Bridges TS, McArdle ME (2004) Comparative risk assessment methods and their applicability to dredged material management decision-making. Hum Ecol Risk Assess 10(3):485–503

DOE (1998) Guidelines for risk-based prioritization of DOE activities. DOE-DP-STD-3023-98. U.S. Department of Energy

Driscoll SBK, Wickwire WT, Cura JJ, Vorhees DJ, Butler CL, Moore DW, Bridges TS (2002) A comparative screening-level ecological and human health risk assessment for dredged material management alternatives in New York/New Jersey Harbor. Hum Ecol Risk Assess 8(3):603–626

Efroymson RA, Nicolette JP, Suter II GW (2003) A framework for net environmental benefit analysis for remediation or restoration of petroleum-contaminated sites. ORNL/TM-2003/17. Oak Ridge National Laboratory, Environmental Sciences Division

Gerlak AK, Heikkila T (2006) Comparing collaborative mechanisms in large-scale ecosystem governance. Nat Res J Summer 46: 657–707

GLDT (1999) Decision making process for dredged material management. Great Lakes Dredging Team, http://www.glc.org/dredging/dmm/decision.pdf

Gregory RS, Keeney RL (2002) Making smarter environmental management decisions. J Am Wat Res Assoc 38(6):1601–1612

Horinko M (2002) Principles for managing contaminated sediment risks at hazardous waste sites. OSWER Directive 9285.6-08. U.S. Environmental Protection Agency, Office of Solid Waste and Emergency Response

Jenni KE, Merkhofer MW, Williams C (1995) The rise and fall of risk-based priority system: lessons from DOE's environmental restoration priority system. Risk Anal 15(3):397–410

Kiker GA, Bridges TS, Varghese A, Seager TP, Linkov I (2005) Application of multicriteria decision analysis in environmental decision making. Integ Environ Assess Manage 1(2):95–108

Kiker GA, Linkov I, Bridges TS (2007) Integrating comparative risk assessment and multi-criteria decision analysis. Working through wicked problems and their impossible solutions, pp. 37–51, In: Environmental Security in Harbors and Coastal Areas, Linkov I, et al. (eds), Springer

Landis WG, Duncan B, Hart Hayes E, Markiewicz AJ, Thomas JF (2004) A regional retrospective assessment of the potential stressors causing the decline of the Cherry Point Pacific Herring run. Hum Ecol Risk Assess 10:271–297

Landis WG, Weigers J (2005) Chapter 2: Introduction to the regional risk assessment using the relative risk model. In: Regional Scale Ecological Risk Assessment Using the Relative Risk Model, Landis WG (ed), CRC Press, Boca Raton, FL

Linkov I, Burmistrov D, Cura J, Bridges TS (2002) Risk-based management of contaminated sediments: Consideration of spatial and temporal patterns in exposure modeling. Environ Sci Technol 36:238–246

Linkov I, Varghese A, Jamil S, Seager TP, Kiker G, Bridges T (2004) Multi-criteria decision analysis: A framework for structuring remedial decisions at contaminated sites, pp. 15–24. In: Comparative Risk Assessment and Environmental Decision Making, Livov I, Ramadan A (eds), Kluwer

Linkov I, Sahay S, Kiker G, Bridges T, Seager TP (2005) Multi-criteria decision analysis: A framework for managing contaminated sediments, pp. 271–297, In: Strategic Management of Marine Ecosystems, Levner E, et al. (eds), Springer

Linkov I, Satterstrom FK, Kiker G, Batchelor C, Bridges T, Ferguson E (2006) From comparative risk assessment to multi-criteria decision analysis and adaptive management: Recent developments and applications. Environ Int 32:1072–1093

Luxon M, Landis WG (eds) (2005) Application of the relative risk model to the upper Willamette River and lower McKenzie River, Oregon, In: Regional Scale Ecological Risk Assessment Using The Relative Risk Model, CRC Press, Boca Raton

Mendoza GA, Martins H (2006) Multi-criteria decision analysis in natural resource management: A critical review of methods and new modeling paradigms. Forest Ecol Manage 230:1–22

Menzie CA, MacDonell MM, Mumtaz M (2007) A phased approach for assessing combined effects from multiple stressors. Environ Health Perspect 115(5):807–816

Morares R, Landis WG, Molander S (2002) Regional risk assessment of a Brazilian rain forest reserve. Hum Ecol Risk Assess 8(7):1779–1803

NRC (1997) Contaminated sediments in ports and waterways. National Research Council, National Academy Press, Washington, DC

NRC (1999) New directions in water resources planning for the U.S. Army. U.S. Army Corps of Engineers. National Research Council, National Academy Press, Washington, DC

Nicolette JP, Hutcheson K (2004) Use of a net environmental benefits analysis (NEBA) approach for remedial decision making at two BRAC sites. CH2M Hill and Marstel-Day Associates. Presentation at the 30th Environmental and Energy Symposium & Exhibition 5–8 April 2004, San Diego, CA, http://www.dtic.mil/ndia/2004enviro/ 2004enviro.html

NRC (2001) A risk-management strategy for PCB-contaminated sediments. National Research Council, National Academy Press, Washington, DC

NRC (2007) Sediment dredging at Superfund megasites: Assessing the effectiveness. National Research Council, Committee on Sediment Dredging at Superfund Megasites. National Academy of Sciences

Ohlson DW, Serveiss VB (2007) The integration of ecological risk assessment and structured decision making into watershed management. Integr Environ Assess Manage 3(1):118–128

Palermo M, Clausner J, Channel M, Averett D (2000) Multiuser disposal sites (MUDS) for contaminated sediments from Puget Sound-subaqueous capping and confined disposal alternatives. U.S. Army Engineer Waterways Experiment Station, Vicksburg, MS

Stahl CH (2003) Multi-criteria integrated resource assessment (MIRA): A new decision analytic approach to inform environmental policy analysis. Vol. 1. Dissertation

Stahl CH, Cimorelli AJ, Chow AH (2002) A new approach to environmental decision analysis: Multi-criteria integrated resource assessment (MIRA). Bull Sci Technol Soc 22(6):443–459

Stahl R (2008) Personal communication concerning the Dupont evaluation of ecological stressors in the Delaware River Estuary. DuPont Corporate Remediation Group, Wilmington, DE

Tetra Tech Inc. (1986) Framework for comparative risk analysis of dredged material disposal options. TC3090-02. Prepared for Puget Sound Disposal Analysis, U.S. Army Corps of Engineers, Seattle District, Seattle, WA

USEPA (1987) Unfinished business: A comparative assessment of environmental problems. U.S. Environmental Protection Agency, Office of Policy Analysis, Washington, DC

USEPA (1993) A guidebook to comparing risks and setting environmental priorities. EPA 230-B-93-003. U.S. Environmental Protection Agency, Office of Policy, Planning, and Evaluation, Washington, DC

USEPA (1998) EPA's contaminated sediment management strategy. EPA-823-R-98-001. U.S. Environmental Protection Agency, Office of Water

USEPA (2000) Toward integrated environmental decision-making. EPA-SAB-EC-10-011. U.S. Environmental Protection Agency, Science Advisory Board, Washington, DC

USEPA (2005) Contaminated sediment remediation guidance for hazardous waste sites. EPA-540-R-05-012, OSWER 9355.0-85. U.S. Environmental Protection Agency, Office of Solid Waste and Emergency Response

Versar and Exponent (2008) Ecological risk assessment related to alternatives for restoration of oysters in Chesapeake Bay. Versar, Inc., Columbia, MD, and Exponent, Inc., Bellevue, WA

Yatsalo BI, Kiker GA, Kim J, Bridges TS, Seager TP, Gardner K, Satterstrom FK, Linkov I (2007) Application of multicriteria decision analysis tools to two contaminated sediment case studies. Integr Environ Assess Manage 3(2):223–233

Yoe C (2002) Trade-off analysis planning and procedures guidebook. U.S. Army Corps of Engineers

Chapter 15
Review of Decision Support Systems Devoted to the Management of Inland and Coastal Waters in the European Union

Paola Agostini, Silvia Torresan, Christian Micheletti and Andrea Critto

Abstract One of the most important policies in the European Union concerns the management of water resources, including rivers, lagoons and coastal areas. Within this strategic and holistic approach to water resources management, contamination issues are relevant aspects of concern.

As an answer to this legislative framework, many different Decision Support Systems are currently available or under development to support river basins and coastal waters management, although very few directly address the problem of contamination.

In fact, the different DSSs have been structured and developed to address specific management needs, often within specific legislative requirements, and to capture specific problematic issues.

This chapter, after an analysis of relevant European policies, reviews some of these systems developed at the European level.

Of the analyzed systems, differences and common aspects are discussed, by focusing on features such as the level of integration among the several considered issues, the scale of application, the GIS potentialities and spatial analysis, the uncertainty and sensitivity analysis performances, the availability and easiness to use for the main stakeholders.

Moreover, the comparison among these systems highlights frameworks, functionalities and interfaces, and focuses on specific technical aspects, such as the structure and the relations among components, the included models, the ways and abilities to facilitate inclusion of different perspectives by decision makers.

The review also puts into evidence the main strengths of the systems in addressing relevant management issues, as well as critical limitations in application and use.

P. Agostini (✉)
Consorzio Venezia Ricerche, via della Libertà 12, 30175 Marghera, Venice, Italy
e-mail: paola.agostini@unive.it

A. Marcomini et al. (eds.), *Decision Support Systems for Risk-Based Management of Contaminated Sites*, DOI 10.1007/978-0-387-09722-0_15,
© Springer Science+Business Media, LLC 2009

15.1 Introduction

Many management problems currently affect fresh waters and coastal waters in Europe, equally concerned with water quality and quantity (EEA, 2005).

Water pollution control is a central concern of European countries. Most rivers have improved across Europe, particularly in formerly badly polluted urban and industrial areas, where point sources of pollution predominated, and where clean-up investment has been concentrated. There is the growing recognition that in an increasing number of water bodies, point sources are no longer the main pollution threat, due to the improvement of their impact achieved by the use of efficient (but costly) waste treatments (EEA, 2005). On the other hand, diffuses sources from agriculture activities are becoming the most dominant sources of pollution. Equally affected by contamination are also sediments (Salomons and Brils, 2004).

In general, it appears quite clear that different challenges are posed to decision makers and politicians in relation to fresh and marine waters in the European Union. First of all, integration is required, and it concerns not only spatial integration between the inland and coastal waters, which are naturally interconnected and mutually dependent, but also disciplinary integration in treating the several aspects (economic, environmental, social, technological, logistical and so on) that compose the assessment and management of these ecosystems. Moreover, it is essential to take into consideration pressures and impacts from agriculture, urbanization, tourism and climate change, and to tackle with the presence of diffuse sources of water pollution. The sustainable management of rivers and coastal environments, altered by engineering activities, and the definition of reliable and cost efficient tools for their assessment and management are other relevant challenges to explore.

This chapter presents methodological and operational issues of Decision Support Systems (DSSs) in relation to inland and coastal waters management in European countries.

In fact, different DSSs are currently available or under development for the assessment and management of fresh waters and coastal waters. An exhaustive review of all these systems would require much more space in the book and should be very much detailed.

The purpose of this review is instead to provide some examples of developed systems in order to highlight the complexity of the problem, also differentiated in several water ecosystems (river basins, lagoons, coastal areas) and how different instruments may respond to the same problems or how they may propose similar approaches. Other tools will be also presented in more specific details in the next chapters of the book.

The intention of the authors is that this preliminary and reduced overview may promote interest in the topic and may put the ground for a more personal exploration of the vast possibilities of tools and instruments for fresh and marine waters management.

The chapter, in Section 15.2, offers a legislative background of relevant EU directives and regulatory frameworks, which are suitable for waters management, highlighting the main requirements, which justify the development of many reviewed Decision Support Systems.

Then in Section 15.3, the six selected DSSs are described and discussed. After a brief presentation of the generalities of the DSS, the section defines firstly how the different DSSs have been evaluated, through the identification of relevant criteria encompassing methodological and technical aspects. Then the same criteria are used to compare these systems, focusing on differences or common approaches.

15.2 Legislative and Management Issues

The main legislative framework for the management of waters in Europe is represented by the EU "Directive 2000/60/EC establishing a framework for Community action in the field of water policy", commonly referred to as Water Framework Directive or WFD (EC, 2000a). The Directive came into force in 2000 and its timetable for implementation extends over 15 years, requiring achievement of good ecological and chemical status by the year 2015.

The aim of WFD is to *establish a Community framework for the protection of inland surface waters, transitional waters, coastal waters and groundwater, in order to prevent and reduce pollution, promote sustainable water use, protect the aquatic environment, improve the status of aquatic ecosystems and mitigate the effects of floods and droughts.* It therefore defines a new, integrated and holistic approach to water protection and management, quality improvement and sustainable use. The WFD applies to all water bodies, including rivers, estuaries, coastal waters out to at least one nautical mile, and man-made water bodies, and has implications for many different industries and activities.

Although contamination issues for water bodies are tackled in this Directive, within the integrated and holistic approach, the WFD does not specifically address sediment management. Other European legislation (EU Landfill Directive or Habitats Directive) may be used in these cases. In fact, the handling of dredged material is complex, because dredged material is at the borderline of water, soil and waste policies. Nevertheless, the WFD is perceived as a tool to tackle the sources of sediment contamination. It offers an opportunity to further improve knowledge about the relation between sediment quality and water quality and to harmonise quality assessment and sediment management on a river-basin scale (Salomons and Brils, 2004).

Public authorities, called to implement the WFD in the whole European territory, are confronted equally with strict and defined activities and with the need to have methodologies and tools that allow the correct implementation of regulations and obligations.

In fact, the decision making process related to the implementation of the Water Framework Directive includes several analytical and management steps, which should be carefully managed by the river basins authorities.

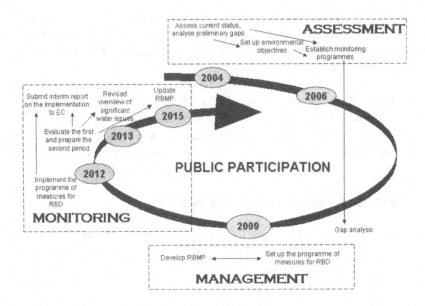

Fig. 15.1 The Management Planning Cycle (modified, European Commission, 2003) (RBD = River Basin District; RBMP = River Basin Management Plan; EC = European Commission)

The River Basin Management Planning Cycle, as shown in Fig. 15.1, can be interpreted as the temporal and practical guide to the implementation of the Directive, and is an iterative process. It includes three main parts: the assessment of the river basin, the management and the monitoring.

The **assessment** part starts with the identification of the River Basin District (RBD), described as *the area of land and sea, made up of one or more neighbouring river basins together with their associated groundwaters and coastal waters, which is identified as the basic management unit of river basins.*

A central aspect of the assessment process is the definition of the ecological and chemical status of the water bodies. The ecological status shall be considered as an expression of the quality and the functioning of aquatic ecosystem associated with surface water; for heavily modified water bodies, resulting from a human physical modification and serving economic activities, the concept of ecological status is translated into that of "ecological potential".

The Directive, in Annex V, defines the quality elements for the classification of ecological status, differentiated for rivers, lakes, transitional waters and coastal waters, encompassing:

- biological elements (e.g. composition and abundance of aquatic flora, benthic invertebrate fauna, fish fauna and phytoplankton);
- hydro-morphological elements supporting the biological elements (e.g. hydrological regime, river continuity, river depth and structure of the river or coastal bed, tidal regime);

– chemical and physico-chemical elements supporting the biological elements (e.g. thermal conditions, salinity, nutrient conditions, priority pollutants and other substances discharged into the body of water).

For each quality element a descriptive definition of a high, good, moderate, poor and bad status is given. In order to assign each element to a quality class, reference conditions are needed. Reference conditions represent the value of quality elements for that surface water body at a high ecological status: they do not adequate necessarily to total undisturbed, pristine conditions and include very minor disturbance which means that human pressure is allowed as long as there are no or only minor ecological effects (EC, 2003). The comparison to reference conditions represents a normalization of the measured value for each parameter and it ensures comparability among different water bodies. If biological elements reveal a class lower than good, hydro-morphological and physico-chemical elements do not need to be evaluated because in this case they are considered as poor or bad by default. Five classes of water quality can be then identified: bad, poor, moderate, good and high.

The ecological status evaluation must be accompanied also by the Chemical Status evaluation. Chemical status is classified only in two categories: "good" and "failing achieving good". A good chemical status is met when all concentrations of pollutants do not exceed the Environmental Quality Standards, established in this Directive as well as in other relevant European community legislation. In particular, priority substances are identified, as chemicals presenting a significant risk to or via aquatic environment, including risks to waters used for the abstraction of drinking waters. Environmental quality standards for priority substances must be set by Member States, which can either agree on these standards at Community level, or be forced to set their own standards within 7 years after the coming into force of the WFD. Once environmental quality standards have been adopted at Community level for the priority substances, these chemicals are considered only in the classification of Chemical Status and no more in the classification of the Ecological Status, within the physico-chemistry category.

In compliance with the integrated approach to water management, during the assessment procedures established to implement the WFD, water managers should also address requirements by the other recent European Directive adopted to reduce and manage the risks of floods in the Community (EC, 2007a). Indeed, the flood risk management directive has complementary objectives and synchronised timetables with the WFD. Therefore, it is required to identify areas with potential significant flood risks and prepare specific flood hazard and risk maps by a description of significant flood events which have occurred in the past, and to assess the potential adverse consequences of future floods, including the estimate of inhabitants, economic activities and environmental resources potentially affected.

The **management** part of the WFD Management Planning Cycle is instead focused on the production and implementation of a River Basin Management Plan (RBMP), including:

- characteristics of the river basin, as described in the previous assessment;
- environmental monitoring data;
- details of the impacts of human activity, in order to identify the pressures on the ecosystem that human activities might be causing;
- analysis of the economic use of water;
- strategic plan for the achievement of "good status", also called the Programme of Measures, including measures to promote an efficient and sustainable water use, to control the abstraction of fresh surface water and groundwater, to regulate the point and diffuse sources of pollution, with prohibition of direct discharges of pollutants in the water.

Therefore, a special concern in the management plan is given to strategies against the pollution of water. The Directive establishes to adopt specific measures against pollution of water by individual pollutants or groups of pollutants presenting significant risk to or via the aquatic environment, including risks to waters used for the abstraction of drinking water. Equally, measures to prevent and control pollution of groundwater shall be adopted.

As in the assessment phase, the implementation of the WFD is connected to the implementation of the parallel flood directive, so that also flood risk management plans and appropriate measures for the reduction of the probability of flooding and of potential adverse consequences to human health, the environment and economic activities should be included in the RBMP.

Finally, it is essential to prepare a good monitoring programme, to mostly provide a coherent and comprehensive overview of ecological and chemical status, thus covering:

- the volume and level or rate of flow, to the extent relevant for ecological and chemical status and ecological potential;
- the ecological and chemical status and ecological potential; and
- the quantitative status of groundwater.

Moreover, the monitoring phase is essential to review and update periodically the assessment and management implementation phases.

The steps established in the WFD cycle have a strict temporal development, as shown in Fig. 15.1, starting form the year 2004 and arriving to the year 2015, when a first full evaluation of measures has to be performed and monitored, after which the cycle can be started again. Within the whole process, public participation, required by Article 14 of the WFD, must be continuously guaranteed. In fact, a key aspect of the river basin management planning process is represented by the public participation and involvement of stakeholders (i.e. all the private, public and non governative associations, that are involved in the management of water bodies and whose interests can be conflicting), which is another effort required to river basin authorities.

Finally, a special consideration must be given to coastal areas. Due the peculiarities of these ecosystems and the addition of other relevant problems (e.g. habitat destruction, coastal erosion, resource depletion and climate change) with respect to those issues regulated by the WFD, in recent years an Integrated Coastal Zone Management (ICZM) strategy has been promoted at European level. The European policy framework regarding the integrated management of coastal areas consists of two communications delineating the ICZM strategy and progresses in Europe (EC, 2000b, 2007b), and of a recommendation (EC, 2002) inviting Member States to implement principles of good coastal zone management and to develop a national ICZM strategy. Nevertheless, a further effort has been made at European level to ensure that EU sectoral policies, including the WFD and the directive on the assessment and management of floods, are compatible with and enable the integrated management of the EU coastal zone.

As in the case of fresh waters management addressed in the WFD, contamination is only one of the main concerns for assessment and management programs of coastal zones, within a holistic and long-term perspective. And equal to the WFD, the ICZM includes assessment, management and monitoring phases, respectively concerned with the analysis of the state of the system and the identification of emerging issues; the formulation or adaptation of plans/programmes/strategies and their formal adoption; the evaluation of the process.

By taking into account this regulatory context, the complexity faced by water managers to accomplish Directives implementation is clearly justified and calls out for support instruments.

In fact, there is the need to assess main impacts by human activities influencing water quality and quantity, as well as to propose prevention, mitigation and monitoring activities on the significant adverse environmental impacts by economic activities on water environments. Integration of policies requires equal consideration of contamination and other relevant problems, and also conjunct assessment and management actions for water and sediments, and for rivers and coastal areas. Contaminated-sediment management issues and problems should be regarded as a common issue within a river basin. Transboundary management is needed for river systems that cross water bodies and national borders. Definition of management measures should be finalised to the identification of the most effective combinations of measures, towards sustainable and long-term objectives.

There is therefore a need for methodologies and tools that bridge the gap between the compliance with legislative requirements and the actual administration of water resources, where contamination may represent a critical issue. Decision Support Systems can offer the proper functionalities and characteristics to bridge this challenging gap. In the next section, some of these Decision Support Systems are going to be presented, in relation to the identified assessment and management problems for the considered water ecosystems.

15.3 Examples of Decision Support Systems for Fresh and Marine Waters

15.3.1 Description of DSSs

Many Decision Support Systems are currently available to tackle decision-making problems for river basins and coastal waters in Europe. Since the European legislation requires integration of policies and management actions, the different DSSs do not address only problems of contamination of rivers, sediments and coastal waters, but rather frame them in the context of the general assessment and management process for water resources.

For the purpose of this introductory chapter to the European state of the art on DSSs for waters resources, a selection of DSSs is proposed, encompassing tools for river basins, coastal waters and lagoons. They are an example of how the management needs of decision makers with respect to water resources are translated into developed tools and softwares. The selected DSSs are briefly described in the next paragraphs and relevant characteristics are also reported in Text Boxes 15.1–15.4.

Text Box 15.1. DSSs comparison: Legislative frameworks and scale of analysis

DITTY: EU Water Framework Directive, for assessment and management phases. It concerns coastal lagoons.

Elbe river DSS: To support the Elbe river management in the light of the EU Water Framework Directive. Scale range from catchment area to river sections.

MODELKEY: Implementation of EU Water Framework Directive, mainly for assessment phase. It concerns river basins

MULINO: In the context of the EU Water Framework Directive, for assessment and management phases, it concerns catchment scale.

River Life: Implementation of EU Water Framework Directive, for assessment and management phases. The scale is river basin.

TRANSCAT: In the context of the EU Water Framework Directive, for assessment and management phases. It concerns transboundary catchments, borderland regions of the European territory

WADBOS: Useful to support the implementation of Integrated Coastal Zone Management process and of the WFD assessment and management phases. Three spatial scales: the whole Wadden sea; 12 large homogeneous basins within the sea; a regular grid consisting of 25 ha cells.

Text Box 15.2. DSSs comparison: Main functionalities

DITTY: Scenario generation and analysis; measures evaluation.

Elbe river DSS: Scenario generation and analysis; measures identification

MODELKEY: Status evaluation; pressures identification; scenarios generation and analysis; indicators production; monitoring programs definition

MULINO: State evaluation; pressures identification; measures evaluation indicators production

River Life: Pressure identification; monitoring program definition; measures identification

TRANSCAT: Status evaluation; scenarios generation and analysis

WADBOS: Scenario generation and analysis; measures identification, measures evaluation

DSSs comparison: Included methodologies

DITTY: DPSIR, MCDA, Social Cost-Benefit Analysis, Integrated system approach, uncertainty analysis

Elbe river DSS: Scenario analysis; end-users involvement

MODELKEY: MCDA, DPSIR, socio-economic and environmental modelling

MULINO: MCDA, DPSIR, socio-economic and environmental modelling, sensitivity analysis

River Life: Environmental Risk Assessment, Environmental impact assessment

TRANSCAT: Water management modelling, MCDA, group decision methodology

WADBOS: Scenario analysis, MCDA, Score Tables, Sensitivity Analysis

DSSs comparison: Output

DITTY: Performance indicators and scenarios' ranking by means of bars graphs

Elbe river DSS: Maps (and related elaborated data) visualizing the state and impacts of Elbe river

MODELKEY: GIS maps and indicators values for river basin and hot-spots analysis

MULINO: Matrix with indicators for the evaluation of management options, linked through the DPSIR chain

River Life: Tables and graphic presentation of impacts, water quality of river basin and changes in land use

TRANSCAT: GIS maps and data of simulation of different conditions, indicators table

WADBOS: Policy indicators required to evaluate the success of scenario and policy options tried out on the system. The outputs are visualized by means of dynamic maps or in the form of text; MS Excel tables and time graphs.

Text Box 15.3.   DSSs comparison: Structural elements, including GIS and Web

DITTY: The model component includes integrated hydrological (water flow and nutrient transport) watershed and hydrodynamic, ecological, biogeochemical and socio-economic models. The user-friendly interface is panel-based. It is GIS-based.

Elbe river DSS: Extensive database containing data of the river Elbe. Models for flow channel hydraulics and morphology; for groundwater dynamics, for river quality and flood risk, for floodplain ecology, for economic analysis, 2D and 3D modelling. Very complex interface composed of many windows opened to answer specific aspects of analysis. It is GIS-based.

MODELKEY: The database component includes one database repository of information about chemical, toxicological, ecological and hydro-morphological data and one database with information about toxicity of (ideally) any type of chemical substance. The model component is wide and includes models for assessing chemical exposure; for assessing effects on populations and ecosystems; for calculation of integrated indicators; for economic analysis of water uses. The clear graphical user interface is similar to the GIS one and is accompanied by a wizard styled application. It is GIS-based and Web-based.

MULINO: Socio-economic and environmental information is stored in appropriate databases and organized according to the DPSIR conceptual framework. Main models include: causal chain models and Multi-attribute decision tools. The user-friendly and clear interface guides within the three different phases, and provides direct links to the DPSIR approach elements. It is GIS-based.

River Life: The GIS is connected to databases of national environmental administrations. The model component includes a hydrological model and ERA model. Very simple graphical interface, that allows to analyse the different aspects of river basin management separately in different windows accessed by suitable buttons.

TRANSCAT: There is a data management system, connected to the GIS. The model component includes models for environmental analysis (to simulate precipitation-run off processes in different kinds of watersheds, to predict water flow in surface and groundwater, to simulate solute transport in surface and groundwater), for socio-economic analysis (to study water supply and sewage water systems) and for GIS-based mapping and manipulation of geographic information. The graphical visualization is similar to the GIS interface, thus providing a familiar interface for spatial analysis. It is GIS-based and can be partly Web-based.

WADBOS: The DSS is based on a stand-alone database. The models variety includes: socio-economic models for economic activities, employment and infrastructures; models to represent the effects of economic activities on the landscape; hydrodynamic models to calculate water flows; transport models for detritus, nutrients and pollutants, ecological models for food chain dynamics. Interactive graphical techniques are applied extensively. It is GIS-based.

Text Box 15.4.   DSSs comparison: Flexibility, case-studies and availability

DITTY: It can be applied to other coastal lagoons with a wide range of social and environmental problems. Case study: Ria Formosa, Portugal; Gera, Greece; Mar Menor, Spain; Etang de Thau, France; Sacca di Goro, Italy. Not available.

Elbe river DSS: Not flexible, developed specifically for the Elbe river. Case study: Elbe river, Germany. Not publicly available, http://elise.bafg.de/servlet/is/3283/.

MODELKEY: Flexible both in inclusion of other analysis parameters and in application to other rives. Case study: Elbe, Germany/Czech republic; Llobregat, Spain; Scheldt, Belgium, Netherlands and France. Available at www.modelkey.org

MULINO: Flexible to application to other rivers. Case study: different cases in Europe, from Romania to Portugal, Italy and Belgium. Available on the web http://siti.feem.it/mulino/

River Life: It has been developed for the Finnish rivers or for boreal rivers. No flexibility for changes in parameters and addition of other models. Case study: Simojoki, Siuruanjoki, Kyrönjoki rivers in Finland. Available on the web www.environment.fi/riverlife

TRANSCAT: The system is flexible to application in other transnational river basins. Pilot areas: Bela/Biala river, Czech/Polish border; Pasvik river, Norwegian/Russian border; Guadiana river, Spanish/Potuguese border; Masta/Nestos river, Bulgarian/Greek border; Sumava catchment, Czech/German border). Prototype available on the web http://transcat.vsb.cz

WADBOS: The DSS was specifically developed for the Wadden sea. It allows the addition of new information and the editing of input data. Case study: Dutch Wadden Sea. Available on the web (http://www.riks.nl/projects/WadBOS)

The **DITTY** Decision Support System was developed within the European project DITTY and its main objective is to support the sustainable management of southern European coastal lagoons affected by the river-basin runoff, taking into account relevant impacts caused by agricultural, urban, and economic (e.g. fish/shellfish farming/fishing, tourism) pressures. In order to achieve this goal, DITTY follows the principles of the Integrated Water Resource Management (IWRM) as a part of the Water Framework Directive (WFD), as well as the DPSIR framework. The DITTY conceptual framework includes three main steps: (i) the decision problem definition, in terms of management options (i.e. policies, strategies, alternative actions) affecting the system behaviour; (ii) the alternatives generation, quantified by means of indicators, simulated by applying ecological, biogeochemical, hydrodynamic, and socio-economic models, taking also into account external factors which can not be manipulated by the decision maker (e.g. climate change); (iii) the alternatives evaluation and ranking by Multi-Criteria Decision Analysis (MCDA), applied to evaluate the system performance under different scenarios taking into account economist, decision makers, and stakeholders. DITTY was applied to five southern Europe coastal lagoons (see Chapter 19 of the book for more details about this DSS).

The **Elbe river DSS** has been developed within the study "Towards a generic tool for river basin management" supported by the German Bundesanstalt für Gewässerkunde (BfG). The principal function of the GIS-based DSS is to address different river problems, such as the improvement of socio-economic use of the river basin, the definition of sustainable level of flood protection and others. The DSS has a modular structure, from a catchment's scale to more detailed river sections, and the different scales are linked through analysis results. At the highest level of analysis (Catchment) there are models describing the impact of landuse and hydrology on diffuse runoff as well as impact of point discharges. At the second level of analysis (River) there are models describing among others navigation conditions, flood risk and water quality. At the third level (River section) there are detailed models describing the impacts of river engineering measures such as dike shifting and the habitat conditions for different species in the river. The system includes a spatial overview model of the Elbe catchment, a network of models that fulfils analysis and communication functions, and 2D or 3D process models (Verbeek et al., 2000; de Kok et al., 2001; Hahn et al., 2002).

The **MODELKEY** DSS aims at interlinking and integrating different analytical tools and exposure/effect models in order to evaluate risks posed by pollution to aquatic ecosystems at river basin scale and to identify areas (hot-spots) in need of management. The system is being developed under the European MODELKEY project (2005–2010) as an open-source, GIS-based system, freely accessible via Web and framed within a risk-based DPSIR approach (Driving Forces, Pressures, State, Impacts, Responses, EEA (1999)). The system is structured on a tiered-procedure for evaluating ecological risks, based on three phases accomplishing different functionalities both at river basin and hot-spot scales. Each phase leads to calculate flexible Integrated Risk Indices (IRI) by means of the Multi-Criteria Decision Analysis (MCDA). Specifically, in order to evaluate and classify the quality status of water bodies five Lines of Evidence are considered: ecology, chemistry, toxicology, physico-chemistry, and hydro-morphology. Moreover, environmental information is integrated with socio-economic factors related to different water uses, in order to define specific environmental objectives and to prioritize hot spots to be managed. Finally, a more detailed risk assessment is carried out at hot-spot scale aiming at evaluating causality and providing support for decision-making (see Chapter 16 of the book for more details on this DSS).

The **MULINO** DSS (MULti-sectoral, Integrated and Operational Decision Support System for Sustainable Use of Water Resources at the Catchment Scale) was developed within the same EU funded project, to be an operational tool which meets the needs of European water management authorities and which facilitates the implementation of the EU WFD, contributing to the quality and transparency of decision making by achieving a truly integrated approach suitable for the development of River Basin Management Plans. The decision process considers alternative options for the use of water resources and integrates multi-disciplinary approaches based on criteria and preferences that are elicited from decision maker and stakeholders, by Multi-criteria Decision Analysis (MCDA) tools. The DSS guides the user to consider the most important social, economic and environmental parameters/indicators that determine changes in water uses and in the state of water resources, organized in the structure of the DPSIR framework. Some of the used criteria concern nutrients concentrations, water quality, energy consumption, land use, recreation. The system is divided into three main phases: the Conceptual phase, with the identification of issues and problem exploration; the Design phase, where possible management options are defined and modelled for the evaluation of their performances; the Choice phase, where all options are judged according to the value functions and the preferences expressed by decision makers, who give weights to the evaluation criteria, through the MCDA methodologies (Fassio et al., 2005; Mysiak et al., 2005).

The **RiverLife** DSS was developed within a project of Finnish institutes. It is an interactive computer-based decision support system, used via Internet, which helps to integrate environmental considerations into land use planning and management practices in river basins. The system contains information on river biota and their habitats, on the effects of land-derived loading on water systems and on the water pollution control measures against non-point pollution. The included methods support the evaluation of the hydrological and ecological status of the rivers and the control of non-point source pollution originating from different forms of land use. RiverLife is characterized by a hydrological watershed model VMOD, which assists in estimating the effects of diffuse and point-source loading on the river flow and water quality; by a GIS tool, for obtaining data on the characteristics of the drainage basin, and examining hydrology and loading in the drainage basin; by a tool focused on the analysis of the ecological status of the river basin area and the river beds, through Ecological Risk Analysis (ERA) (Karjalainen and Heikkinen, 2005).

The **TRANSCAT** DSS was developed, within the EU funded project with the same name, for the Integrated Water Management of transboundary catchments. The system, built on a GIS platform, should provide the basis for the water management in the borderland regions in the contexts of the EU WFD. Therefore the DSS is built upon modules that allow simulation of the different climatic, topographic, environmental and socio-economic conditions of various transboundary catchments. The system is composed of four main subsystems: the mapping subsystem with the manipulation and visualization of geographic information; the modelling subsystem which encapsulates various models, including runoff models, precipitation models, river system analysis and groundwater flow and solute transport models; the data management subsystem which concentrates on editing and management of data needed in other subsystem; and the DSS subsystem which specifies decision alternatives, through one of the Multi-Criteria which Decision Analysis (MCDA) methodologies available or through group decision devices, and helps in choosing between them. A TRANSCAT prototype, based on Open Source solutions, which are freely available, is available and has been applied to 5 selected pilot areas in Central-Eastern EU countries (Horak and Owsinski, 2004).

The **WadBOS** DSS was developed in the Netherlands specifically for the integrated management and policy preparation in the Dutch part of the Waddensea, an important estuarine system in the north of the Netherlands. In fact, it links ecological and economic knowledge and information about the Wadden Sea in order to facilitate the planning and decision making process. Its main purpose is therefore to design and analyse potential policy measures and to be useful for all those involved in the management of the system. Accordingly, the system provides different functionalities covering all phases in a typical decision making process. First of all, WadBOS allows to gather, order and link knowledge about issues and problems; then, it allows to deepen the understanding about particular topics and linkages, providing a holistic representation of the Wadden system; and finally, it allows to evaluate the effects of different policy interventions onto the system and facilitates the discussions among policy makers, stakeholders and the public. The core element of WadBOS is an integrated dynamic model of the Wadden system representing strongly coupled social, economic, ecological, biological, physical and chemical processes. The output of the DSS includes summarised information and policy indicators required to evaluate the success of scenarios and policy options tried out on the system (Engelen, 2000; Van Buuren et al., 2002)

15.3.2 Comparison Discussion

The selected Decision Support Systems have been reviewed in consideration of methodological as well as structural issues.

As already stated in other parts of this book, a DSS can be characterized by two main elements: a framework and a structure. The first one refers to the assessment and management issues to which the DSS responds and for which it offers specific functionalities. The structure describes instead the main components of the system in terms of databases, models and graphical interface.

In consideration of these general characteristics and other relevant aspects of evaluation for Decision Support Systems, the following criteria have been identified and reported in Text Boxes 15.1–15.4, in order to present the different DSSs and propose their preliminary review:

- Legislative framework. It specifies to which legislation the DSS refers to and to which phase of the decision process it provides support.
- Scale of analysis. It specifies if the system is applicable to watershed, river basin, or coastal waters and to local, regional or global scale.
- Functionalities. Relevant functionalities of the systems are here reported. They include, according to the decision process: status evaluation, relevant pressures identification, scenarios generation and analysis, measures identification, measures evaluation, indicators production, monitoring programs definition.
- Included methodologies. It refers to the methodologies included in the system and used in the elaborations, such as risk assessment, the Driving Forces/Pressure/State/Impact/Response (DPSIR) framework, the Multicriteria Decision Analysis (MCDA), scenario analysis, socio-economic analysis.
- Structural elements. The three main elements of the DSS are detailed: models, such as economic, morphological, hydrological or ecological; database with specification of its nature; interface, addressing if the system is user-friendly and what kind of visual facilitations are provided to the user.
- Output. The main results of the system are here reported, in addition to their format.
- GIS-based. It is specified if the system is built within the Geographical Information System environment.
- Web-based. This feature assesses if the system is partly or totally accessible and usable via Web, and it is not in the form of a downloadable software to be installed in the user PC.
- Flexibility. It is the characteristic of the system to be adaptable, in terms of change of input parameters or addition of new models and functionalities. It is also linked to the possibility to be adaptable to different coasts or basins than those of the case-studies.
- Case-studies. The European river basins or coastal areas where the DSS has been tested and applied are listed.
- Status and availability. It is specified if the system is under development or ready for use, and the website where it is available.

In the following paragraphs, the Decision Support Systems will be discussed and compared based on the information reported in Text Boxes 15.1–15.4.

As far as the river basins are concerned, all the reviewed DSSs reflect assessment and management aspects required or proposed in the decision making process for the implementation of the WFD. In fact, the majority of them has been developed within EU funded projects, thus with the clear objective of supporting implementation of EU regulatory frameworks. This common objective may justify the inclusion in the reviewed systems of similar functionalities and approaches to tackle general problems.

Many of them, for instance, support very well the assessment part of the decision process, providing tools for analysis of characteristics of river basins, including definition of the water quality (RiverLife, Elbe River DSS, MODELKEY). RiverLife includes specifically a tool for environmental risk assessment and for the River basin analysis, while the Elbe River DSS provides models for describing water quality, navigation and other aspects of the river conditions. MODELKEY analyses in details assessment issues, providing tools for river basins and hot-spots analysis, and a wide range of models and integrated indicators for the definition of water quality in compliance with legislative requirements which cover chemical as well as ecological indications.

Equally, with respect to the WFD objective of a more comprehensive approach to water management, the tools also provide useful models for economic analysis of the impacts of human activities and use of water. In this context, with RiverLife is possible to analyse in details problems such as ditch maintenance. MULINO and MODELKEY allow to clearly study the causal chain between the human activities and impacts on water resources through the DPSIR approach.

However many DSSs are mostly concerned with the definition of management options and their evaluation, which represents a critical step in the WFD implementation. To this purpose, tools such as MULINO, Elbe River DSS, TRANSCAT or DITTY include models for scenario generation and simulation, as well as Multi-criteria Decision Analysis (MCDA) tools, which allow the comparison of the management alternatives as well as an active involvement of stakeholders in the decision process (another specific requirement of the WFD).

The European legislative and policy framework concerning the management of coastal zones encourages the further development of Decision Support Systems as appropriate analytical tools to cope with the many problems in coastal systems.

Accordingly, the coastal DSS reported in the present review (i.e. WadBOS) support the management of different aspects related to different EU policy targets.

In fact, WadBOS provides a broad framework to implement the strategic European principles under the Recommendation for ICZM. In fact, it allows experimentation with different combinations of policy measures, including limitations on economic activities such as fishing and mining, taxes and subsides, and to assess costs-benefits related to different economic and environmental scenarios and policy options. Moreover, it can facilitate the implementation of the WFD for coastal waters, supporting the classification of their ecological quality and the description of the economic pressures on the system.

Numerical models often represent the operative nucleus of these systems. While some DSSs employ mostly analytical models aimed at the characterization of the state and main processes of the examined system (e.g. Elbe river DSS, RiverLife, TRANSCAT, MODELKEY), others use simulation models to predict future environmental and socio-economic situations (e.g. DITTY, WadBOS, MODELKEY).

In the same way, while some DSSs are based on models mostly representing natural processes (e.g. RiverLife, TRANSCAT), many DSS (e.g. MULINO, DITTY, WadBOS, MODELKEY) make use of both environmental models and socio-economic ones, in order to achieve the integrated management of inland and coastal aquatic resources. Accordingly, they do not consider only key environmental aspects of the analyzed problem, but also allow the generation and evaluation of alternative management options, thus supporting the assessment and management phases of the decision-making processes.

Several DSS (e.g. WadBOS, Elbe river DSS, MODELKEY) utilize also more complex integrated models in order to represent the interrelations among different categories of processes (e.g. physical, morphological, ecological, chemical and socio-economic), and provide more comprehensive information about linked environmental and socio-economic phenomena.

In addition to the kind of models employed, reviewed DSS are also associated by the set of methodologies selected to grant their main functionalities. For instance, several DSS (e.g. DITTY, Elbe river DSS and WadBOS) include a scenario generation/analysis algorithm, which allow to make hypothesis about external influences not under the control of the policy maker (e.g. climate variability, economic prosperity and decline, population growth). Moreover, many DSSs make use of Multi-Criteria Decision Analysis (MCDA) in order to rank different management options, taking into account experts' knowledge and decision makers/stakeholders preferences.

Many of the analysed DSSs are GIS-based. In fact, in addition to automatic map production, GIS tools allow a better information management and a higher quality analysis and visualization of the study area. Moreover, GIS functionalities facilitate the storing, checking, manipulating, and displaying of data and allow the integration of environmental, economic and social factors into a shared platform. Finally, the use of geographically referenced data and the concise communication of complex spatial patterns are required by legislative frameworks and useful for environmental reporting.

Otherwise, only RiverLife, and partly MODELKEY and TRANSCAT, are developed on a web-based technology supporting a large group of users in a networked system. This is probably due to the additional effort required to develop this complex systems and to the necessity of update and handle continuously a computer server.

An important aspect that emerges from Text Box 15.4 concerns the flexibility of the proposed systems. The flexibility should be taken into high consideration, since the possibility to use a tool in other contexts allows a common approach to decision making process and also the possibility to obtain more comparable results.

In general, more flexible and adaptable systems are MODELKEY and MULINO, which are generic systems applicable to different water systems. Nevertheless, the fact that some of the reviewed systems, as the majority of worldwide available DSSs, have been developed specifically to address assessment and management issues in well-defined river basins or coastal areas must be kept in mind. Therefore, the tools usually include models and functionalities that respond to common legislative requirements, but at the same time they are also narrowed to specific contexts, for which they provide suitable parameters or problem solving instruments.

If a generic tool may be preferable due to its adaptability to other contexts, it must be recognized that the development of generic tools is a complicated task and brings the disadvantage of providing a comprehensive but also very complex and often unpractical tool, which may be discarded by decision makers. On the other hand, a tool developed for a specific geographical context may be more appealing to local authorities, which may feel more comfortable in its application.

The analysed systems are still applied in few cases, usually within the respective development projects (see case-study in Text Box 15.4). Most of the DSSs support the implementation of EU directives, but they are not compulsory instruments, and they are not routinely used in water systems management. Except for cases where the DSS is specifically devoted to the management of a river system, e.g. the Elbe River DSS, the presented DSSs still lack of a wider applicability in regular management practices.

15.4 Conclusions

The review has provided only a partial presentation of the various Decision Support Systems that were developed in Europe and during European projects to tackle water issues, including water pollution.

Water managers and decision makers are therefore confronted with a wide choice of adoptable systems. Which may be the most suitable tool, or which criteria should be used to select one or another tool?

Uran and Janssen (2003) suggest for instance to look at how management alternatives are specified, how outputs are presented and if the DSSs actually do what they promise to.

In fact, the primary criteria that drive the selection of one tool rather than another from a decision maker's point of view is the applicability to the specific problem and the environmental system he is confronted to.

But also other considerations may be perceived important, particularly when similar systems can be used, e.g. how simple the system appears, how the running of the tool is technically elementary and straightforward, how the included models, although accomplishing their analytical tasks, are equally understandable and not perceived as black boxes. Linked to this, the inclusion in a DSS of widely known tools (such as GIS) or models, already familiar to decision makers and end-users, can be an important advantage, as well as the match of the language and results of the DSS elaborations with the regulatory language and requirements.

Another aspect that could be considered is that some systems are mainly focused and have greater performances on specific phases of the decision process, such as the management options generation and evaluation. Therefore, their application is subordinated to the questions the users have to address.

Finally another aspect to consider is the characteristic of time and cost consuming of the DSSs: complex systems may require longer times of operation, and may be built on sophisticated but very costly technologies, not to mention the massive data management required. To avoid this disadvantage, some recent DSSs, for instance, are built on open source softwares, and linked to wide information sources.

The field of Decision Support Systems for contaminated waters is still open to different accomplishments and improvements.

References

de Kok, J.L., Wind, H.G., van Delden, H. and M. Verbeek (2001) Towards a generic tool for river basin management. Feasibility assessment for a prototype DSS for the Elbe Final report. Bundesanstalt für gewasserkunde.

EC (2000b) Communication from the Commission to the Council and the European Parliament on Integrated Coastal Zone Management: A Strategy for Europe. COM(2000) 547, Brussels.

EC (2000a) Directive 2000/60/EC "Directive 2000/60/EC of the European Parliament and of the Council Establishing a Framework for the Community Action in the Field of Water Policy" Official Journal (OJ L 327) on 22 December 2000.

EC (2002) Recommendation of the European Parliament and of the Council of 30 May 2002 Concerning the Implementation of Integrated Coastal Zone Management in Europe, 2002/413/EC.

EC (2003) Common Implementation Strategy for the Water Framework Directive (2000/60/CE). Guidance Document n. 11. Planning process. Office for Official Publications of the European Communities, Luxembourg.

EC (2007a) Directive 2007/60/EC Directive of the European Parliament and of the Council on the Assessment and Management of Flood Risks. Official Journal on 6 November 2007.

EC (2007b) Communication from the Commission. Report to the European Parliament and the Council: An Evaluation of Integrated Coastal Zone Management (ICZM) in Europe.

EEA (European Environment Agency) (1999) Environmental indicators: typology and overview. Technical report n. 25. Copenaghen, Denmark.

EEA (European Environment Agency) (2005) The European Environment. State and outlook 2005. Copenhagen, Denmark. p. 584.

Engelen G. (2000) The WADBOS Policy Support System: Information Technology to Bridge Knowledge and Choice. Technical paper prepared for the National Institute for Coastal and Marine Management/RIKZ. The Hague, the Netherlands.

Fassio, A., Giupponi, C., Hiederer, R. and C. Simota (2005) A decision support tool for simulating the effects of alternative policies affecting water resources: an application at the European scale. Journal of Hydrology 304: 462–476.

Hahn, B., Engelen, G., Berlekamp, J. and M. Matthies (2002) Towards a generic tool for river basin management. IT Framework Final report. Bundesanstalt für gewasserkunde.

Horak, J. and J.W. Owsinski (2004) Transcat project and prototype of DSS. 10th EC GI &GIS Workshop, ESDI State of the Art, Warsaw, Poland, 23–25 June 2004.

Karjalainen Satu, M. and K. Heikkinen (2005) The RiverLife project and implementation of Water Framework Directive. Environmental Science & Policy 8: 263–265.

MULINO consortium (2004). MULINO project Final report, part 1 and 2, MULINO EVK1-2000-000082, www.mulino.com.

Mysiak, J. Giupponi, C. and P. Rosato (2005) Towards the development of a decision support system for water resource management. Environmental Modelling & Software 20: 203–214.

Salomons, W. and J. Brils (eds). (2004) Contaminated Sediments in European River Basin. A report from the European Sediment Research Network (SedNet). p. 79.

Uran, O. and R. Janssen (2003) Why are spatial decision support systems not used? Some experiences from the Netherlands. Computers, Environment and Urban Systems 27: 511–526.

Van Buuren, J., Englelen G.K., van de Ven, 2002. The DSS WadBOS and EU Policies Implementation. Littoral 2002, The Changing Coast. Ed. EUROCOAST-Portugal, ISBN 972-8558-09-0.

Verbeek, M., van Delden, H., Wind, H.G. and J.L. de Kok (2000) Towards a generic tool for river basin management. Problem definition report. Bundesanstalt für gewasserkunde.

Chapter 16
MODELKEY: A Decision Support System for the Assessment and Evaluation of Impacts on Aquatic Ecosystems

Stefania Gottardo, Elena Semenzin, Alex Zabeo, and Antonio Marcomini

Abstract The MODELKEY DSS aims at interlinking and integrating different analytical tools and exposure/effect models in order to evaluate risks posed by pollution to aquatic ecosystems at river basin scale and to identify areas (hot spots) in need of management. In particular, the system helps decision makers and water managers in fulfilling the European Water Framework Directive requirements (EU WFD 2000/60/CE), which establishes a framework for Community action in the field of water policy and promotes the achievement of the "good" quality status in all surface waters (rivers, lakes, coastal and transitional waters) by 2015.

Currently, the system is under development but a dedicated risk-based DPSIR framework schematizing objectives, outputs and methodologies as well as the overall technical structure of the DSS are already defined.

In general, the system is characterized by an "open configuration" able to manage and integrate different types of data, parameters and models and freeing end users to include their own specific tools. The tiered procedure for evaluating ecological risks is based on two phases accomplishing multiple functions at both river basin and site-specific scales: data exploration and evaluation, quality status classification, identification of causes, economic analysis of water uses, hot spot prioritization, and provision of monitoring recommendations. Each phase leads to calculation of flexible Integrated Risk Indices (IRI) by means of Multi Criteria Decision Analysis (MCDA). Specifically, five Lines of Evidence (LOE) for grouping different environmental information are considered: biology, chemistry, toxicology, physico-chemistry, and hydromorphology. Moreover, environmental information is integrated with socio-economic factors related to different water uses in order to prioritize hot spots to be managed. The results will support decision makers in targeting future management actions on the most critical ecological endpoints, stressors and hot spots.

S. Gottardo (✉)
Center IDEAS, University Ca' Foscari of Venice, San Giobbe 873, 30121, Venezia, Italy
e-mail: Stefania.gottardo@unive.it

A. Marcomini et al. (eds.), *Decision Support Systems for Risk-Based Management of Contaminated Sites*, DOI 10.1007/978-0-387-09722-0_16,
© Springer Science+Business Media, LLC 2009

The MODELKEY DSS will be implemented in an open-source GIS environment and will be freely accessible via the Web. The structure will also allow the interaction with external tools, models and data sources (e.g. existing European databases, DSSs or other tools).

The system is being developed under the European MODELKEY project (2005–2010), and it is being applied to three case-studies: the Elbe, Scheldt and Llobregat river basins.

16.1 Introduction

The main piece of legislation for the management of river basins and water quality in Europe is the European Directive 2000/60/CE (EC 2000), also known as Water Framework Directive (WFD), which establishes a framework for community action in the field of water policy. It entered into force in 2000 and its timetable for implementation extends over 15 years.

The WFD represents a milestone in European water legislation since "integration" becomes a key concept providing a common and coherent framework within which the previous European directives regarding water policy can be reformulated or coordinated (EC 2003a). Moreover, as stated by Borja (2005), for the first time water management is: (i) based mainly upon biological and ecological elements with ecosystems at the centre of the management decisions; (ii) applied to all European water bodies, including inland surface waters (rivers and lakes), transitional and coastal waters and groundwater; and (iii) based upon the whole river basin including adjacent coastal area.

The WFD sets new goals for the European water management and introduces innovative means and processes for achieving them (Kallis and Butler 2001). The main environmental objectives related to surface waters are (Art. 4):

- to prevent further deterioration of the surface water body's conditions;
- to protect, enhance and restore all surface water bodies with the ultimate aim of achieving at least the "good ecological status" and the "good chemical status" by 2015;
- to protect and enhance all artificial and heavily modified water bodies with the aim of achieving the "good ecological potential" and the "good chemical status" by 2015;
- to reduce, cease or phase out emissions, discharges and losses of "priority pollutants";
- to promote sustainable use of water.

The management measures to achieve the WFD environmental objectives should be coordinated at the geographical/administrative level of the "river basin district" (Art. 3) while individual "water bodies" represent the classification and management units of the Directive (Borja 2005). For each river basin district a "competent authority" (Art. 3) should be designated, which is responsible for preparing and implementing a River Basin Management Plan (RBMP; Art. 13), reporting a summary of the river basin environmental and economic characterizations (Art. 5), and providing a description of the "programme of measures" (Art. 9) to be implemented to achieve the goals. The RBMP relies heavily on monitoring (Art. 8) to provide information for classifying water quality status and to address additional measures in response to non compliance with the environmental objectives (Dworak et al. 2005). In particular, the Directive describes (Annex V, paragraph 1.3) three different designs of monitoring programs (i.e. surveillance, operational and investigative monitoring) and specifies in which cases they are requested.

In order to achieve the main environmental objectives (i.e. good status and no-deterioration) by 2015, a planning process (Management Planning Cycle, MPC) was established that includes a series of tasks to be accomplished by prescribed deadlines (e.g. characterisation of pressures and impacts as well as economic analysis of water uses by 2004, setting up of monitoring programs by 2006, definition of management measures by 2009) (see EC 2003a). After the first MPC ending in 2009, information and results will be refined and updated during further 6-years management cycles. Stakeholders participation as well as public consultation should be assured throughout the whole process (EC 2003a).

A main innovative concept introduced by the WFD is that of "ecological status" for the evaluation of water quality. As explained by Vighi et al. (2006), the WFD overcomes the use of traditional chemically-based water quality criteria by emphasizing the site-specific evaluation of ecological effects. This means that classification systems for the ecological status should evaluate how the structure of the biological communities and the overall ecosystem functioning are altered by multiple anthropogenic stressors (Heiskanen et al. 2004).

The overall evaluation of surface waters quality involves the ecological status on the one hand and the chemical status on the other hand: both are required equally to reach the good status (for more details see EC 2005; Heiskanen et al. 2004). WFD Annex V explicitly defines which "quality elements" must be evaluated in order to assess the overall status with separate lists for each surface water category (i.e. rivers, lakes, transitional waters and coastal waters). The quality elements are subdivided into three groups:

- biological elements;
- hydro-morphological elements;
- chemical and physico-chemical elements.

The ecological status evaluation is based upon biological quality elements with hydromorphological and physico-chemical quality elements as support. Each water body has to be assigned to a quality class (i.e. "high", "good", "moderate", "poor" and "bad") by comparison with a reference condition (see EC 2003b). A good chemical status is encountered when no concentrations of priority pollutants exceed the Environmental Quality Standards (EQS) to be set at Community level (Art. 16, Annex V, par. 1.2.6). In the recently published proposal for a Daughter Directive on priority substances COM(2006) n. 397 (EC 2006), EQS for surface waters are preliminary derived according to the risk assessment approach (see Lepper 2005 Crane and Babut 2007).

Since WFD requests often are complex and difficult to correctly interpret and apply, supporting activities have been carried out by both EU Commission and scientific community. In fact, in order to assure the effective implementation of the WFD by Member States, a "Common Implementation Strategy" (CIS) was agreed in 2001. Several CIS working groups have been created dealing with specific topics and issues to provide guidance documents and promote harmonization. The EU Commission has recently published the first WFD implementation report, i.e. COM(2007) n. 128 (EC 2007), describing achievements but also highlighting gaps and giving recommendations for the years 2007–2009 (see Quevauviller 2006, 2007). To date one of the main activities carried out is the "intercalibration exercise" (Annex V, par. 1.4.1) which aims at making comparable the classification results obtained by applying each Member State's monitoring and assessment system for the biological quality elements, therefore assuring a common interpretation of good ecological status across the Europe (Birk and Hering 2006; Borja et al. 2007; Buffagni et al. 2007; EC 2005; Heiskanen et al. 2004). In addition to the CIS, the EU, single member states, and national research institutes have initiated many research projects to develop and apply new methodologies and tools supporting implementation of the WFD (e.g. Allan et al. 2006; Borja et al. 2004, 2006; Casazza et al. 2004; Dodkins et al. 2005; Henoque and Andral 2003; Verdonshot and Moog 2006).

In this context, the EU MODELKEY project (Models for Assessing and Forecasting the Impact of Environmental Key Pollutants on Marine and Freshwater Ecosystems and Biodiversity) was set up gathering 26 partners from 14 countries (see www.modelkey.org). The 5-years project funded by the Sixth Framework Program started in 2005 with the ultimate aim of developing a risk-based Decision Support System (i.e. MODELKEY DSS) interlinking and integrating a set of predictive/diagnostic models and analytical tools to assist decision makers and water managers in evaluating and managing impacts on European river basins as well as in protecting freshwater and marine ecosystems' biodiversity in compliance with the WFD regulations.

Currently, the MODELKEY project is in progress. For this reason, this chapter gives an overview of the first results obtained in developing the DSS. In particular, it describes the decisional framework and the conceptual framework of the DSS and explains the technical structure of the system and the preliminary choices related to its implementation. Moreover, the main decisional issues and the process of end users involvement are briefly clarified.

16.2 The MODELKEY Decisional Framework

Within the MODELKEY project, a decisional framework outlining general phases and objectives of the river basins assessment and management process under WFD was derived. It was defined by taking into account and partially

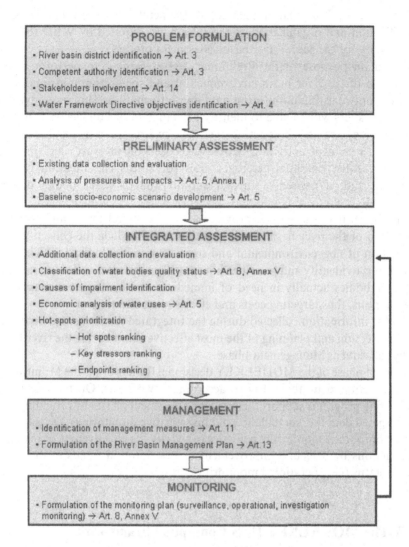

Fig. 16.1 The MODELKEY decisional framework for assessment and management of European river basins under the WFD (*arrows* designate links to the WFD articles and annexes)

fulfilling regulations and goals of the Directive. The resulting decisional framework is visualized in Fig. 16.1 which shows the different phases of the whole decisional process specifying assessment and management objectives of each phase. Moreover, it shows how the objectives are directly linked to the WFD articles.

The phases of MODELKEY decisional framework are: Problem Formulation, Preliminary Assessment, Integrated Assessment, Management, and Monitoring.

The Problem Formulation includes all the activities concerning the initial identification and organization of the river basin required by WFD that are preparatory for the actual river basin assessment process.

The main purpose of the Preliminary Assessment is to perform a first evaluation of the river basin environmental and socio-economic conditions by using only existing monitoring data and information. This way, involved decision makers will be able to identify gaps in data or knowledge as well as driving forces and pressures acting over their river basin to be focused during the next assessment activities. Moreover, based on pressures and impacts analysis results, potential reference sites for status classification can be identified and, if needed, a further sub-division of water bodies can be performed.

The Integrated Assessment provides a more comprehensive and complex evaluation of the river basin conditions that could include the collection and integration of new environmental and socio-economic information. The ultimate aim is to identify and prioritize hot spots throughout river basins, i.e. sites or water bodies actually in need of immediate and consistent management interventions, thus targeting costs and efforts in an effective way.

All the information collected during the Integrated Assessment is necessary for the selection and planning of the most effective solutions for the river basin in the subsequent Management phase.

The last phase of the MODELKEY decisional framework is the Monitoring, which is required by the WFD to accomplish two goals. On the one hand, monitoring programs support the previous assessment phases by providing new and targeted data (i.e. surveillance and investigative monitoring). On the other hand, appropriate monitoring activities can verify the effectiveness of management actions by detecting improvement or deterioration trends in the water body's status (i.e. operational monitoring).

16.3 The MODELKEY DSS Conceptual Framework

According to Rekolainen et al. (2003) the DPSIR (Driving forces, Pressures, State, Impacts, Responses) framework developed by the European Environment Agency (EEA 2003) is suitable for WFD implementation since many of the tasks required by the Directive refer directly to the elements of the DPSIR. Specifically, Rekolainen et al. (2003) proposed a modified framework for the implementation of the WFD, called DPCER, where the "State" and "Impacts" indicators are substituted with the "Chemical state" and the "Ecological state", respectively. As shown in Fig. 16.2, such framework specifically addresses the assessment phase of the MPC by identifying "Driving forces", "Pressures", "Chemical state" and "Ecological state", while the production of the RBMP required in the management phase corresponds to the identification of the "Responses".

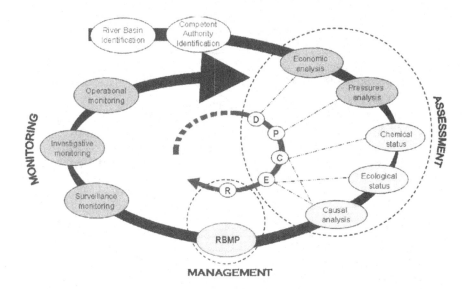

Fig. 16.2 Integration of WFD Management Planning Cycle and DPCER framework D = Driving forces; P = Pressures; C = Chemical state; E = Ecological state; R = Responses; RBMP = River Basin Management Plan

Considering both the DPSIR framework adopted by the EEA (2003) and the DPCER scheme outlined by Rekolainen et al. (2003), a risk-based DPSIR conceptual framework was proposed for the MODELKEY DSS fulfilling each element by means of risk-based methods and tools. It is visualized in Fig. 16.3 and aims at describing in detail functionalities (i.e. squares), tools (i.e. parallelograms) and outputs (i.e. circles) supplemented by the DSS, illustrating where the system provides support within the overall WFD-based decisional framework previously depicted and outlined in Fig. 16.1. Specifically, the MODELKEY DSS encompasses the whole assessment process, including both the Preliminary Assessment and the Integrated Assessment phases.

The ultimate goal of the assessment process supported by the MODELKEY DSS is to assist decision makers in targeting future management actions on river basins by providing three main outputs: (i) classification of the quality status of sites and water bodies, (ii), evaluation of the most responsible causes of impairment, (iii) identification of the most critical hot spots. In order to accomplish this task, a tiered risk-based procedure composed of the two assessment phases outlined in the decisional framework (i.e. Preliminary and Integrated) was implemented allowing end users to make an effective use of available data at site-specific and basin scales and to refine evaluations by improving the dataset when a lack of knowledge is highlighted. For each assessment phase one or more Integrated Risk Indices (IRI) are calculated by a risk-based integration of heterogeneous information coming from different areas of investigation (i.e. economy, ecology, ecotoxicology, chemistry, physico-chemistry, hydromorphology).

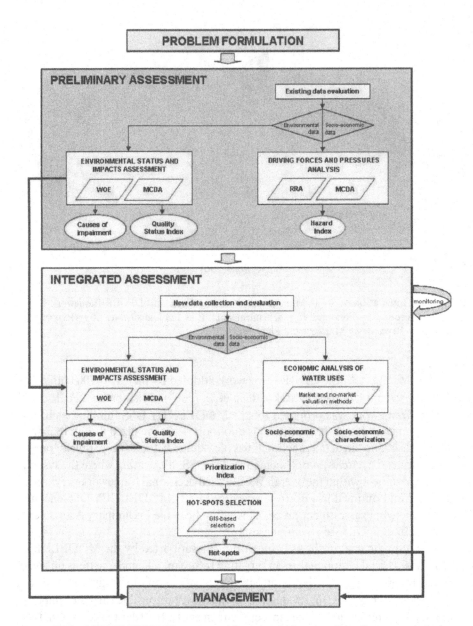

Fig. 16.3 The MODELKEY DSS conceptual framework: functionalities (i.e. *squares*), tools (i.e. *parallelograms*) and outputs (i.e. *circles*) of both the Preliminary Assessment and the Integrated Assessment phases WOE = Weight of Evidence; MCDA = Multi Criteria Decision Analysis; RRA = Regional Risk Assessment

Each Member State has over time carried out different monitoring programs and partly developed its own tools for fulfilling WFD requirements. In order to properly manage such diversity, the MODELKEY DSS is characterised by an "open configuration" which can use any type of relevant data, parameters and models. It provides end users with "default options", but also allows them to include their own specific tools. For this purpose, the MODELKEY DSS assessment process is based on a set of "flexible" IRIs allowing applications on every river basin to take into account specific environmental characteristics and existing data and tools availability. The final outcome is a procedure to normalize and integrate the computational and experimental data with the ultimate aim of developing an easy-to-use DSS supporting decision makers in assessing and managing river basins in compliance with the WFD regulations.

The Driving forces and Pressures elements of the risk-based DPSIR framework are addressed by the Preliminary Assessment (see Fig. 16.3) of the conceptual framework by using only existing data on the basin of interest. Practically, in the Preliminary Assessment, the system first helps end users to collect and explore existing datasets in terms of typology, richness, spatial and temporal distribution in order to reveal needs for additional data.

Subsequently, in order to identify significant driving forces and related pressures causing potential impacts (hazard) on river basins, socio-economic information on key economic drivers and sources as well as environmental data on vulnerability of water bodies are integrated according to a Regional Risk Assessment approach (RRA; Landis 2005) and by means of Multi Criteria Decision Analysis methods (MCDA; Kiker et al. 2005). RRA is applied since it is able to provide a relative ranking of areas, stressors and receptors at regional scale (e.g. a river basin) by integrating the magnitude and spatial distributions of pollution sources and stressors with vulnerability assessments of receptors. The main output of this stage is the Hazard Index (HI) which highlights the most relevant pressures, the water bodies that are of greatest concern and those water bodies that might be references sites.

Status and Impacts elements of the risk-based DPSIR framework are tackled by both phases: by relying only on existing data and by using sites (i.e. sampling stations) as assessment units in the Preliminary Assessment; and by enlarging the datasets as needed and by considering both sites and water bodies as assessment units in the Integrated Assessment phase. In fact, if a lack of knowledge or an excessive discordance among results is highlighted in the Preliminary Assessment, the DSS will require the acquisition of additional monitoring data on the sites of concern elaborating them in the IA phase in order to refine previous evaluations.

Both phases of the MODELKEY DSS conceptual framework support decision makers in evaluating and classifying the overall quality status of sites and water bodies according to the five quality classes proposed by the WFD: high, good, moderate, poor and bad. The main output is the Quality Status Index (QSI): taking into account the TRIAD scheme (Long and Chapman 1985; Chapman and Hollert 2006) all available environmental data and indicators are grouped into five Lines of Evidence (LOE), i.e. biology, chemistry, ecotoxicology, physico-chemistry and hydromorphology, and aggregated according to a Weight of Evidence approach (WOE; Burton et al. 2002; Suter 2003) and by means of MCDA methods.

Moreover, the separate aggregation and evaluation of environmental data and indicators as a function of specific stressors leads to identify the most responsible causes of impairment (e.g. eutrophication, acidification, toxic pressure) (see for details Suter et al. 2002; US EPA 2000); otherwise the separate aggregation of environmental data and indicators related to individual biological communities allows to identify the most damaged ecological endpoints (e.g. macroinvertebrates rather than fish).

The ultimate aim of the Integrated Assessment (see Fig. 16.3) and of the overall assessment procedure supported by the MODELKEY DSS is the prioritisation of hot spots in need of immediate and consistent management interventions by using both environmental and socio-economic information. The system carries out the economic analysis of water uses by providing a socio-economic characterization of the basin of interest and by calculating a set of Socio-Economic Indices (SEI) related to different water uses (e.g. agricultural, industrial, residential, recreational or fish-farming use). The SEIs are developed by applying appropriate market and non-market valuation methods to purposely collected socio-economic data with the aim of estimating in monetary terms the socio-economic loss over the basin due to inadequate environmental quality conditions that compromise water uses. Finally, hot spots on the basin of interest are ranked by means of the Prioritization Index (PI) integrating the QSI results with the SEI results. The MODELKEY DSS hot spots are visualized by means of GIS-generated maps.

The last element of the risk-based DPSIR framework (R) is directly linked to the Management phase of the decisional process, as it aims to identify and select adequate responses. An adequate response would be a technical measure, mitigation measure or policy instrument that would protect or improve water quality of a river basin, so as to maintain or restore the good ecological status by 2015. This management phase needs decision support tools to guide water managers in making decisions on intervention alternatives, to assure stakeholders' involvement and participation, and to communicate results in a transparent and simple way. The risk-based MODELKEY DSS interlinking different assessment methodologies and tools in a comprehensive structure is able to guide management actions and to make the decision process flexible, repeatable, changeable, traceable and transparent. The assessment process implemented by the MODELKEY DSS can be used not only for analyzing existing conditions, but also for developing scenarios to evaluate different management alternatives. For example, some input parameters values (e.g. chemicals concentrations) could be modified according to the abatement efficiency of a set of restoration measures, to determine whether the final results will change, i.e. if the quality status of the water body of concern will actually improve.

The management process is cyclic: after defining the RBMP and applying the program of measures, the overall assessment procedure begins again. The purpose of iterating the process is to determine whether goals are being met after the prior round of assessment and management and, if not, to guide additional management actions. In this case all derived outputs and collected data will become new input for the first assessment phase of the MODELKEY DSS conceptual framework.

16.4 The MODELKEY DSS Technical Structure

A Decision Support System (DSS) can be defined as a computer-based system that aids the process of decision making. It should be interactive, flexible, and adaptable and should support the solution of non-structured management problems. It should utilize available data, provide an easy-to-use interface, and allow for the decision maker's own insights (Druzdzel and Flynn 1999; Finlay 1994; Turban 1995).

The MODELKEY DSS achieves these goals by using a variety of environmental and socio-economic data in a set of mathematical models to obtain reliable river basin status snapshots and predictions of future conditions.

The MODELKEY DSS is characterized by an "open configuration" helping the user in managing both the different supplied (internal) models and external models, in order to obtain results of interest by using the available data on any type of river basin. The system can be seen as a sort of wizard that takes the user by the hand and guides him through the whole assessment process.

A DSS can be classified by different perspectives. To classify the MODELKEY DSS, "taxonomy" and "application" perspectives are used.

From the taxonomic point of view (Power 2002) the MODELKEY DSS can be seen as a model-driven application. Since the project concerns the investigation and evolution of physical processes related to aquatic environments, in the MODELKEY DSS all the involved models, both internal and external, are simulation models. Therefore, because of the large amount of environmental data that are computed during the assessment process, the MODELKEY DSS can be defined as a model-driven, data intensive DSS.

From the application point of view (Holsapple and Whinston 1996), the MODELKEY DSS can be seen as a Compound DSS, a hybrid between the Database-oriented and Solver-oriented basic DSS application types. That is, it is designed to deal with huge amounts of data and to be easy to use.

The technical structure of a DSS has 3 basic components (Sprague and Carlson 1982):

- DataBase Management Systems (DBMS) to manage, retrieve and store data in specific data structures;
- Model-Base Management Systems (MBMS) to simplify the application and management of heterogeneous models;
- Dialog Generation and Management System (DGMS), the graphical user interface systems used to interact with the users.

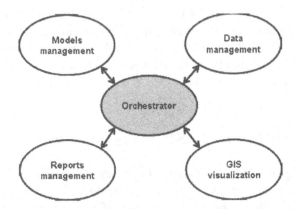

Fig. 16.4 The star connection architecture of the MODELKEY DSS modules

The technical structure of the MODELKEY DSS contains all of these basic components.

As visualized in Fig. 16.4, the MODELKEY DSS technical structure is composed of five main interactive modules: Orchestrator module, Data management module (i.e. the DBMS), Models management module (i.e. the MBMS), GIS visualization module (i.e. the DGMS), and Reports management module.

As depicted in Fig. 16.4 the DSS's internal modules collaborate with each other by using a logical star connection network. This means that a central unit, the Orchestrator module, manages all communications between the peripheral objects, i.e. the other modules.

The Orchestrator module is a resilient entity waiting for input through the user interface (the user interface is not represented in the figure because it is contained in the Orchestrator module); though, it is an always active module. Conversely, other modules are never active unless they are directly involved in the procedure by the Orchestrator.

The Data management module guarantees communication between different databases and the Orchestrator and simplifies the acquisition of information. This separation between data and data consumer is useful to simplify the modification of the Database structure and the eventual addition or substitution of databases.

The Models management module is intended to work as an interface between different types of models and the Orchestrator allowing a standardized flow of information. This way, the addition of external models as well as the changes in internal models can be easy and secure. All problems related with transferring data to models are handled by this module.

The GIS visualization module is intended to show results to users by providing effective GIS layers.

The Reports management module is used to create printable reports summarizing the executed assessments and their results. Reports will have specific formats and will contain information organized differently than on the screen;

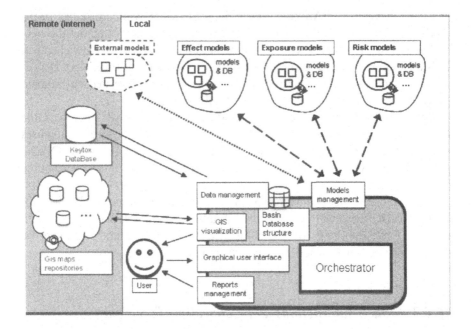

Fig. 16.5 Extended representation of the whole MODELKEY DSS structure

the usual contents of reports are tables and graphs. Reports will help decision makers to examine data and understand model results. As stated before for the GIS visualization module, the Orchestrator transfers only raw data to the reports management module that formats it in the proper way.

In Fig. 16.5, in addition to the five modules presented before, the connections with the external resources are shown, including: Databases, Models, and GIS map repositories. Databases (DB) are the repositories for data required for the operations included in the decision process. Models should not be considered strictly part of the core DSS application. In fact each model is an independent application supplying specific functionalities for the assessment process. GIS map repositories are web-available servers containing free downloadable maps to be used for the GIS visualization.

16.4.1 Databases Organization

Databases are fundamental components of a DSS, because they contain the data used for all assessments. In the MODELKEY DSS data are organized in two main databases, BASIN and KEYTOX.

The BASIN database is a database structure aimed at being a repository of chemical, physico-chemical, ecotoxicological, ecological and hydro-morphological data. Within its structure the users will be able to include their own data about any basin of interest. BASIN is a local database which will be installed in the user's machine, in order to avoid spreading his information throughout the web.

The KEYTOX database contains information about the toxicity of chemical substances that is intended to be a constantly growing repository. KEYTOX is an on-line resource, to which the DSS connects to retrieve needed data on demand, thereby avoiding the need to download updates.

Both databases are currently developed using Microsoft Access as the DBMS. The DBMS will probably be changed later, since Access databases are limited in space and performance in respect to other solutions (like Oracle, mySql, PostgreSql, etc.) and the MODELKEY project should use open source software.

Since models developed within the project need data not included in BASIN or KEYTOX, there will be additional databases in the DSS that are model specific. These databases will be embedded in the models and will never be directly accessed by the DSS's other components.

16.4.2 Models Interactions and Connection Techniques

Although models are in some ways external to the DSS, their role is fundamental in the achieving any result. Without models, no assessment could be performed on the data.

Some models are delivered embedded in the DSS (internal models). This does not imply that the DSS and the models constitute a monolithic structure, in fact the system keeps models as self dependent as possible using the models managing module to interact with them.

As depicted in Fig. 16.5, three types of models are embedded in the DSS:

- exposure models simulating fate and transport of contaminated sediments and waters at basin scale;
- models for prediction and diagnosis of effects on populations;
- risk models calculating indices for quality evaluation and hot spots prioritization.

It is important to note that the MODELKEY DSS is characterized by an "open configuration", so that it is possible to connect new external models providing additional functionalities as long as they adhere to the established connection protocol. External models can be linked to the DSS from the user's machine or from the net.

As it can be seen in Fig. 16.5 data interchange between models and the Models management module is based on the use of XML sheets (eXtensible Markup Language sheets) and, as a preliminary choice, by means of the web services paradigm (W3C 2007). This connection and data transfer protocol will be used for models residing both inside and outside the user's computer (internal and external models).

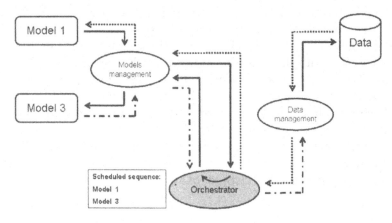

Fig. 16.6 Models interactions and data flow in the MODELKEY DSS

Sometimes data could flow from one model to another. As can be seen in Fig. 16.6, the Orchestrator creates a schedule of subsequent models calls and then executes it. The execution of the schedule presumes that the Orchestrator initially gains data from the database (via the Data management module, dotted arrow) and passes it to the first executed module, then the results obtained from the first module flow to the Orchestrator and from there to the second module and so on (continuous line arrow). When the executions are all completed the modelling results are stored in the database by the Orchestrator (semi-dotted arrow). There is never a direct flow of data between databases and models or between models and models.

16.4.3 GIS Interfaces

We intend that the MODELKEY DSS will use open source GIS software that offers a clear graphical user interface and a reliable object model in an extensible framework. This means that the MODELKEY DSS needs a basic set of well documented objects organized in a GIS generic framework that the system can extend and specialize in order to generate an application specific for the problem at hand, i.e. the WFD implementation. UDig (User-friendly Desktop Internet GIS) (UDig 2007) was identified as a product that met our criteria.

UDig is a Java/Eclipse (Eclipse 2007) based application available for all kinds of operative systems; in fact it is both a GeoSpatial application and a platform through which developers can create new derived applications. UDig has all the required characteristics: its user interface is built on a reliable GIS object model and it is expandable via the usual Eclipse plug-in procedure.

The MODELKEY DSS infrastructure will be based on the UDig framework. More precisely, MODELKEY DSS will be a plug-in for UDig, extending the base system with all the assessment functions needed to evaluate river basin quality status according to the WFD regulations.

After installing the MODELKEY plug-in, the standard UDig user interface will be enriched with the new features given by the application. After data insertion, GIS maps will be provided to the user by the UDig framework. In addition, when a specific spatial analysis is needed, the MODELKEY DSS will use UDig's GIS objects to obtain information. This means that UDig plays the role of the GIS visualization module (see Fig. 16.4).

The MODELKEY DSS will be a wizard styled application, i.e. will ask to the user some questions related to data and assessment needs in order to address the correct sequence of functional steps to follow.

16.4.4 Users

The MODELKEY DSS application is designed for several kinds of users who are enable to select which steps of analysis they want to perform. In particular, four types of users have been elicited: average user, expert user, specialized user, decision maker.

The average user is supposed to be a technician who can operate the DSS without being an expert in any aspect of the system. Expert users have the specialized user's skills in every aspect of the assessment process and they are asked for providing their judgment. The specialized user is an expert in some specific components of the application and he can directly focus on the model/ functionality of his interest. The decision maker is not intended to directly interact with the application, but to express his opinions/preferences and to use only reports and on screen results presentations in order to take decisions.

16.5 Decisional Issues and Involvement of End Users

As explained in the previous paragraphs, the MODELKEY DSS aims to provide decision makers (i.e. end users) with outputs specifically related to two main decisional issues: (1) which are the priority hot spots to be selected and (2) which are the most appropriate management actions to be undertaken. Outputs include the quality status classification of sites and rivers, the most responsible causes of impairment, the most damaged biological communities, and the ranking and GIS-based visualization of hot spots along the river basin. In addition, the DSS highlights missing information and helps end users to improve monitoring programs by providing recommendations about sites to be added or further investigated, missing indicators, and biological communities or stressors to be considered because of discordance among results.

To effectively assist decision makers, they must be informed and involved from the beginning of the assessment procedure in order to express their opinions and therefore to influence specific choices. As regards the MODELKEY DSS, decision makers are initially involved in defining boundaries of the assessment procedure, i.e. the steps and functionalities they want to perform and the scale of analysis. In fact the system provides different entry points to the assessment procedure (i.e. complete evaluation, environmental evaluation, socio-economic evaluation) and the most appropriate one can be chosen for the specific management objectives. Moreover, as the DSS includes analytical functionalities for providing both site-specific and river-specific results, end users can select the part of the river basin to be investigated (e.g. a river segment) and check the quality status, stressors and hot spots on that part. In addition, thanks to the "open configuration" of the MODELKEY DSS, end users can select, weigh or include specific environmental and socio-economic indicators or compare results obtained by applying different assessment scenarios before taking decisions.

In order to develop a DSS effectively addressing management needs and expectations, end users from European and not European countries have been involved in the overall assessment procedure as well as in the software system design. Moreover, in the course of the DSS software system development, intermediate prototypes will be tested by means of applications to the three case studies of the MODELKEY project, i.e. Elbe, Scheldt and Llobregat river basins. End users for each case study will evaluate the DSS functionalities and outputs to provide feedback for programming the definitive prototype.

16.6 Conclusions

The EU WFD sets strict objectives and multiple tasks for the assessment and management of river basins resulting in a considerable workload for water managers. In this context, the MODELKEY DSS is an innovative software system that combines several tools addressing all major aspects of river basin assessment according to a dedicated risk-based DPSIR framework. It will allow the identification of driving forces and pressures acting on the basin of interest to assess the ecological and chemical status of water bodies, to identify relevant stressors and key endpoints, to prioritize hot spots by integrating environmental and socio-economic information and to support the design of additional monitoring activities. All these outputs are needed in order to effectively allocate management costs and efforts. However, collaboration with end users in designing and developing a DSS is extremely important in order to effectively fulfill needs and expectations of decision makers and to assure DSS application and updating over time.

Acknowledgments This work was financially supported by the Sixth Framework Program (Sub-Priority 1.3.6.3, Global Change and Ecosystems) of the European Commission within the MODELKEY project (Models for Assessing and Forecasting Impacts of Key Pollutants on Marine and Freshwater Ecosystems and Biodiversity, Contract n. 511237-2) coordinated by UFZ (Germany). We acknowledge all MODELKEY partners for their contribution and assistance, specifically Dick de Zwart (RIVM, The Netherland).

References

Allan IJ, Mills GA, Vrana B, Holmberg A, Guigues N, Laschi S, Fouillac AM, Greenwood R (2006) Strategic monitoring for the European Water Framework Directive. Trends in Analytical Chemistry 25:704–715

Birk S, Hering D (2006) Direct comparison of assessment methods using benthic macro-invertebrates: a contribution to the EU Water Framework Directive intercalibration exercise. Hydrobiologia 556:401–415

Borja A, Franco J, Valencia V, Bald J, Muxika I, Belzunce MJ, Solaun O (2004) Implementation of the European water framework directive from the Basque country (northern Spain): a methodological approach. Marine Pollution Bulletin 48:209–218

Borja A, Galparsoro I, Solaun I, Muxika I, Tello EM, Uriarte A, Valencia V (2006) The European Water Framework Directive and the DPSIR, a methodological approach to assess the risk of failing to achieve good ecological status. Estuarine, Coastal and Shelf Science 66:84–96

Borja A (2005) The European water framework directive: A challenge for nearshore, coastal and continental shelf research. Continental Shelf Research 25:1768–1783.

Borja A, Josefson AB, Miles A, Muxika I, Olsgard F, Phillips G, Rodríguez JG, Rygg B (2007) An approach to the intercalibration of benthic ecological status assessment in the North Atlantic ecoregion, according to the European Water Framework Directive. Marine Pollution Bulletin 55:42–52.

Buffagni A, Erba S, Furse MT (2007) A simple procedure to harmonize class boundaries of assessment systems at the pan-European scale. Environmental Science & Policy 10:709–724.

Burton GA, Chapman PM, Smith EP (2002) Weight-of-Evidence approaches for assessing ecosystem impairment. Human and Ecological Risk Assessment 8(7):1657–1673.

Casazza G, Lopez y Royo C, Silvestri C (2004) Implementation of the Water Framework Directive for coastal waters in the Mediterranean ecoregion: the importance of biological elements and of ecoregional co-shared application. Biologia Marina Mediterranea 11:12–24

Chapman M, Hollert H (2006) Should the sediment quality triad become a tetrad, a pentad, or possibly even a hexad? Journal of Soils and Sediments 6(1):4–8.

Crane M, Babut M (2007) Environmental quality standards for water framework directive priority substances: challenges and opportunities. Integrated Environmental Assessment and Management 3(2):290–296.

Dodkins I, Rippey B, Harrington TJ, Bradley C, Ni Chathain B, Kelly-Quinn M, McGarrigle M, Hodge S, Trigg D (2005) Developing an optimal river typology for biological elements within the Water Framework Directive. Water Research and Policy 8:301–306

Druzdzel MJ, Flynn RR (1999) Decision support systems. Encyclopedia of library and information science. A Kent, Marcel Dekker, Inc.

Dworak T, Gonzalez C, Laaser C, Interwies E (2005) The need for new monitoring tools to implement the WFD. Environmental Science and Policy 8:301–306.

Environment European Agency (2003) Environmental Indicators: Typology and Use in Reporting. EEA internal working paper.

EC (2000) Directive 2000/60/EC of the European Parliament and of the Council of 23 October 2000. Establishing a framework for Community action in the field of water policy. European Commission. Official Journal of the European Communities L 327/72-22.12. 2000.

EC (2003a) Common Implementation Strategy for the Water Framework Directive (2000/60/EC). Guidance Document n.°11. Planning Process. European Commission, Working Group 2.9 on Planning Processes. Office for Official Publications of the European Communities, Luxembourg.

EC (2003b) Common Implementation Strategy for the Water Framework Directive (2000/60/EC). Guidance Document n.°10. River and lakes – typology, reference conditions and classification systems. European Commission, Working Group 2.3 – REFCOND. Office for Official Publications of the European Communities, Luxembourg.

EC (2005) Common Implementation Strategy for the Water Framework Directive (2000/60/EC). Guidance Document n.°13. Overall approach to the classification of ecological status and ecological potential. European Commission, Working Group ECOSTAT 2.A on Ecological Status. Office for Official Publications of the European Communities, Luxembourg.

EC (2006) Directive 2006/118/EC of the European Parliament and of the Council of 12 December 2006 on the protection of groundwater against pollution and deterioration. Official Journal of the European Union. L 372/19-27.12.2006.

EC (2007) Communication from the Commission to the European Parliament and the Council. Towards sustainable water management in the European Union – First stage in the implementation of the Water Framework Directive 2000/60/EC. COM(2007) 128 final. European Commission, 22.3.2007, Brussels.

Eclipse (2007) Eclipse Official web site: www.eclipse.org/

Finlay PN (1994) Introducing decision support systems. Oxford, UK Cambridge, Mass., NCC Blackwell, Blackwell Publishers.

Heiskanen AS, van de Bund W, Cardoso AC, Nöges P (2004) Towards good ecological status of surface waters in Europe-interpretation and harmonisation of the concept. Water Science and Technology 49:169–177.

Henocque Y, Andral B (2003) The French approach to managing water resources in the Mediterranean and the new European Water Framework Directive. Marine Pollution Bulletin 47:155–161

Holsapple CW, Whinston AB (1996) Decision support systems: a knowledge-based approach. West Publishing, St. Paul.

Kallis G, Butler D (2001) The EU Water Framework Directive: measures and implications. Water Policy 3:125–142.

Kiker GA, Bridges TS, Varghese A, Seager TP, Linkov I (2005) Application of multicriteria decision analysis in environmental decision making. Integrated Environmental Assessment and Management 1(2):95–108.

Landis WG (2005) Regional scale ecological risk assessment. Using the relative risk model. CRC Press.

Lepper P (2005) Manual on the Methodological Framework to Derive Environmental Quality Standards for Priority Substances in accordance with Article 16 of the Water Framework Directive (2000/60/EC). (Report by the Fraunhofer-Institute Molecular Biology and Applied Ecology, Schmallenberg, Germany.)

Long ER, Chapman PM (1985) A sediment quality triad: measures of sediment contamination, toxicity and infaunal community composition in Puget Sound. Marine Pollution Bulletin 16:405–415.

Power DJ (2002) Decision support systems: concepts and resources for managers. Westport, Conn., Quorum Books.

Quevauviller P (2006) News from Water Front. Water Framework Directive. Journal of Soils and Sediments 7(2):111–116.

Quevauviller P (2007) EU Policy. News from the Water Front. Journal of Soils and Sediments 6(4):255–258.

Rekolainen S, Kümüri J, Hiltunen M, Saloranta TM (2003) A conceptual framework for identifying the need ands role of models in the implementation of the Water Framework Directive. International Journal of River Basin Management 1:347–352.

Sprague RH, Carlson ED (1982) Building effective decision support systems. Englewood Cliffs, NJ Prentice-Hall.

Suter GW II, Norton SB, Cormier SM (2002) A methodology for inferring the causes of observed impairments in aquatic ecosystem. Environmental Toxicology and Chemistry 21:1101–1111.

Suter GW II (2003) Definitive Risk Characterization by Weighing the Evidence. In: Suter GW II (ed) Ecological Risk Assessment. Second Edition. CRC Press, Taylor and Francis Group, Boca Raton, Fla.

Turban E (1995) Decision support and expert systems: management support systems. Englewood Cliffs, NJ Prentice Hall.

UDig (2007) UDig official web site: udig.refractions.net/confluence/display/UDIG/Home

US EPA (2000) Stressor Identification Guidance Document. EPA/822/B-00/025 (Report by United States Environmental Protection Agency, Office of Water, Office of Research and Development, Washington, DC.)

Verdonschot PFM, Moog O (2006) Tools for assessing European streams with macroinvertebrates: major results and conclusions from the STAR project. Hydrobiologia 566:299–309

Vighi M, Finizio A, Villa S (2006) The evolution of the Environmental Quality Concept: from the US-EPA Red Book to the European Water Framework Directive. Environmental Science and Pollution Research 13(1):9–14.

W3C (2007) W3C Web Services Activity home page: www.w3.org/2002/ws/

Chapter 17
CADDIS: The Causal Analysis/Diagnosis Decision Information System

Susan B. Norton, Susan M. Cormier, Glenn W. Suter II, Kate Schofield,
Lester Yuan, Patricia Shaw-Allen, and C. Richard Ziegler

Abstract Biological monitoring and assessment methods have become indispensable tools for evaluating the condition of aquatic and terrestrial ecosystems. When an undesirable biological condition is observed (e.g., a depauperate fish assemblage), its cause (e.g., toxic substances, excess fine sediments, or nutrients) must be determined in order to design appropriate remedial management actions. Causal analysis challenges environmental scientists to bring together, analyze, and synthesize a broad variety of information from monitoring studies, models, and experiments to determine the probable cause of ecological effects. Decision-support systems can play an important role in improving the efficiency, quality and transparency of causal analyses.

CADDIS (http://www.epa.gov/caddis) is an on-line decision framework for identifying the stressors responsible for undesirable biological conditions in aquatic systems. CADDIS was developed in response to requirements under the U.S. Clean Water Act to develop plans for restoring impaired aquatic systems. CADDIS is based on U.S. EPA's 2000 Stressor Identification Guidance document, and draws from multiple types of eco-epidemiological evidence. A major update in 2007 added summaries of commonly encountered causes of biological impairment: metals, sediments, nutrients, flow alteration, temperature, ionic strength, low dissolved oxygen, and toxic chemicals. These reviews are designed to help practitioners choose which causes to consider, based on sources, site information, and observed biological effects. A series of conceptual models illustrates connections between sources, stressors and effects. Another major new section provides advice and tools for analyzing data and interpreting results as causal evidence; these tools help quantify associations between any cause and any biological impairment using innovative methods such as species-sensitivity distributions, biological inferences, conditional probability analysis, and quantile regression analysis.

S.B. Norton (✉)
National Center for Environmental Assessment U.S. Environmental Protection
Agency, Washington, DC, USA
e-mail: norton.susan@epa.gov

A. Marcomini et al. (eds.), *Decision Support Systems for Risk-Based Management*
of Contaminated Sites, DOI 10.1007/978-0-387-09722-0_17,
© Springer Science+Business Media LLC 2009

An essential part of the development strategy for CADDIS has been the use of case studies to test the process and tools in different regions, and with different causal factors. Case studies have been conducted in streams on the urbanized east coast and the agriculturally-dominated mid-west to the arid west, and have considered causes including low dissolved oxygen, increased temperature, toxic substances, altered food resources and fine sediments. Lessons learned from the case studies include the importance of a structure for organizing the large variety of evidence that is often available, the need for well-matched reference sites for comparison, the benefits of iterative and directed data collection, and the frequency of surprising results. The case studies illustrate the promise of CADDIS: by building on the foundation of biological monitoring, we can provide a powerful means for improving the health of our aquatic systems.

17.1 Introduction

The Causal Analysis/Diagnosis Decision Information System (CADDIS: http://www.epa.gov/caddis) is an on-line decision support system to help scientists identify the stressors responsible for undesirable biological conditions in aquatic systems.

The development of CADDIS was motivated by the increased use of biological monitoring and survey methods to evaluate aquatic ecosystems (Davis 1995; Ohio Environmental Protection Agency 1987; Plafkin et al. 1989). Fifty-seven U.S. states and tribes currently use biological assessments in water resource management (U.S. Environmental Protection Agency 2002). Biological assessments have also become an integral part of programs in the United Kingdom, Europe, Australia, Canada, New Zealand and South Africa (Marchant 1997; Metcalfe-Smith 1994; Stark and Maxted 2007). Biological assessments often reveal impairments previously overlooked by water quality measurements and that were not necessarily resolved by controlling point source emissions. However, biological assessments do not identify the cause of impairment; they only indicate where conditions are unacceptable. So, when biological assessments indicate that a water body is impaired, the cause of the change in the biological community needs to be identified before it can be rectified.

CADDIS aims to improve the practice of causal analysis of biological effects, by providing a formal inferential methodology and technical content useful for implementing the method. A formal method for making decisions about causation has many benefits; it provides a structure for organizing data and thinking when a situation is complex and provides transparency when a situation is contentious. When remedial alternatives are costly, a formal method can increase confidence that a proposed remedy will truly improve environmental condition.

Although the primary application of CADDIS has been to lotic systems, it is based on principles that are applicable to any ecosystem; lakes, estuaries, and terrestrial systems. CADDIS is designed to be prompted by the results of biological monitoring programs, but the principles can be applied to assessments prompted by concerns over sources, such as a non-point source inputs, a particular industrial outfall or a hazardous waste site, or over other types of observed effects, such as mass mortalities. In particular, the principles have been applied to assessments of fish kills and impaired populations and communities on contaminated sites.

17.2 Content

CADDIS is designed to help practitioners find, analyze and use information to produce causal evaluations in aquatic systems. It contains an inferential process and information needed to apply that process. The inferential process is based on U.S. EPA's 2000 Stressor Identification Guidance document and draws from multiple types of eco-epidemiological evidence (Section 17.2.1). A major update in 2007 provides tools and reviews that make information useful for causal analysis more accessible (Section 17.2.2).

17.2.1 The Step-by-Step Guide to Stressor Identification

The Step-by-Step Guide to Stressor Identification provides a formal process for making decisions about causation at specific sites (Fig. 17.1). It is a general

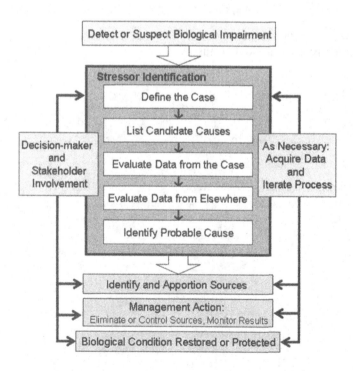

Fig. 17.1 The Stressor Identification Process (shown in the *darker gray box*), within the broader management context

framework that can be applied to the great range of causal scenarios and data availability that investigators encounter.

A formal process for causal analysis can mitigate many of the cognitive shortcomings that arise when we try to make decisions about complex subjects. Common errors include clinging to a favorite hypothesis when it should be doubted; using default rules of thumb that are inappropriate for a particular situation; and favoring data that are conspicuous (Kahneman et al. 1982; Nisbett and Ross 1980; Norton et al. 2003). The general attributes of a good decision process have been the subject of study for several decades. A good decision process provides a means for:

1. Choosing the most appropriate frame or scope for the analysis
2. Collecting the right information for the analysis
3. Organizing the information
4. Reaching conclusions
5. Obtaining feedback on the effectiveness of the decision (Russo and Schoemaker 1989).

The sections below discuss the five steps of the Step-by-Step guide in the context of these characteristics.

17.2.1.1 Choosing the Most Appropriate Frame or Scope for the Analysis

Step 1: Defining the Case, and
Step 2: Identifying Candidate Causes

Text Box 17.1. The Little Scioto River, OH, USA The Little Scioto River case study was developed to illustrate the application of the Stressor Identification process (Cormier et al. 2002, Norton et al. 2002, Cormier and Ferster 2007). The case study involves a 15-km reach of a river in north-central Ohio (Fig. 17.2).

Many point and non-point sources of pollutants are associated with the Little Scioto River. Point sources include a wastewater treatment plant and combined sewer overflows that enter between 9.5 and 10.5 km, respectively, upstream of its confluence with the Scioto River. Non-point sources include runoff from agricultural land uses and from the city of Marion. Releases may also originate from several contaminated industrial areas, including an abandoned wood treatment plant, a landfill, an appliance plant, and a rail facility (Ohio Environmental Protection Agency 1994). Finally, the stream was channelized in the early 1900s starting from river kilometer 15 and continuing downstream to the confluence with the Scioto River.

Several sites in the Little Scioto Case study area were considered to be biologically impaired based on the results of fish and macroinvertebrate surveys. Specifically, the values of two multimetric indices, the Index of Biotic Integrity (IBI) and the Invertebrate Community Index (ICI) were below criteria set by the Ohio Environmental Protection Agency.

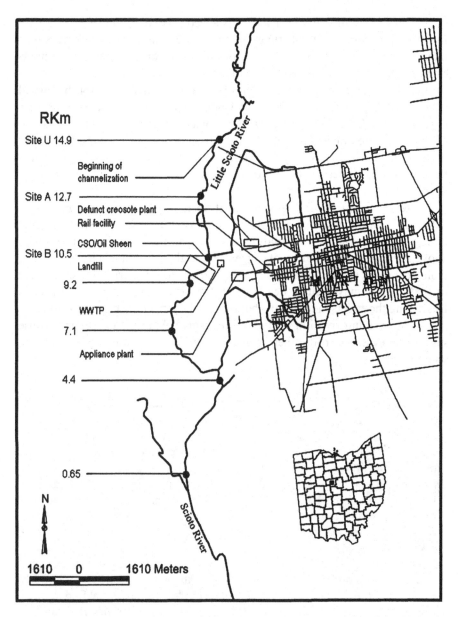

RKm

Site U 14.9

Beginning of
channelization

Site A 12.7

Defunct creosote plant
Rail facility

CSO/Oil Sheen

Site B 10.5

Landfill

9.2

WWTP

7.1

Appliance plant

4.4

0.65

N

1610   0   1610 Meters

Fig. 17.2 Map of the Little Scioto River case study area (adapted from Cormier et al. 2002)

Methods for causal analysis have frequently considered a very general spatial frame, but a targeted candidate cause frame. Examples include: Does smoking cause cancer (Hill 1965)? Can chlorinated dioxins, furans and biphenyls cause deformities in wildlife in the Great Lakes region of the United States (Fox 1991)? The types of causal analyses addressed by CADDIS reverse this scope.

They focus instead on a more localized spatial frame (e.g., a stream reaches), but consider a full range of candidate causes. They ask questions like: Did excess fine sediments, low dissolved oxygen, or chemical contaminants cause the loss of mayflies in this stream reach?

A causal analysis is prompted by the observation of an undesirable biological effect; a fish kill, a decline in a biological index, or a high incidence of anomalies (e.g., Text Box 17.1). The evaluation of ecological condition, including selection of appropriate biological indicators and sampling designs is a complex subject in itself and not addressed by CADDIS. Rather, causal analysis begins when these conditions assessments or other observations indicate that something is amiss.

In Step 1 of the CADDIS guide, practitioners begin scoping the analysis by defining the case that will be investigated (e.g., Text Box 17.2). First, the specific biological effects that will be analyzed are defined. For example, as mentioned above, the biological impairment triggering causal analysis may be a decline in a biological index score. The specific biological effects or metrics that contribute to that decline may include decreases in the abundance of larval stoneflies. Describing the effects in terms of what is actually happening biologically makes it easier to use information on the mechanisms behind the cause, and to use supporting evidence from other areas (e.g., similar situations in other locales, scientific literature, etc.).

Text Box 17.2. Defining the Case: The Little Scioto River The IBI and ICI indices were disaggregated to gain additional insights into the changes occurring in the Little Scioto River. A subset of individual metrics were identified that indicated distinctive changes in the assemblage at different points along the stream reach: the weight of fish normalized to 1 km distance (relative fish weight); the percent of fish having deformities, eroded fins, lesions or tumors (DELT anomalies); percent of macroinvertebrate individuals that were mayflies (percent mayflies); and the percent of macroinvertebrates that were taxa considered to be tolerant of stress (percent tolerant invertebrates).

The pattern of effects changed at different locations on the Little Scioto, indicating that different causes may be operating (Fig. 17.3). For example, the weight of fish normalized to 1 km distance (relative fish weight) increased at Site A, whereas the percent of fish having deformities, eroded fins, lesions or tumors (DELT anomalies) did not increase dramatically until Site B. For this reason, separate causal analyses were performed for each site. The reference sites used for comparison moved incrementally down stream. That is, the furthest upstream site (Site U Rkm 14.9) was used as a baseline for comparison for Site A (Rkm 12.5); both Sites U and Site A were used as baselines for comparison to Site B (Rkm 10.5).

Second, the geographic scope of the analysis is defined. The geographic scope of the case has two parts: (1) the impaired stream reach (or similar

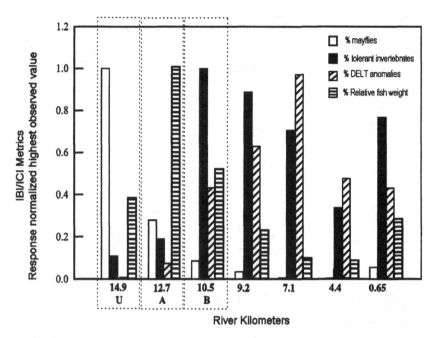

Fig. 17.3 Patterns of selected fish and macroinvertebrate metrics at different locations within the Little Scioto River case study; percent of macroinvertebrate individuals from mayfly taxa (% mayflies); percent of macroinvertebrate individuals from taxa considered to be tolerant of stress (percent tolerant invertebrates); the percent of fish having deformities, eroded fins, lesions or tumors (DELT anomalies); the weight of fish normalized to 1 km distance (relative fish weight) (adapted from Cormier et al. 2002)

stream part), and (2) other sites within the same aquatic system (e.g., the same stream, watershed, bay, or reservoir) that are either unimpaired or impaired in a different way, for use as a reference for comparison. Whenever possible all locations within a case should be part of the same system. They also should be located relatively close together, and aside from anthropogenic effects, should be as similar as possible physically, chemically, and biologically.

In Step 2, the scope of the analysis is further defined in terms of the candidate causes that will be analyzed (e.g., Fig. 17.4). Rather than trying to prove or disprove a particular candidate cause, CADDIS instead identifies the most probable cause from a list of candidates. Candidate causes are the stressors the organisms either contact (e.g., increased metals) or co-occur with (e.g., lack of suitable habitat). The list of candidate causes is compiled by reviewing available information from the site and from the region. People who have an interest in the assessment may have insights and opinions on candidate causes; these candidate causes should be included in the list so that they can be appropriately addressed.

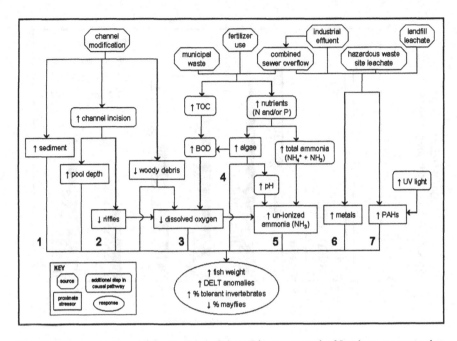

Fig. 17.4 Conceptual model for the Little Scioto River case study. Numbers correspond to candidate causes: **1.** Increased fine sediments. Channel modification (i.e., narrowing, deepening, and straightening of channel) leads to increased deposition of fine sediments. **2.** Altered pools and riffles. Channel modification (i.e., narrowing, deepening, and straightening of channel) leads to deeper pools and fewer riffle habitats. **3.** Low dissolved oxygen (DO) concentration. Several pathways may contribute to low DO levels: decreases in large woody debris and loss of riffle habitats can decrease aeration; increased organic carbon loading (e.g., from wastewater inputs) can increase biological oxygen demand (BOD); and/or increases in nutrients (nitrogen and/or phosphorus) can stimulate algal production and thus BOD (due to respiration of living plants and/or decomposition of algal detritus). **4.** Increased algal biomass. Nutrient inputs lead to moderate increases in algal biomass. These increases are insufficient to significantly reduce DO, but sufficient to stimulate secondary production in the system (i.e., lead to increased fish weights) and to alter invertebrate community structure. **5.** Ammonia toxicity. Increases in nitrogen loading lead to increases in total ammonia within the system; this ammonia dissociates into un-ionized ammonia, which is toxic to aquatic organisms. This candidate cause can be significantly affected by pH levels and DO concentrations, as the relative abundance of unionized ammonia increases with increasing pH and decreasing DO (due to reduction of nitrate to ammonium). Because increases in photosynthesis can raise pH, increased nutrients can both directly and indirectly increase unionized ammonia concentrations. **6.** Metal toxicity. **7.** Polycyclic Aromatic Hydrocarbon (PAH) toxicity

An important part of describing candidate causes is the construction of a conceptual model that describes the linkages between potential sources, stressors or candidate causes, and biological effects in the case (Fig. 17.4). The models show in graphical and narrative form the working hypotheses and assumptions about how and why effects are occurring. They also provide a framework for keeping track of what information is available and relevant to each candidate cause, setting the stage for the next steps of the analysis.

17.2.1.2 Collecting and Organizing Information

Step 3: Evaluating Data from the Case
Step 4: Evaluating Data from Elsewhere

A wide variety of arguments and data analyses can be used to support causal analyses. The objective is to show that fundamental characteristics of a causal relationship are indeed present at the case under investigation; for example, that the effect is associated with a sequential chain or chains of events; that the organisms are exposed to the causes at sufficient levels to produce the effect; that manipulating or otherwise altering the cause will change the effect; and that the proposed cause-effect relationship is consistent with general knowledge of causation in ecological systems .

The Step-by-Step guide walks practitioners through fifteen different types of evidence (Tables 17.1 and 17.2). Confidence in conclusions increases as more types of evidence are evaluated for more candidate causes. Although

Table 17.1 Types of evidence that use data from the case

Type of evidence	The concept
Spatial/Temporal Co-occurrence	The biological effect must be observed where and when the cause is observed, and must not be observed where and when the cause is absent.
Evidence of Exposure or Biological Mechanism	Measurements of the biota show that relevant exposure to the cause has occurred, or that other biological mechanisms linking the cause to the effect have occurred.
Causal Pathway	Steps in the pathways linking sources to the cause can serve as supplementary or surrogate indicators that the cause and the biological effect are likely to have co-occurred.
Stressor-Response Relationships from the Field	As exposure to the cause increases, intensity or frequency of the biological effect increases; as exposure to the cause decreases, intensity or frequency of the biological effect decreases.
Manipulation of Exposure	Field experiments or management actions that increase or decrease exposure to a cause must increase or decrease the biological effect.
Laboratory Tests of Site Media	Controlled exposure in laboratory tests to causes (usually toxic substances) present in site media should induce biological effects consistent with the effects observed in the field.
Temporal Sequence	The cause must precede the biological effect.
Verified Predictions	Knowledge of a cause'7;s mode of action permits prediction and subsequent confirmation of previously unobserved effects.
Symptoms	Biological measurements (often at lower levels of biological organization than the effect) can be characteristic of one or a few specific causes.

Table 17.2 Types of evidence the use data from elsewhere

Type of evidence	The concept
Stressor-Response Relationships from Other Field Studies	At the impaired sites, the cause must be at levels sufficient to cause similar biological effects in other field studies.
Stressor-Response Relationships from Laboratory Studies	Within the case, the cause must be at levels associated with related biological effects in laboratory studies.
Stressor-Response Relationships from Ecological Simulation Models	Within the case, the cause must be at levels associated with effects in mathematical models simulating ecological processes.
Mechanistically Plausible Cause	The relationship between the cause and biological effect must be consistent with known principles of biology, chemistry and physics, as well as properties of the affected organisms and the receiving environment.
Manipulation of Exposure at Other Sites	At similarly impacted locations outside the case sites, field experiments or management actions that increase or decrease exposure to a cause must increase or decrease the biological effect.
Analogous Stressors	Agents similar to the causal agent at the impaired site should lead to similar effects at other sites.

most assessments will have data for only some of the types of evidence, a ready guide to all of the types of evidence may lead practitioners to seek additional evidence.

The fifteen types of evidence provide a system for organizing the data and information relevant to a causal analysis. Each relevant analysis can be isolated, which is helpful when so much information is being evaluated. Human minds can only retain and process about seven pieces of information at time. Breaking up or chunking information into pieces is a more effective way to manage complex tasks (Nisbett and Ross 1980). Isolating each analysis helps prevent cognitive overloading and our tendencies to give undue weight to information that is easily obtained.

The types of evidence are organized into two sets: those that utilize data from the case itself (Step 3) and those that bring in information from other situations, or biological knowledge (Step 4). Causal analyses often begin with an examination of data from the case at hand (Table 17.1). For example, a field biologist might observe that effects occur when a particular candidate cause is present, but do not occur when it is absent (evidence of spatial co-occurrence). Such associations provide the core of information used for characterizing causes. It is beneficial to evaluate associations from the case first, because they can be powerful enough to eliminate candidate causes from further consideration.

Data from elsewhere may include information from other sites within the region; stressor-response relationships derived from field or laboratory studies; studies of similar situations in other streams, and numerous other kinds of information (Table 17.2). After assembling the information, it must then be related to observations from the case.

CADDIS includes a scoring system, adapted from the system by Susser (Susser 1986), that can be used to summarize the degree to each type of evidence that is available strengthens or weakens the case for a candidate cause. A consistent system for scoring the evidence facilitates the synthesis of the information into a final conclusion. The number of plusses and minuses increases with the degree to which the evidence either supports or weakens the argument for a candidate cause. Evidence can score up to three plusses ($+ + +$) or three minuses ($---$).

There are two other types of scores:

- Refute (R) is used for indisputable evidence that disproves that the candidate cause is responsible for the specific effects.
- Diagnose (D) is used when a set of symptoms for a particular causal agent or class of agents is, by definition, sufficient evidence of causation, even without the support of other types of evidence.

For example, the scoring table for spatial-temporal co-occurrence is shown in Table 17.3, and is applied in the Little Scioto Case Study in Text Box 17.3.

Table 17.3 Scoring system for spatial/temporal co-occurrence

Finding	Interpretation	Score
The effect occurs where or when the candidate cause occurs, OR the effect does not occur where or when the candidate cause does not occur.	This finding somewhat supports the case for the candidate cause, but is not strongly supportive because the association could be coincidental.	+
It is uncertain whether the candidate cause and the effect co-occur.	This finding neither supports nor weakens the case for the candidate cause, because the evidence is ambiguous.	0
The effect does not occur where or when the candidate cause occurs, OR the effect occurs where or when the candidate cause does not occur.	This finding convincingly weakens the case for the candidate cause, because causes must co-occur with their effects.	- - -
The effect does not occur where and when the candidate cause occurs, OR the effect occurs where or when the candidate cause does not occur, and the evidence is indisputable.	This finding refutes the case for the candidate cause, because causes must co-occur with their effects.	R

17.2.1.3 Reaching Conclusions

Step 5: Identifying the Probable Cause

After the evidence has been assembled and analyzed, the probable cause may be obvious. However, in many cases, a more systematic approach to synthesizing the evidence is useful for reaching and communicating conclusions. CADDIS provides advice on using the evidence and scores developed in Steps 3 and 4 to identify the probable cause.

Alternative approaches to using a system to reaching causal conclusions include relying on an expert's knowledge base of patterns and intuition, for example, it "feels" like toxic substances are the cause. In studies of medical diagnoses, intuitive approaches have been shown to yield results that are inconsistent and difficult to replicate (Russo and Schoemaker 1989). Diagnostic accuracy increased when physicians were given the results of probabilistic rules, suggesting that experts can profit from formal analyses (Dawes 2001). Rules of thumb are another alternative; for example, one might apply a rule that any chemical that is above its Ambient Water Quality Criterion is a probable cause. Rules of thumb are most useful when developed and applied to a particular subset of questions; they can be inaccurate and insensitive when applied generally.

Text Box 17.3. Example Analysis of Spatial Co-Occurrence from the Little Scioto River Case Study Spatial co-occurrence was evaluated by comparing stressor levels at sites upstream (Site U) of the impaired reach with those downstream. At the first site where the impairment was observed (Site A), concentrations of metals, biochemical oxygen demand (BOD), and nutrients were higher and dissolved oxygen was lower than at the upstream site (where no impairment was observed), so these could not be eliminated a potential causes. Ammonia was not detected at the site, but water column measurements of ammonia are highly variable. The concentrations of polycyclic aromatic hydrocarbons (PAHs) in sediments at Site A were not greater than the upstream site (in fact PAHs were not detected at either location), so PAHs were eliminated as a cause. A summary of the scores is shown in the following table:

Candidate cause	Result	Score
Sediment	Elevated sediment co-occurs with impairment	+
Pool/riffle	Poor pool/riffle condition co-occurs with impairment	+
Dissolved oxygen	Reduced DO co-occurs with impairment	+
Ammonia	Elevated ammonia not detected at site	−
Metals	Elevated metals concentrations observed at impairment	+
PAHs	Elevated PAHs not detected at Site A	R

Text Box 17.4. Identifying the Probable Cause in the Little Scioto River The team found that the Little Scioto River could be divided into three general geographical segments—upper, middle, and lower—based on the biological conditions and causal analysis.

- *Upper (Above Site U)* Biologically unimpaired, the upper regions of the river had a diverse community of fish and invertebrates.
- *Middle (Site A to Site U)* The middle segment exhibited less biological diversity, with a fish community dominated by the presence of large carp. The probable cause for that impairment was attributed to channelization and deepening of the stream, as well as alterations of stream habitat and water quality.
- *Lower (Site B to the Confluence with the Scioto)* The lower reaches of the Little Scioto were the most impaired. The fish community had decreased diversity and fewer fish. In addition, individual fish were smaller, and showed an increased incidence of external anomalies and lesions. The biological impairments observed here were consistent with toxicological effects associated with exposure to PAH, metals, and ammonia. (Heavy creosote contamination began approximately 270 m upstream from the confluence of North Rockswale Ditch.) Although the chemical contamination was sufficient to cause the biological impairments, the contaminated portions of the Little Scioto was also affected by the same channelization problems as the middle segment of the river.

CADDIS uses a strength of evidence approach. Evidence for each candidate cause is weighed, then the evidence is compared across all of the candidate causes. The evidence and scores developed in Steps 3 and 4 provide the basis for the conclusions. The scoring approach is advantageous because it incorporates a wide array of information, and the basis for the scoring can be clearly documented and presented.

One of the challenges commonly faced by causal analyses of stream impairments is that evidence is sparse or uneven. Because information is rarely complete across all of the candidate causes, CADDIS does not employ direct comparison or a quantitative Multi-Criteria Decision Analysis approach. The scores are not added. Rather the scores are used to gain an overall sense of the robustness of the underlying body of evidence and to identify the most compelling arguments for or against a candidate cause (Text Box 17.4).

In the best case, the analysis points clearly to a probable cause or causes. In most cases, it is possible to reduce the number of possibilities. At the least, the analysis identifies data gaps that need to be filled to increase confidence in conclusions.

17.2.1.4 Obtaining Feedback on the Decision

Mechanisms for obtaining feedback on the effectiveness of decisions are an important part of improving decision making processes. However, it is also very difficult to do so. Causal analysis is only one of several activities required to improve and protect biological condition. In some cases, the most effective management action will be obvious after the probable cause has been identified. In many cases, however, the investigation must identify sources and apportion responsibility among them. This can be even more difficult than identifying the stress in the first place (e.g., quantifying the sources of sediment in a large watershed), and may require environmental process models. The identification and implementation of management alternatives can also be a complex process that requires additional analyses (e.g., economic comparisons, engineering feasibility) and stakeholder involvement (Text Box 17.5). In the best case, a causal analysis is compelling enough to prompt management action, and follow-up monitoring confirms that that the management action improved biological condition (Text Box 17.6).

> **Text Box 17.5. Management Actions in the Little Scioto River** Findings from the case study clarified remediation options and likely environmental outcomes. However, the results of the causal analysis were unknown to the U.S. EPA and Ohio EPA resource managers negotiating the remediation of the Little Scioto. A decision was made to focus remediation on removal of the contaminated sediments.
>
> Managing expectations of recovery potential is critical for evaluating remediation success. For the lower segment of the Little Scioto, removing sediment alone, through dredging could reduce exposure to chemical contamination; however, the biological condition would be expected to improve to levels similar to those in the middle segment of the river. Achieving conditions of the most upstream segment of the river would also require extensive, habitat restoration efforts.

17.2.2 Information and Tools

A second major objective of the CADDIS project is to make information relevant to causal arguments more available. Studies of how people form judgments have shown that we are overly influenced by data that are conspicuous and easy to find (Nisbett and Ross 1980). Using readily available information is not a problem in itself: our goal is to expand the range of readily available information and ensure that it is science-based. Our approach is three-fold. First we provide basic information on commonly encountered causes that is applicable to a broad range of assessments. Second, we provide downloadable tools that make it easier for scientists to analyze their own data. Third, we provide databases of stressor-response relationships that may be difficult to find or generate. An advantage of a Web-based system is that this information can be cross-linked with the causal analysis framework described in Section 17.2.1.

> **Text Box 17.6. Management Actions and Environmental Outcomes in the Willimantic River Connecticut, USA**
>
> Causal Assessment
>   Monitoring by the Connecticut Department of Environmental Protection in the autumn of 1999 identified biological impairment in the Willimantic River in northeastern Connecticut, USA. The specific biological impairment defined as low numbers of Ephemeroptera, Plecoptera and Trichoptera (EPT) taxa at a site on the Middle River, and low numbers of EPT and non-EPT taxa downstream on the Willimantic River. The assessors suggested that unreported episodic, acute exposures were causing the impairment because the magnitude of the measured candidate causes were deemed insufficient to cause the severe impairment observed at the site. They recommended sampling in a way to localize the area of the impairment.
>
> Management Action
>   New biological sampling localized the upper bounds of the impairment near a raceway previously obscured by vegetation. A grey discharge was discovered and was traced to a broken pipe under a loading dock of a textile mill. The episodic toxic discharge was confirmed as the probable cause after rerouting of the illicit discharge and observing an increase in number of EPT and non-EPT taxa at two impaired locations.
> Environmental Outcome
>   Three years after rerouting the illicit discharge, the impaired sites reached acceptable biological conditions as defined by the State's Department of Environmental Protection. These findings have given confidence to the state agency to apply causal analysis to other rivers and we have been able to demonstrate that scientific information can be presented in a way that results in management action that improves the environment.

17.2.2.1 Candidate Causes

CADDIS provides information on eight commonly encountered candidate causes: metals, sediments, nutrients, flow alteration, temperature, ionic strength, and low dissolved oxygen, and toxic chemicals. Currently these reviews are designed to help practitioners choose which causes to consider, based on sources, site information, and observed biological effects.

A series of generic conceptual models illustrates connections between sources, stressors and effects (Fig. 17.4). Conceptual models are graphic representations of the potential links among sources, stressors, and biological responses. These diagrams support decision-making in several ways. Initial development of conceptual model diagrams provides a framework for brainstorming and prioritizing possible stressors and causal pathways. As the causal assessment progresses, these diagrams can help investigators identify data gaps, track the likelihood of different candidate causes, and, perhaps most importantly, clearly and efficiently communicate the logic of the causal analysis to stakeholders.

Conceptual models also can be powerful tools for organizing and providing access to information relevant to stream impairment. CADDIS includes a prototype Interactive Conceptual Models (ICMs) for phosphorus that builds on the conceptual model diagram. The diagram serves as a structural framework, or scaffold, for organizing stressor-specific information. Users can query the diagram (via hyperlinks) to access detailed information relevant to their own decision-making processes; currently, users can select two or more shapes in the diagram to retrieve literature citations that support the hypothesized relationship between those shapes. In the future, other types of information, such as quantitative stressor-response relationships and measurement techniques, may be built onto the framework in different layers.

17.2.2.2 Analyzing Data

Another section of the site provides advice and tools for analyzing data and interpreting results as causal evidence. CADDIS presents selected methods and describes how to apply them in a causal assessment. The methods range from well-established exploratory data analysis methods such as scatter plots, box plots and correlations, to statistical modeling methods like regression, conditional probability analysis and species sensitivity distributions. The Analyzing Data section also reviews fundamental concepts and best practices for data handling. Advice is provided that discuss how the source, quality and structure of data (e.g., timing, variability) influence how it should be organized and analyzed for causal analysis. Best practices for interpreting statistical outputs are also provided.

The application of some of these methods to causal analysis has required adaptation and extension. For example, paleolimnological methods for inferring environmental concentrations from algal species occurrences have been adapted for use in analyzing macroinvertebrate species occurrence data that are often available in causal analyses (U.S. Environmental Protection Agency 2006a; Yuan 2007). The use of these approaches to calculate tolerance values and to predict environmental conditions from biological assemblage information is discussed in detail.

CADDIS provides downloadable tools that can quantify associations between any cause and any biological impairment, including a Species Sensitivity Distribution generator and a downloadable statistical package (CADStat) that provides a graphical user interface that makes a variety of exploratory and statistical methods easier to use. As of the writing of this chapter, methods in CADStat include scatter plots, box plots, correlation, linear regression,

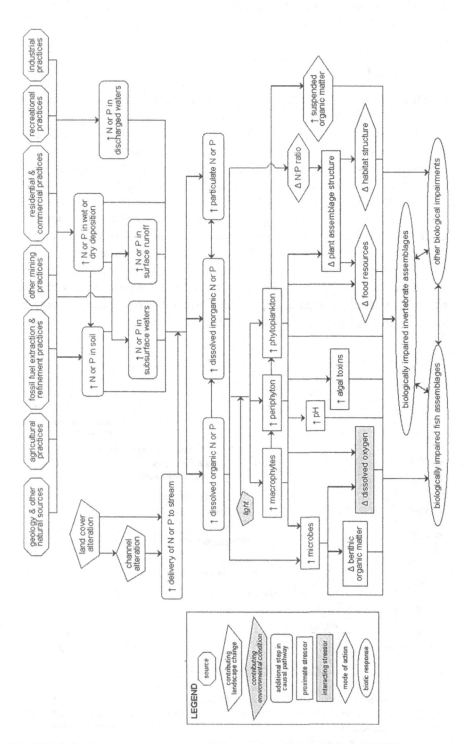

Fig. 17.5 A simple generic conceptual model for nutrients (Source, Schofield 2007.)

quantile regression, conditional probability analysis and classification and regression tree (CART) analysis.

17.2.2.3 Databases

CADDIS provides a series of databases that house information that users can modify and use in their analyses. A library of conceptual models includes both simple and complex generic models for each of the commonly encountered candidate causes described above (e.g., Fig. 17.5). It also contains conceptual models from case studies. Models are provided in downloadable form in Adobe Acrobat ® (.pdf) and Microsoft Power Point ® (.ppt) formats.

The Databases section also contains three databases that contain quantitative stressor-response information. Two of the databases synthesize laboratory test results for metals, yielding concentration-response curves and species sensitivity distributions for metals (U.S. Environmental Protection Agency 2005) (Fig. 17.6). A third database compiles stressor-response associations from regional data (e.g., Fig. 17.7). Practitioners can use these materials with their site data to develop evidence, in particular, stressor-response from laboratory studies, and stressor-response from regional data.

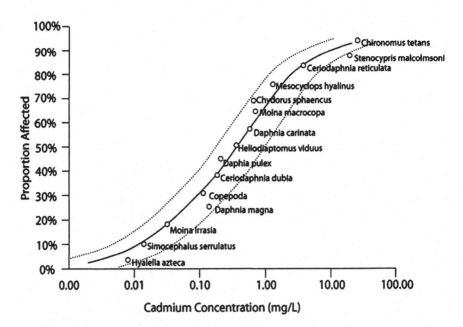

Fig. 17.6 The Species Sensitivity Distribution Gallery provides a collection of generic SSDs for metals. The example shown plots the proportion of species that have LC50 s less than a given concentration of cadmium. Data from U.S. EPA's ECOTOX database, following the methods described in U.S. EPA 2005

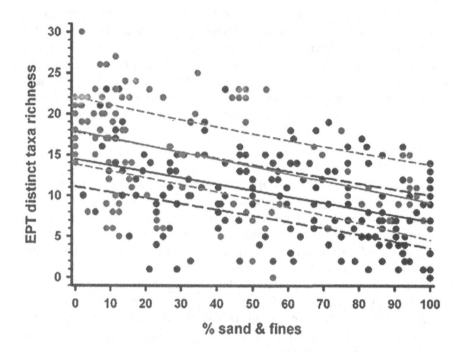

Fig. 17.7 An example linear regression plot from the CADDIS Stressor-Response Association Gallery, showing the taxa richness of Ephemeroptera, Plecoptera and Trichoptera (EPT) taxa richness vs. percent sands and fines. Data from Minnesota Pollution Control Association, plot generated by Michael Griffith, U.S.EPA. *symbols* = observations, *solid line* = mean line, *dashed line* = 95% prediction limit. *Black lines and symbols*: Ecoregions 46, 47 and 51, *gray lines and symbols*: Ecoregions 50 and 52

17.3 Structure

CADDIS is implemented primarily using an Adobe ColdFusion ® front end to an Oracle ® database. The downloadable tools are provided in a variety of formats, from Microsoft Excel ® spreadsheets to programs that can be used with the R statistical package (http://cran.r-project.org/index.html). CADStat was implemented using the Graphical User Interface for R (http://rosuda.org/JGR/). The interactive conceptual model is implemented in Macromedia Flash ®. The application includes a password-protected editing function that allows U.S. EPA to keep content up to date.

17.4 Case Studies

An essential part of the development strategy for CADDIS has been the use of case studies to test the process and tools in different regions, and with different causal factors. To date, case studies have been conducted to determine the causes of macroinvertebrate or fish community attributes in the states of Ohio, Connecticut, Maine, Iowa, Washington, Mississippi, Maine, West Virginia and Virginia. Case studies conducted early in the development cycle, such as the Little Scioto case study discussed above, led us to make the process more linear, and influenced the direction of tool and information development. We expect that testing the analytical tools and information included in the 2007 release of CADDIS will yield additional insights into making causal analyses more defensible and practicable.

17.5 Future Directions

Our overall objective for CADDIS is to provide an on-line destination for information, methods, and experiences relevant to conducting causal analysis in aquatic systems. Toward that end we intend to continue adding content that can be used to evaluate additional candidate causes, for example altered habitat. Analytical methods of interest include those that address multivariate and spatial issues. The methodology, information and tools useful for causal analysis also have applications in risk assessment and setting of benchmarks and criteria; we are working to take advantage of this nexus. Finally, we intend to make CADDIS a user-supported system, building a community of scientists that share information useful for causal and risk assessment.

17.5.1 Multiple Stressors

The stressor-specific information and statistical methods currently presented in CADDIS have emphasized the analysis of individual candidate causes to determine whether they are sufficient to induce the observed effects. However, interactions among causal agents should also be analyzed. Future editions of CADDIS will contain information, tools and guidance for such analyses. Examples include information on frequency and strength of co-occurrence of different stressors under different watershed land-use scenarios, and stressor-response relationships for combinations of stressors that produce response interactively, i.e., with greater- or less-than response or concentration addition.

17.5.2 CADDIS and Risk Assessment

Although CADDIS was developed to determine causes of impairments identified by biological monitoring programs, it can play two roles in risk assessments for contaminated sites. First, it can determine the cause of impairments identified during biological sampling for a site risk assessment. For example, fish community sampling for the Clinch River unit of the Oak Ridge, Tennessee,

Superfund site revealed that fish abundances were low in a contaminated embayment and the fish that were present displayed physical and physiological impairments (Environmental Sciences Division and Jacobs Engineering Group 1996). An ad hoc analysis of the evidence found that contaminants were likely causes of the impairments, but CADDIS would have eased the analysis and made the conclusions more convincing (Suter et al. 1999). CADDIS has been applied to a stream contaminated by a Superfund site (Cormier et al. 2002; Norton et al. 2002). It has also been applied to as yet unpublished assessments of contaminated sites in California, Colorado, Delaware and Tennessee. CADDIS's distinction between evidence of causal relationships at the site and evidence from elsewhere, its standard types of evidence and its scoring system are all directly applicable to risk assessments.

Second, the statistical modeling tools in CADDIS and the galleries of exposure-response relationships are useful for risk assessments. CADDIS is designed to support inferences from identified effects to uncertain causes, while risk assessment makes inferences from identified causes to uncertain effects, but the models are the same for both. If biological response data are available for the contaminated site and reference locations, the statistical tools such as regression can be used to develop site-specific exposure-response models for the contaminants of concern. Models of exposure-response relationships from regional monitoring data can be used to determine the credibility of potentially causal relationships at the site. In particular, regional monitoring data are commonly available for common stressors of aquatic communities such as sediment, temperature and dissolved oxygen that are alternatives to contaminants as causes of impairment at contaminated sites. These models can be used to evaluate whether the observations of candidate causes and effects at the site are consistent with regional patterns.

17.5.3 CADDIS and Criteria Setting

Some environmental stressors do not lend themselves to the traditional methods of deriving environmental quality criteria, because their effects are not readily tested in the laboratory. The U.S. EPA (U.S. Environmental Protection Agency 2006b) has developed an alternative approach for such stressors that is based on the analysis of multiple types of evidence concerning causal relationships that was inspired by CADDIS. That is, different types of field and laboratory data are analyzed using different statistical methods and the results are compared to arrive at a protective level for a particular location or region. The statistical methods for analysis of field data in CADDIS are useful for developing and evaluating alternative approaches to criteria development.

17.5.4 CADDIS as a Platform for Collaborative Information Sharing

The wide variety of information relevant to causal analysis and the need for region and ecosystem specific data argue for evolving CADDIS towards a collaborative platform. In this way the entire community of investigators can share results and advances. CADDIS currently provides a framework and functional context for collecting information on ecological causal relationships. Increasing the availability of such relationships would make CADDIS a more effective decision support tool. What if any individual or group could input— and felt compelled to do so—details about an ecological causal relationship into an online CADDIS platform, that information was assessed, by peers or otherwise, for accuracy to an appropriate degree, and then harnessed to enhance causal assessment efforts?

Recent advancements in technology and innovative online paradigms bring the potential achievement of this task into view. Information gathering mechanisms have recently gained momentum in today's information technology world. Various terms for these and related efforts include "peer production" or "commons-based peer production," "collective intelligence," "crowdsourcing" and "massively distributed collaboration." Platforms that currently employ such mechanisms include, for example: Wikipedia, (http://www.wikipedia.org), and the Encyclopedia of Life (http://www.eol.org), a project that aims to allow scientists from around world and from different sectors to develop a Web page that holds information specific to each species on Earth.

The CADDIS project team is at the fledgling stages of moving in the direction of a collaborative platform. We see potential in allowing collaborators to enter information through our conceptual model diagrams, given the intuitive nature of these cause and effect illustrations, improvements in graphical user interface design tools (for example, Macromedia's Flash ®, and Java ™ technology), and their online aesthetic quality. Whether or not a diagrammatic approach is taken, issues related to informatics offer challenges when, for example, users wish to enter different types of stressor-response relationships, with varying levels of accuracy, different naming conventions, varying sample size, and dataset inconsistencies. As such, flexibility of the user-interface and underlying database will be a critical component of this endeavor.

Open participation often spurs concern about accuracy of information. Research scientists have become accustomed to and comfortable with classic peer-reviewed journal articles, whereby articles are submitted for publication and academically reviewed by an expert panel of, say, three peers, for content and accuracy. Models for the scrutiny of internet-based information submissions, whereby anyone can contribute, are evolving; Wikipedia and the Encyclopedia of Life both address such concerns on their Web sites. The CADDIS team suspects that the potential benefits of a collaborative platform— for example, an exponential increase in available causal relationship information—far outweigh potential challenges associated with the introduction of innovative review processes. A CADDIS-based mass collaboration platform may in fact outperform more traditional academic peer reviewed mechanisms for amassing causal knowledge in terms of speed, usability, accuracy, and cost.

17.6 Conclusions

Investigating causes of adverse effects observed in aquatic systems poses many challenges. Investigators must have a deep understanding of the many ways that physical, chemical and biological stressors impact aquatic systems. They must be able to organize many types of evidence into a coherent whole. In many investigations, the initial information that is available is quite sparse and investigators must be able to decide whether the information is sufficient to support a decision, and describe the value of additional data collection.

CADDIS is designed to make the formal process of causal analysis more accessible and feasible. A formal process forces assessors to confront unexpected and counterintuitive findings. Our experiences have shown a high frequency of surprising results. For example, in one case study, upstream sites that were expected to serve as reference locations were just as biological degraded as downstream sites. In the Willimantic River, the source was intermittent and spatially disjunct from the effects that were initially observed, making detection difficult.

The identification of the stressors responsible for biological degradation is only one step in a management approach that begins with the detection of a biological impairment and ends with an effective management action that restores desired condition. By making causal analyses more defensible and transparent, we hope to increase the application and utility of biological assessment methods, improve the scientific basis for sound management action and contribute to improving the condition of the world's waters.

References

Cormier SM, Norton SB, Suter GW II, Altfater D, Counts B (2002) Determining probable causes of ecological impairment in the Little Scioto River, Ohio, USA: Part 2. Characterization of causes. Environ Toxicol Chem 21:1125–1137

Davis WS (1995) Biological Assessment and Criteria: Building on the Past. In: Davis WS, Simon TP (eds). Biological Assessment and Criteria. Lewis Publishers, Boca Raton, pp 7–14

Dawes RM (2001) Everyday Irrationality. Westfield Press, Boulder, CO, USA

Environmental Sciences Division, Jacobs Engineering Group (1996) Remedial Investigation/ Feasibility Study of the Clinch River/Poplar Creek Operable Unit.ORNL/ER-315/ V1&D3. U.S. Department of Energy, Oak Ridge, TN

Fox GA (1991) Practical Causal Inference for Ecoepidemiologists. J Toxicol Environ Health 33(4):359–373

Hill AB (1965) The Environment and Disease: Association or Causation? Proc Roy Soc Med 58:295–300

Kahneman D, Slovic P, Tversky A (1982) Judgment under uncertainty: Hueristics and biases. Cambridge University Press, New York

Marchant R (1997) Classification and prediction of macroinvertebrate assemblages from running water in Victoria, Australia. J North Am Benthol Soc 16:644–681

Metcalfe-Smith J (1994) Biological water-quality assessment of rivers: Use of macroinvertebrate communities. In: Calow P, Petts G (eds). The Rivers Handbook, Hydrological and Ecological Principles. Blackwell Science, Cambridge, pp 144–170

Nisbett R, Ross L (1980) Human Inference: Strategies and Shortcomings of Social Judgment. Prentice-Hall, Inc., Englewood Cliffs NJ, USA

Norton SB, Cormier SM, Suter GW II, Subramanian B, Lin E, Altfater D et al. (2002) Determining probable causes of ecological impairment in the Little Scioto River, Ohio, USA: Part I. Listing candidate causes and analyzing evidence. Environ Toxicol Chem 21:1112–1124

Norton SB, Rao L, Suter GW (2003) Minimizing cognitive errors in site-specific causal assessments. Human Ecol Risk Assess 9:213–229

Ohio Environmental Protection Agency (1987) Biological Criteria for the Protection of Aquatic Life: Volume II: Users Manual for Biological Assessment of Ohio Surface Waters.WQMA-SWS-6. Division of Water Quality Planning and Assessment, Ecological Assessment Section, Columbus, OH

Plafkin JL, Barbour MT, Porter KD (1989) Rapid Bioassessment Protocols for Use in Rivers and Streams: Benthic Macroinvertebrates and Fish.EPA-440-4-89-001. Office of Water Regulations and Standards, Washington DC, USA

Russo JE, Schoemaker PJH (1989) Decision Traps: The Ten Barriers to Brilliant Decision-Making and How to Overcome Them. Simon and Schuster, New York

Schofield K (2007) Simple Conceptual Model for Nutrients. In: U.S. Environmental Protection Agency. Causal Analysis/Diagnosis Decision Information System (CADDIS) http://www.epa.gov/caddis. Accessed October 9, 2008.

Stark TB, Maxted JR (2007) A biotic index for New Zealand's soft-bottomed streams. NZ J Mar Freshwater Res 41:43–61

Susser M (1986) Rules of Inference in Epidemiology. Regul Toxicol Pharmacol 6:116–128

Suter GW II, Barnthouse LW, Efroymson RE, Jager H (1999) Ecological risk assessment of a large river-reservoir: 2. Fish community. Environ Toxicol Chem 18(4):589–598

U.S. Environmental Protection Agency (2002) Consolidated Assessment Programs and Bio-criteria Development for States, Tribes, Territories, and Interstate Commissions: Streams and Wadeable Rivers.EPA-822-R-02-048. Office of Water, Washington DC, USA

U.S. Environmental Protection Agency (2005) Methods/Indicators for Determining when Metals are the Cause of Biological Impairments of Rivers and Streams: Species Sensitivity Distributions and Chronic Exposure-Response Relationships from Laboratory Data. NCEA-1494. Office of Research and Development, National Center for Environmental Assessment, Cincinnati, OH

U.S. Environmental Protection Agency (2006a) Estimation and application of macroinvertebrate tolerance values. EPA/600/P-04/116A. U.S. Environmental Protection Agency, Office of Research and Development, Washington, DC

U.S. Environmental Protection Agency (2006b) Framework for developing suspended and bedded sediments water quality criteria. EPA-822-R-06-001. Washington, DC

Yuan LL (2007) Using biological assemblage composition to infer the values of covarying environmental factors. Freshw Biol 52:1159–1175

Chapter 18
BASINS: Better Assessment Science Integrating Point and Nonpoint Sources

Russell S. Kinerson, John L. Kittle, and Paul B. Duda

Abstract The U.S. Environmental Protection Agency's (EPA's) Better Assessment Science Integrating point and Nonpoint Sources (BASINS) is a Decision Support System for multipurpose environmental analysis by regional, state, and local agencies performing watershed and water quality-based studies. It was developed by the EPA's Office of Water to meet the needs of the Total Maximum Daily Load (TMDL) process as specified under Section 303(d) of the Clean Water Act of 1977. BASINS integrates environmental data, analytical tools, and modeling programs to support development of cost-effective approaches to watershed management and environmental protection, making it possible to quickly assess large amounts of data in a format that is easy to use and understand. The BASINS system is configured to support environmental and ecological studies in a watershed context, with flexibility to support analysis at a variety of scales using tools that range from simple to sophisticated. BASINS' analytical utility has resulted in a framework for examining management alternatives far beyond the original objectives.

All versions of BASINS contain a suite of Geographic Information System (GIS) based tools and operate in a GIS environment, using the graphical user interface as the front end. The current release of BASINS, version 4.0, is the first to be primarily based on a non-proprietary, open-source GIS foundation. As a result, the core of BASINS is now independent of any proprietary GIS platform. BASINS users are no longer limited by expensive proprietary GIS software, and BASINS has greater stability and transparency, as the source code for all components is available to developers and end users alike.

BASINS is designed around an extensible architecture that allows the addition of new data types and new tools. This flexibility enables BASINS to continue evolving to meet the changing needs of the watershed management community.

R.S. Kinerson (✉)
Retired U. S. Environmental Protection Agency, 6527 El Reno Lane, Joplin,
MO 64804, USA
e-mail: rskinerson@gotsky.com

A. Marcomini et al. (eds.), *Decision Support Systems for Risk-Based Management of Contaminated Sites*, DOI 10.1007/978-0-387-09722-0_18,
© Springer Science+Business Media LLC 2009

18.1 Introduction

The U.S. Environmental Protection Agency's (EPA's) Office of Water developed BASINS as a multipurpose environmental analysis system (US EPA, 2007). As a multipurpose system, BASINS (www.epa.gov/waterscience/basins/) was designed to support watershed and water quality-based studies by facilitating examination of environmental information, by supporting analysis of environmental systems, and by providing a means to examine the consequences of management alternatives. BASINS encompasses a suite of watershed models, from sophisticated broad-spectrum watershed models to agricultural models to planning and management level models, plus supporting tools and data, all within one package.

State and local agencies have found that water quality standards cannot be met merely by controlling the point source discharges into that waterbody. Therefore agencies must use a watershed-based approach to meet water quality standards. BASINS is configured to support environmental studies by including information and tools applicable to the entire watershed. The system is designed to be flexible by including a wide range of tools so that it can support analyses for study areas of widely varying size and composition. The user has the flexibility to choose the model and tools best suited for the requirements of the study, for example from a screening-level tool to a full continuous simulation watershed model.

A major driving forces for watershed-based analyses is the legal requirement of Section 303(d) of the U.S. Clean Water Act of 1977, which requires states to develop Total Maximum Daily Loads (TMDLs) for waterbodies that do not meet water quality standards. TMDLs are developed by assessing both point and nonpoint sources of pollutants in a waterbody to determine how much of a given pollutant may be assimilated without violating water quality standards. BASINS was originally designed to support the TMDL process, by analyzing the joint impacts of point and nonpoint sources in a watershed. This approach provides the technical basis for a manager to determine, for an impaired waterbody, how much loading of a pollutant may be allowed and to allocate the load among sources. Thus the system allows users to explore and research different techniques for reducing the impacts of those pollutants, while assessing alternative management scenarios.

The capabilities that make BASINS useful for TMDLs, especially differentiating and quantifying the impacts of point and nonpoint sources, also make it useful for other types of decisions. BASINS is commonly used for purposes as diverse as understanding the impacts of land and reservoir management operations, performing future conditions analyses for both hydrology and water quality, aiding in the process of prioritizing watersheds for implementation strategies, and recommending management practices that will best enhance water quality. All of these decisions are guided through quantifying the loading contributions from different sources and tracking these constituents of concern throughout the watershed and associated waterbodies.

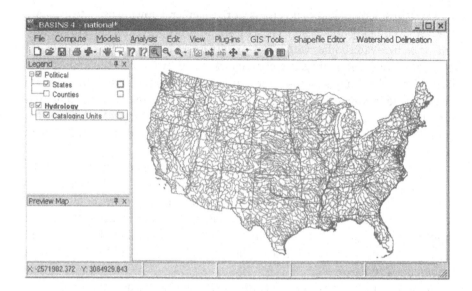

Fig. 18.1 BASINS geographic information system interface

The main interface to BASINS is provided through a Geographic Information System (GIS) (Fig. 18.1). GIS provides tools to display and analyze spatial information. Because GIS combines mapping tools with a database management system, it provides the integrated framework necessary to bring modeling tools together with environmental spatial and tabular data. Through this GIS foundation, BASINS has the flexibility to display and analyze diverse data at a user-chosen scale. That scale can range from one or more of the standard hydrologic cataloging units designated by the U.S. Geological Survey down to a site of only a few acres. BASINS includes tools that operate on large or small watersheds, and thus BASINS is flexible in its support for a broad user community. Adding locally developed, high-resolution data sources to existing data layers is an additional option that expands the local-scale evaluation capabilities.

BASINS brings together a suite of interrelated components for performing a complete watershed analysis, from data compilation and source assessment, through model construction and alternatives analysis. The components include:

- national databases
- utilities to organize and evaluate data
- watershed delineation tools
- assessment tools for watershed characterization based on observed data
- a simple GIS-based model that estimates nonpoint loads on an annual average basis
- a GIS-based hydrologic modeling system geared toward the arid southwestern United States
- and two watershed loading and transport models

The assessment and modeling tools work together, allowing users to evaluate study areas quickly and easily. The assessment tools provide means to identify and prioritize waterbodies with water-quality issues based on observed data. As point and nonpoint sources are characterized and evaluated, the appropriate level of modeling may be considered. Once a model has been used to simulate loadings and in-stream processes, potential control strategies can be compared for effectiveness. At each step of the process the tools within BASINS provide graphics and tabular results useful for communicating and explaining results and recommendations to stakeholders.

The latest release of BASINS is Version 4.0. This version provides significant enhancements and functions beyond those provided by the earlier releases of BASINS in an open-source environment. The continuing modification and enhancement of the system reflect the extensive comments and input provided by the user community. Other enhancements have included adding additional types of data, higher-resolution data, additional models and analysis tools.

Version 4.0 of BASINS is the first to be primarily based on a non-proprietary, open-source GIS foundation. A careful analysis of BASINS' needs revealed a relatively small number of critical core GIS functions, all of which could be provided through publicly available algorithms and source code. By using open-source GIS tools and non-proprietary data formats, the core of BASINS becomes independent of any proprietary GIS platform while still accommodating users of several different GIS software platforms. The underlying software architecture provides a clear separation between interface components, general GIS functions, and GIS platform-specific functions. Separating these components and functions provides a future migration path for using core GIS functions from other GIS packages or for accommodating future updates to the already-supported GIS packages.

18.2 History and Context

Through the early 1990s researchers at institutions around the world recognized the potential for linking watershed models with GIS systems and databases. Concurrently, awareness of water quality issues and needs was growing as the Clean Water Act was enforced through TMDLs. BASINS was conceived during that time as a system that could combine models and data through GIS to ease the burden of developing TMDLs.

BASINS was developed to be a fully comprehensive watershed management tool, assisting the user through all of the steps typically involved in a watershed assessment. The system greatly reduces effort needed to prepare data, summarize information, and develop maps and tables. To that end, the GIS interface provides a conduit through which data can pass, setting up models of varying scopes. The system not only helps the user apply the model, but it helps interpret the model output as well. Instead of performing each step of a watershed assessment using a series of independent and informally connected computer programs, BASINS coordinates and integrates those steps, resulting in improved efficiency and greatly reduced effort. As the analysis time is reduced, costs are reduced and the user is freed to answer a greater variety of questions in more detail.

BASINS has always been a dynamic system, with increased capabilities added as technology and needs demand. The first release of BASINS was in May of 1996, and the current version (4.0) was released in March 2007. With each new version additional data and tools were added along with new modeling tools expanding the spectrum of models available from simple to sophisticated.

One of the most significant changes to the BASINS system throughout its history was a shift in data distribution. Older version of BASINS had been distributed as a set of CDs containing data for each EPA region. With unlimited data available on the World Wide Web, static data on CDs was no longer adequate. As datasets are maintained and updated by various agencies, BASINS takes advantage of the power of Internet connections to provide the users with current data.

Prior to version 4.0, BASINS was dependent upon proprietary software from Environmental Systems Research Institute (ESRI), as most of the interface was built using the scripting components of the ArcView 3.x desktop GIS. In order to make BASINS system components most reuseable in later releases, components were gradually migrated away from use of proprietary software tools. The core GIS functionality was separated from the rest of the BASINS system components, which led to a smoother evolution to another GIS platform as well as managing changes in ESRI's software.

BASINS 4.0 is based upon an open source GIS package known as MapWindow. Moving to a non-proprietary GIS platform makes BASINS available to many new BASINS users who previously could not use this federally funded tool because of inability to purchase expensive proprietary GIS software. Additionally, the use of open source software provides BASINS with greater stability and transparency because the source code for all components—including the foundational GIS software—will always be available to end users and the federal government.

18.3 Design Considerations

The design of BASINS has evolved in concert with the capabilities of desktop GIS and with identified user needs. Consistent throughout the history of the system has been the design decision to develop interfaces to the more significant modeling tools separate from the core GIS system. For example the Hydrological Simulation Program - Fortran (HSPF) (Bicknell et al., 2005) is integrated through programming code that builds input files and then invokes the model. The model runs in the native language of development. In the case of HSPF, the original FORTRAN code base is maintained. The sophisticated watershed models are fully integrated, and yet they remain separate from the core system thus ensuring that the models are not inadvertently altered while facilitating maintenance and development of the system.

One of the most significant design achievements of the BASINS system is the extension architecture that was engineered for version 3.0. Prior versions of BASINS had all customized components of the GIS interface combined into one project file. A number of serious consequences arose from that design decision. The project file was quite large, and it was slow to load. Perhaps more importantly, the original design required extensive coordination among BASINS developers, and it restricted the ability to provide updates to existing BASINS projects. Starting with version 3.0, all customized components of BASINS were developed as independent extensions, loaded through an extension manager. One BASINS tool could be developed independently of another BASINS tool, greatly increasing the potential for independent groups to develop compatible BASINS extensions simultaneously. Another important implication is that users then had the capability to load only a subset of the BASINS extensions, so they can load only those needed for their BASINS project. This extension architecture also allows the BASINS system to operate at several levels of hardware and software sophistication.

Another major benefit of the BASINS extension architecture is that this design allows other groups not directly affiliated with the BASINS development team to develop tools for the BASINS system. An example of a model extension added to BASINS through the benefits of the extension architecture is the AQUATOX model (US EPA, 2004b). This model is distributed independently of BASINS, yet, if a user has both BASINS and AQUATOX installed, the user can proceed from the BASINS GIS directly into AQUATOX.

The current development approach within BASINS is a component-based architecture. BASINS system components are designed to be reusable and independent of GIS platform. BASINS GIS functions are separate from the rest of the BASINS system components, allowing for possible migration between different foundational GIS platforms.

Throughout recent BASINS development efforts, a design goal has been to use the GIS platform only when performing GIS functions, not as an environment for all BASINS functions. Following this design decision, utility tools and model interfaces have been created to be independent of the GIS platform. While these components are invoked seamlessly, the component code is not dependent upon the GIS environment. This design decision facilitates implementation of BASINS in any GIS environment, and has been particularly advantageous in the recent migration to an open source GIS platform.

Beginning in 2004, BASINS development efforts focused on a new version of BASINS, known as BASINS 4.0. The major design consideration governing the development of BASINS 4.0 was the issue of the changing underlying GIS platform. The desktop GIS from ESRI was moving from ArcView 3.x to ArcGIS, and the BASINS development team recognized that as GIS users made this move BASINS would have to somehow accommodate users of both GIS platforms. The issue is particularly complicated considering that each current BASINS user would be making decisions regarding that switch on their own schedule, as organizational budgets allowed.

The BASINS 4.0 development presented the challenge of building a system that would accommodate both ArcView 3.x and ArcGIS as GIS analysis tools. Early BASINS 4.0 prototypes included what was called the System Application to handle transfer of BASINS projects between the ArcView 3.x platform and the ArcGIS platform. The system application, while including a mapping interface, did not use any proprietary mapping tools, so this application would not require any run-time licensing.

A key advantage to this approach was the removal of ArcView 3.x as a prerequisite to the use of BASINS, though allowing for its continued use according to the desires of end-users. Through the System Application, the BASINS system was to be available with very limited GIS functionality to a user without either ArcView 3 or ArcGIS. All of the functionality from the BASINS 3 ArcView interface would still be available, while components for ArcGIS were being developed and rolled out to the user community. The System Application would identify which (if any) GIS software products are available on the user's computer, and thus indicate the GIS-based functionality available to the user. In this way the design provided a migration path from the ArcView 3.x components to the ArcGIS components.

While the BASINS System Application was being designed, the BASINS development team created a list of all GIS related functionality needed in BASINS. This list consisted of the specific GIS functions that were needed for BASINS, including such functions as determining which polygon contains a given point, identifying which feature of a layer was selected, and overlaying one polygon layer with another. At the time this list was developed it was thought it would help the development team decide which basic GIS functionality should reside in the system application and which should be left to be done by the GIS foundation.

The final list of core required GIS functionality was more limited than originally expected. Some of the items on the list were fairly trivial, others could be written with very modest effort, and for others there were already established open-source solutions available. With this realization it followed then that BASINS could be developed completely independently of any proprietary GIS software, making use of open-source GIS tools and non-proprietary data formats. These observations together with an interest in providing BASINS users a fully functional tool with no third party software purchase requirements (except for Microsoft Windows) drove a decision to migrate BASINS 4.0 to a non-proprietary, open-source GIS foundation (Fig. 18.2).

A major benefit of being independent of any proprietary GIS software is that the BASINS system can now be available to any user without cost. No

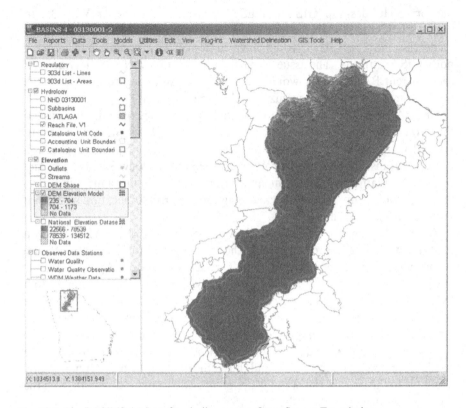

Fig. 18.2 The BASINS 4.0 Interface built upon an Open Source Foundation

prerequisite commercial software is required, so there are no financial hurdles
to impede use of the BASINS system by anyone who wants to use it. But
perhaps more importantly, the move away from proprietary GIS means that
all source code is open and freely available to the federal government and end
users. With the source code freely available, EPA now has the ability to maintain and/or upgrade core GIS functions as needs and budgets permit, not as
dictated by the commercial GIS market.

While not being dependent upon any proprietary GIS platform, the core of
BASINS 4.0 is designed to complement and interoperate with enterprise and
full-featured GIS systems. BASINS 4.0 can import and export projects from
ArcView 3.x and ArcGIS 9.0. This interoperability allows access to GIS features available in these systems that are not built into BASINS 4.0.

Learning from the challenges posed in migrating from one foundational
GIS to another, the BASINS development team was able to institute a strict
architectural standard for BASINS 4.0. Through this standard, general GIS
functions are separated from GIS platform specific functions. The component-
based architecture requires the programmer to use an intermediate generic class
for GIS functions, which are then implemented through a specific GIS

platform. For instance every time the programmer intends to overlay one GIS layer with another, all BASINS code uses one specific method in a class. The specific method then accesses the GIS foundational algorithm to do that overlay task. The major implication of this design is that in the future any change in the foundational GIS will have to be implemented in only one place in the BASINS source code, drastically simplifying maintenance and minimizing the cost of future enhancements. Following this design standard, a future migration path is provided for using core GIS functions from other GIS packages or for accommodating future updates to the already-supported GIS packages.

18.4 Using a Lightweight, Open-Source GIS Foundation

With the realization that BASINS could be written to be completely independent of proprietary GIS software, the BASINS development team examined existing open-source GIS tools. MapWindow GIS (www.MapWindow.org) was identified as a product that met the criteria of being a lightweight open source GIS with the necessary BASINS functionality already built-in, or as enhancements planned shortly thereafter.

MapWindow provides BASINS with a fully functional GIS foundation, including a complete GIS application programming interface (API) for both vector (shapefile) and raster (grid) data. MapWindow is a component based GIS platform that includes a core standalone library of GIS functions and an end-user graphical user interface with a plug-in architecture. As an open source end user GIS tool, MapWindow builds upon and takes advantage of several underlying GIS data and geoprocessing libraries including GDAL, GPC, PROJ4 and others, allowing it support both raster and vector data manipulation in most common file formats. MapWindow includes standard GIS data visualization features (zoom, pan, layer management, etc.) as well as DBF attribute table editing, shapefile editing, and grid importing and conversion. Also, because of its open source distribution, a worldwide development community is contributing to the already wide feature set contained in MapWindow.

By building on existing open source libraries, MapWindow supports over 3000 mapping projections, can be used internationally with multiple languages supported, and includes a scripting interface for running scripts written in VB.NET or C#. Its functionality has been extended to support GeoTiff as a grid file format, and it includes tools for clipping and merging raster and vector data. The platform has been adopted by several private companies, government agencies and universities as a GIS foundation for distributing data, models and research tools.

The extensibility of MapWindow is one reason why it was identified as an excellent candidate GIS foundation for BASINS. MapWindow can be extended with plug-in components written in any Microsoft .NET language. The plug-in interface operates much like the extension interface in ArcView, allowing third-party

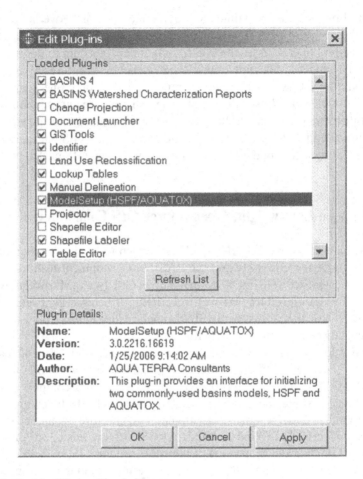

Fig. 18.3 The MapWindow Plug-in Manager

developers to create plug-ins that become fully integrated into the BASINS interface (Fig. 18.3). This means that third parties can write plug-ins to add additional functionality (models, special viewers, hot-link handlers, data editors, etc.) to BASINS and pass these tools along to other clients and cooperators. The MapWindow interface not only operates very similarly to the extension interface in BASINS, but each BASINS GIS component has now been converted into a parallel component for MapWindow.

Following the architectural design described in the previous section, the BASINS GIS components have been re-factored for MapWindow. The GIS-related functionality has been separated from the user-interface and data management functionality, providing for easier maintenance and upgrades in the future. While some advanced features of BASINS 4.0 may require proprietary products, the base BASINS 4.0 system does not require any run-time licensing or commercial software purchases.

18.5 BASINS Data

The BASINS system includes a tool, known as the BASINS Data Download tool, for downloading and extracting databases for watershed analysis and modeling. Some of the data downloaded using this tool have been preprocessed for use in BASINS. These prepared data are known collectively as the BASINS data holdings. Other data that can be downloaded using the Data Download tool have not been preprocessed and are extracted directly from the agency responsible for collecting the data.

18.5.1 The BASINS Data Holdings

These national databases, hosted on an EPA web server, were compiled from a wide range of federal sources and selected for inclusion in BASINS based on their relevance to environmental analysis. The data prepared for BASINS provide a starting point for watershed analysis, but users are encouraged to add additional data sets where locally derived data may be at a higher resolution or compiled more recently.

The BASINS databases are compiled into compressed files according to geographic location, according to the 8-digit Hydrologic Unit Codes (HUCs) established for the United States by the United States Geological Survey (USGS). A BASINS user begins a project by specifying one or more HUCs of interest, and data for those HUCs are downloaded and extracted for the project.

The types of data prepared and hosted for BASINS can be grouped into one of four classifications. They are base cartographic data, environmental background data, environmental monitoring data, and point source or loading data. Each type is described briefly below.

18.5.2 Base Cartographic Data

The base cartographic data in BASINS includes political and administrative boundaries (such as states and counties), hydrologic features and drainage boundaries, and major roads. These data are useful for base mapping to give the user a frame of reference for the rest of the data. Through these data the user can define and locate study areas and begin to further define watershed drainage areas.

18.5.3 Environmental Background Data

Environmental background data provide spatially distributed information to support watershed characterization and environmental analyses. They include information on soil characteristics, land uses, topography, and stream hydrography. The BASINS tools use this information in performing technical assessments of watershed conditions and loading characteristics. To cite a common example, the BASINS topographic data are often used to determine subwatershed drainage boundaries using the BASINS delineation tools. Those boundaries are then combined with the land use layers to determine how much of each land use type is in each subwatershed, which is critical information for watershed modeling.

18.5.4 Environmental Monitoring Data

Several existing national databases of environmental observations were adapted and converted into BASINS data sets. These databases were converted into spatial data layers so the user can see where these observation stations are located in a given watershed, which facilitates the assessment of water quality conditions. Examples of these types of databases include the water quality monitoring data summaries, the water quality observation data from the EPA STORET system, and USGS flow and water quality data from the National Water Information System (NWIS). The water quality datasets can also be used to prioritize and target water bodies and watersheds for remediation, as well as to assess the current status and historical trends.

One of the most recent enhancements to BASINS is the extension and expansion of the national database of meteorological data in BASINS. BASINS contains a national database of meteorological data that are essential to the successful application of BASINS assessment models. To be effective, these data need to be of high quality, have thorough spatial coverage, and be current. An updated version of the meteorological database has been compiled using data from NOAA's National Climatic Data Center (NCDC). This updated database brings all data up to the currently available time period and greatly expands the number of stations for which data are available. For instance, while BASINS previously provided access to roughly 500 hourly precipitation stations, that number has now expanded to over 2100.

18.5.5 Point Source/Loading Data

BASINS also includes data related to direct pollutant loading from point source discharges. The estimated loadings are provided along with the location and type of facility. These data were extracted from the EPA PCS database. The primary purpose of this loading data is to provide input for watershed models to represent the point source component, or point load allocation, of pollutants.

18.5.6 Dynamically Downloaded Data

Since version 3.1 the BASINS system has included a tool for dynamically downloading data from an additional set of sources. In addition to downloading the BASINS data from the EPA web server, the Data Download tool (Fig. 18.4) provides links to the federal agencies where certain data types are hosted, as well as tools to download the data and convert them into forms usable by BASINS. Since data available on the web are not static, this tool allows a user to check for more recent data and update the BASINS project data as appropriate.

When the Data Download tool is started, a window appears listing all of the available data types that the tool may add or update. The list of data types is determined at run-time, so this list may expand as new data-type components are created. The user chooses as many of the data types as desired, and the tool accesses the specified data through the World Wide Web and adds the data to the BASINS project.

Data types that are available for dynamic download include USGS flow data, the EPA Permit Compliance System (PCS) discharge data, the modernized EPA STORET system, the USGS water quality data, the National Hydrography Dataset, and the National Land Cover Database. Other data types will be added for dynamic download over time as resources allow.

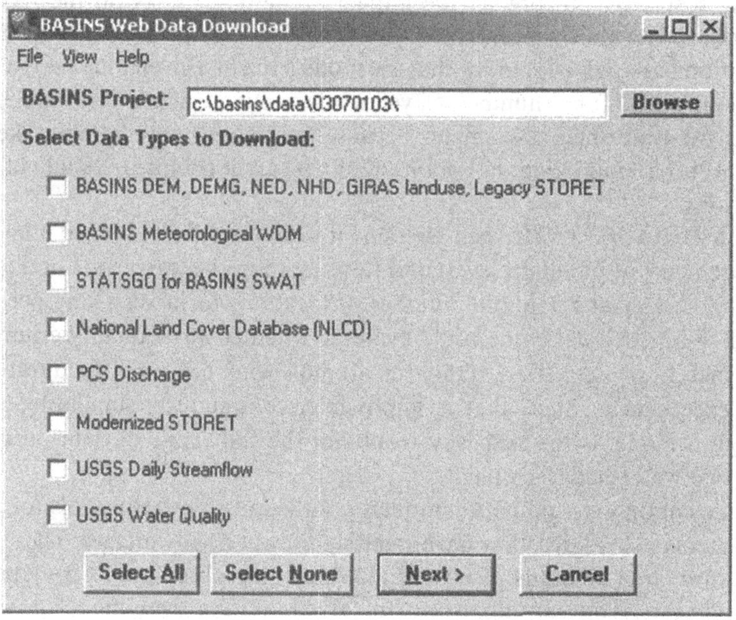

Fig. 18.4 The BASINS data download tool

A key feature of BASINS Web Data Download tool's architecture is the separation of the list of data types into individual components. For each data type available for downloading, there is a unique Dynamic Link Library (DLL). This design allows the list of data types to be populated at runtime, but it also greatly enhances the maintainability of the Web Data Download program. Very often the way the data are hosted on a web site changes over time. With this design, if a data type's web storage is changed, only the DLL for that data type will need to be updated and distributed, not the whole Web Data Download program.

This tool provides great flexibility in pulling data from a variety of sources. Instead of distributing all BASINS data through a specially compiled BASINS data holding, the data can be retrieved from the source of the data directly. This design makes the BASINS system easier and less expensive to maintain. In addition, updates to the data are available as soon as the agency producing the data makes the update available, making the most updated data available directly to the user.

18.6 BASINS Models

Watershed models predict loadings into surface waterbodies. Through the use of watershed models, one can simulate various point and nonpoint source loading scenarios and predict the impact of these loadings on the receiving waterbody. The most sophisticated watershed models operate on a continuous simulation basis, which is to say that the models run at a given time step (usually hourly or daily) for a number of years. Continuous simulation modeling is critical for watershed assessment, because continuous simulations take into account both point and nonpoint loadings at a complete range of flow conditions.

In an ideal world a watershed assessment would have an unlimited number of measurements of pollutant levels and flows both in the waterbody and exiting the land surface at an infinite number of locations for a very long period of record. With such data, one could evaluate whether water quality criteria are being met, how often those standards are not being met, and the duration of those exceedances. Since that is impossible, continuously simulating values through a model is the best way to obtain the full range of data needed to perform a watershed assessment.

While continuous simulation models are the most powerful tools for assessing watershed loadings, they have some significant disadvantages. These models require large amounts of input data, including observations over periods of many years. The learning process involved in using these models is significant, and uncertainty is inherent in input data, algorithms, and modeling assumptions.

BASINS reduces the disadvantages of using continuous simulation models by addressing each of these issues. BASINS provides a tremendous amount of input data so that the data gathering process is much less daunting. BASINS includes graphical user interfaces to the models to make the models easier to use, as well as analysis tools to help make model output easier to understand. BASINS also provides a suite of watershed models with a broad range of sophistication and complexity, so that the user can choose the model most appropriate for a given study or assessment.

Three models are integrated into BASINS to allow the user to simulate the loading of pollutants and nutrients from the land surface. These three models are spatially distributed, lumped parameter models, or in other words they may be used to analyze watersheds by subdividing the study area into homogenous parts. When deciding which model to use, one should consider factors such as the amount of data available, the processes to be modeled, the spatial and temporal resolution required, and the how the output results will be used. The integration of each model into BASINS is discussed individually below.

18.6.1 HSPF

The Hydrological Simulation Program Fortran (HSPF) (Bicknell et al., 2005) is a continuous simulation watershed model that simulates nonpoint source run-off and pollutant loadings for a watershed and performs flow and water quality routing in stream reaches. HSPF can be used to estimate nonpoint source loads from various land uses, as well as fate and transport processes in streams and lakes.

The Windows interface to HSPF, known as WinHSPF (Duda et al., 2001), was created for BASINS (Fig. 18.5). BASINS contains an extension that allows the user to open WinHSPF directly from the BASINS user interface, extracting appropriate information for the preparation of HSPF input files.

WinHSPF is designed to interact with the BASINS utilities and data sets, including the BASINS watershed delineation tools. HSPF requires land use data, reach data, meteorological data, and information on the pollutants to be modeled. The reach network is automatically developed based on the subwatershed delineations. Users can modify and enhance input files based on land use, meteorological data, and other locally derived data sources through WinHSPF. WinHSPF works with postprocessing tools to facilitate calibration as well as display and interpretation of output data. The HSPF User's Manual is available for reference as a Help file.

While HSPF is fully integrated into BASINS through the WinHSPF interface, the code base of HSPF is maintained separately. This separation is accomplished by compiling the HSPF model as a dynamic link library (DLL), called by WinHSPF for running a simulation. Maintaining HSPF as a separate DLL means that it can be enhanced independently of WinHSPF and BASINS.

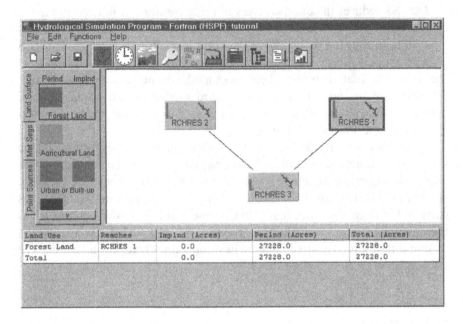

Fig. 18.5 The WinHSPF interface

A revised DLL can be copied into place on the user's computer and the user will have access to the latest HSPF features.

18.6.2 SWAT

The Soil and Water Assessment Tool (SWAT) model version 2000, (Arnold et al., 1998), is a physically based continuous simulation watershed model developed by the USDA Agriculture Research Service (ARS). With its agricultural roots, it is most often used to predict the impact of land management practices on water, sediment, and agricultural chemical yields. The model can be used on complex watersheds with varying soils, land use, and management conditions. The model combines these loadings with point source contributions and performs flow and water quality routing in stream reaches.

The SWAT Extension in version 3.1 is an ArcView extension. but a SWAT Plug-in for BASINS 4.0 is currently under development. The SWAT interface in BASINS is designed to set up SWAT input files using BASINS watershed delineations and data sets. BASINS data including land use, soils, reach data, meteorologic data, and pollutant characteristics can be used, or the user can provide custom data. SWAT input files can be modified through BASINS to facilitate the calibration of the model based on site-specific conditions and data sources. The BASINS SWAT interface works with postprocessing tools to facilitate display and interpretation of output data.

18.6.3 Pollutant Loading Estimator (PLOAD)

The BASINS Pollutant Loading Estimator (PLOAD) is a simplified GIS based model originally developed by CH2M HILL for calculating pollutant loads from watersheds. PLOAD estimates nonpoint source loads (NPS) of pollution on an annual average basis for any pollutant specified by the user. The NPS loads may be calculated using either the export coefficient or the EPA's Simple Method approach. Best management practices (BMPs) and point source inputs may also be included in computing total watershed loads. PLOAD results can be displayed as maps and tabular lists, and the model facilitates comparison of multiple scenarios.

PLOAD was designed to be simple so that it can be applied as a screening tool in typical watershed assessment or reservoir protection projects. As it operates on an average annual basis, it is not a continuous simulation model.

The PLOAD application requires spatial landuse data, subwatersheds, pollutant loading rate tables, impervious terrain factor tables, and optional spatial and tabular BMP and point source data. Landuse and point source data are provided with BASINS, and subwatersheds can be provided using the BASINS watershed delineation tools.

18.6.4 AGWA

The Automated Geospatial Watershed Assessment (AGWA) tool (Semmens et al., 2004), developed by the U.S. Agricultural Research Service (ARS), is a multipurpose hydrologic analysis system for performing studies ranging from watershed to basin scale. This tool was designed by ARS for use by watershed, water resource, land use, and biological resource managers and scientists. It provides the functionality to conduct a watershed assessment using SWAT and another model geared toward the arid southwest known as KINEROS2.

The BASINS AGWA extension was designed to interact with the BASINS utilities and data sets to provide the data needed by AGWA to parameterize either the KINEROS2 or SWAT model. AGWA was implemented as an Arc-View extension in BASINS 3.1, which facilitates the transfer of data from BASINS to the core models. As with the HSPF and SWAT implementations in BASINS, AGWA is kept separate from the ArcView extension for maintenance and enhancement.

The incorporation of the AGWA extension demonstrated the strengths of the flexible design of the underlying BASINS architecture. Recognizing the power and convenience of the large databases provided through BASINS, ARS decided to adapt AGWA to be a BASINS extension so that AGWA users would have convenient access to BASINS data. With very limited support from the BASINS development team, the AGWA developers were able to adapt AGWA to be fully incorporated into the BASINS system, making that convenient access possible.

18.7 Postprocessing and Analysis

For postprocessing and analysis of time-series data, BASINS includes the program *Generation of Scenarios GenScn* (Kittle et al., 1998) originally developed by the U.S. Geological Survey. GenScn is included in BASINS because of its excellent functionality for analyzing model simulation results including multiple model scenarios.

GenScn facilitates the display and interpretation of output data derived from model applications (Fig. 18.6). This tool allows users to select time periods and locations of interest and display results in graphical and tabular forms. GenScn handles a broad range of data formats, including HSPF simulation output, BASINS water quality observation data, USGS flow data, and SWAT output data. It also performs statistical functions and data comparisons. Due to its ability to display and compare observed and modeled data, this postprocessor is a useful tool in model calibration as well as environmental systems analysis.

GenScn is distributed with BASINS, and may be invoked through the BASINS GIS interface, as well as through the WinHSPF interface. While it is

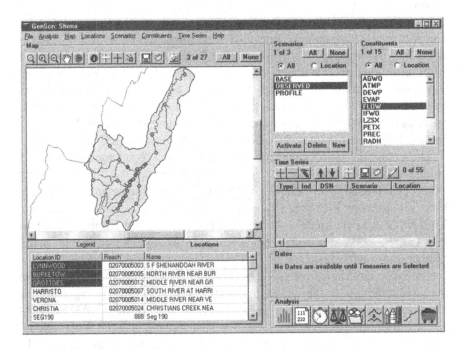

Fig. 18.6 The GenScn postprocessor

fully integrated into BASINS, it is kept as its own separate executable program for maintenance and enhancement.

The time-series analysis features of GenScn are being migrated to direct access through the BASINS 4.0 MapWindow interface. The BASINS 4.0 MapWindow interface includes capabilities to list and plot time-series data from a variety of data formats. New time-series analysis tools added for BASINS 4.0 include the ability to generate and analyze time-series on a seasonal basis, create subsets of timeseries by date, and perform mathematical computations on time-series data.

One of the more recently added analysis tools in BASINS is known as the BASINS Climate Assessment Tool (CAT). CAT provides a flexible set of capabilities for representing and exploring climate change and its relationship to watershed science. Tools have been integrated into the BASINS system allowing users to create climate change scenarios by modifying historical weather data, and use these data as the meteorological input to the Hydrological Simulation Program - FORTRAN (HSPF) watershed model. A capability is also provided to calculate specific hydrologic and water quality endpoints important to watershed management based on HSPF model output (e.g. the 100-year flood or 7Q10 low flow event). Finally, the CAT can be used to assess the outcomes of a single climate change scenario or to automate multiple HSPF runs to determine the sensitivity or general pattern of watershed response to different types and amounts of climate change.

Users can modify historical climate data using standard arithmetic operators applied monthly, seasonally or over any other increment of time. Increases or decreases in a climate variable (precipitation, air temperature) can be applied uniformly, or they can be selectively imposed on only those historical events that exceed (or fall below) a specified magnitude. This capability allows changes to be imposed only on events within user-defined size classes and can be used to represent the projected effects of 'intensification' of the hydrologic cycle, whereby larger precipitation events intensify, instead of events becoming more frequent. In addition, users are able to create time series that contain more frequent precipitation events. These capabilities provide users with an ability to represent and assess the impacts of a wide range of potential future climatic conditions and events. An example of the CAT window is shown in Fig. 18.7.

BASINS CAT provides a capability for quickly creating and running climate change scenarios within the BASINS system. Diverse sources of information such as records of historical and paleo-extreme events, observed trends, and projections based on global or regional scale climate models can be used to guide scenario development. Data requirements will vary depending on assessment goals. BASINS CAT provides capabilities to support a range of assessment goals, including simple screening analysis, systematic sensitivity analysis, detailed scenario analyses based on climate model projections.

Other advanced data analysis tools provided through BASINS include select functions of the USGS SWSTAT statistical software. The USGS Office of

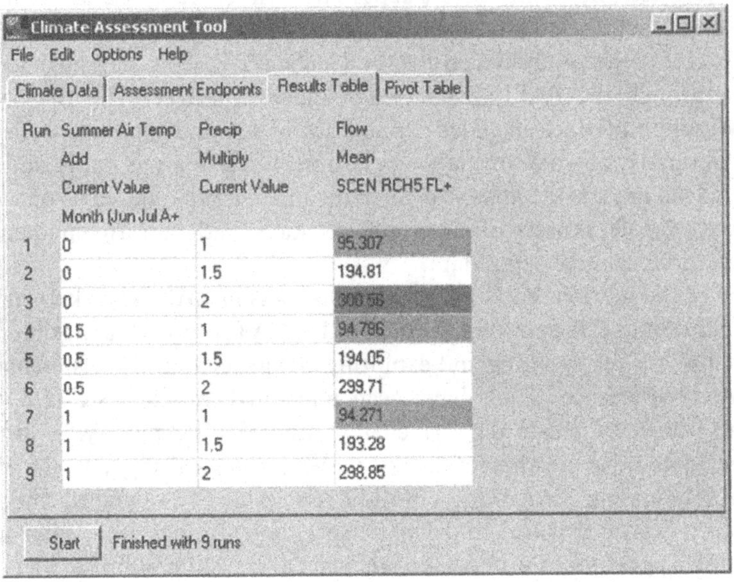

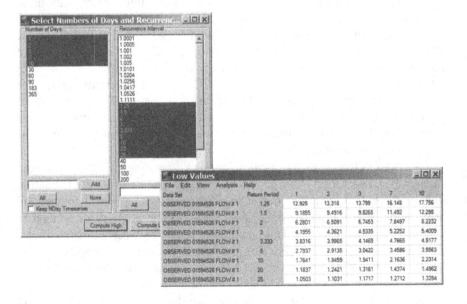

Fig. 18.7 The BASINS Climate Assessment Tool (CAT)

Surface Water's SWSTAT is a software package for statistically analyzing time-series data. A user interface to some functions of SWSTAT has been written as a BASINS plug-in (Fig. 18.8). The available functions include frequency distribution, trend analysis, and n-day annual time series.

Fig. 18.8 SWSTAT frequency analysis in BASINS

The BASINS system also includes tools designed to assist in preparing reports. A standard set of report scripts is included for inventorying and characterizing both point and nonpoint sources at the watershed and subwatershed scales. These watershed characterization reports can be used to evaluate the watershed condition, while providing the necessary information to assess monitoring programs, identify data gaps, and develop watershed-water quality modeling strategies.

The BASINS Watershed Characterization Reports operate within a user-defined area of interest. The reports are generated using .NET scripts, which the user can modify to provide customized reports. Output from the reports is displayed on the computer screen and is written to tab-delimited files. The scripting capabilities of BASINS allow users to generate reports based on any type of data, including GIS data and time-series data. Other types of reports commonly produced through BASINS include model results compiled as comparison statistics, watershed summaries, and constituent balances.

18.8 Users Support and Training

User support and training for BASINS is provided by EPA, often through a qualified contractor. Users receive responsive support related to all aspects of BASINS, including models, GIS, and BASINS utility programs. Web-based Frequently Asked Questions (FAQs) and BASINS Technical Notes are available on the BASINS web page (www.epa.gov/waterscience/basins/).

EPA sponsors BASINS workshops periodically, in various locations throughout the United States. The scopes of the workshops vary, with different workshops placing varying amounts of emphasis on BASINS components. Some of the workshops are general BASINS workshops, while many others focus on use of a particular model within BASINS. Hands-on experience with BASINS and its component models is provided during the extensive computer work sessions.

18.9 Case Studies

Many examples of the use of BASINS are available. The BASINS website (www.epa.gov/waterscience/basins/) provides a case study on the Cottonwood Creek Watershed in Idaho County, Idaho, illustrating use of BASINS as a decision support system for TMDL development.

At the request of the Idaho Department of Environmental Quality (IDEQ) and EPA's Region 10 Idaho Office, the EPA Office of Water Headquarters conducted this study to model the loading of fecal coliform bacteria to creeks in the Cottonwood watershed, and to evaluate the level and types of controls required to reduce bacteria loading to acceptable levels. Data from a variety of sources, both within BASINS and outside of BASINS, were used to inventory and quantify point and nonpoint sources in the watershed. The Cottonwood wastewater treatment plant is the sole point source in this largely agricultural watershed. A fraction of the septic systems in this largely rural population were believed to be failing. Livestock and wildlife were known to be numerous in the area as well.

The HSPF model was used to represent the Cottonwood watershed's hydrology and fecal coliform loads to the creek. The Cottonwood model hydrology was calibrated against measured Lower Cottonwood Creek flow gage data. A spreadsheet was used to calculate bacteria related HSPF input parameters, and then the model was calibrated against fecal coliform monitoring data. The model was re-run with a percent reduction to nonpoint source loads to creeks, in addition to cattle-in-stream and faulty septic system "point source" reductions, to determine nonpoint and point load reductions required to achieve the state water quality standard. Required nonpoint source load reductions ranged from 23% to 88%, when "cattle-in-stream" and faulty septic system loads were reduced by 80–100%.

Additional control scenarios were run to evaluate the level of impact from individual sources, and the study arrived at the following key conclusions:

- The Cottonwood wastewater treatment plant is not a significant source of fecal coliform loadings in the subject creeks;
- The cattle-in-streams (or other) point source in the subject creeks, in late Spring, is a significant source of fecal coliform loadings during periods of dry weather;
- Accumulation of fecal coliform on land surfaces, due to both grazing/pasturing of cattle and manure spreading from hog and dairy operations, appears to be a significant source of fecal coliform loading to creeks, particularly during wet weather events; and
- Faulty septic systems appear to be a significant contributor to exceedances of the fecal coliform criteria in the watershed.

An implementation plan was developed to reduce bacteria loading to acceptable levels and the State of Idaho was able to issue the draft TMDL (US EPA, 2000).

Other representative uses of BASINS include the following:

- HSPF Model of the Streamflow Simulation for the Lower Flint River Watershed (Wen, 2007). Agricultural irrigation in southwestern Georgia is one of the most important water uses in the region. In this study BASINS/ HSPF was used along with a groundwater model (MODFE) to understand the impacts of irrigation operations and the interactions between groundwater and surface water in the Lower Flint Watershed. The modeling results were used in the development of water management strategies.
- Calibration of a Watershed Model for Metropolitan Atlanta (Hummel et al., 2003). The Metropolitan North Georgia Water Planning District was

created to provide a mechanism for regional coordination on water supply, wastewater treatment, and stormwater management. The District's first mandate included the development of a Watershed Management Plan (WMP) to provide for compliance with water quality standards, while allowing for continued sustainable growth in the region. BASINS/HSPF was used to provide a consistent modeling approach throughout the district. Along with the data layers provided in BASINS, local data were collected and compiled for meteorology, land cover, hydrology, water quality, point source, and water withdrawals. The model provides the ability to perform future conditions analyses for hydrology and water quality, and it aides in the process of prioritizing watersheds for implementation strategies such that maximum benefit is achieved with limited financial resources.

- Modeling of Nonpoint Sources in Tickfaw River Watershed (Gala, 2006). The Tickfaw River watershed in the Lake Pontchartrain basin of southeastern Louisiana is especially challenged from rapid population growth, industrial activities, and agricultural use. Fish and wildlife propagation and primary contact recreation are not supported. There are many suspected sources of impairment, including agriculture, construction, forest management, and industrial sources. The specific objective of this study was to model the Tickfaw river watershed in order to quantify and differentiate the sources of pollution from agriculture, forestry, urban storm water runoff, and other sources, making use of BASINS. An assessment can then be performed to enhance the water quality within the watershed by recommending best management practices.

18.10 Conclusions

Several decisions along the BASINS development path have been critical in making BASINS a leading system for watershed analysis decision support.

The design of a watershed analysis system must support the addition of new data and new techniques for analyzing that data. BASINS, through its extensible component-based architecture, is a dynamic system whose capabilities have increased as technology has allowed and needs have arisen. Another implication of the extensible architecture is that each BASINS tool can be developed independently of each other BASINS tool, greatly increasing the potential for independent groups to develop compatible BASINS extensions simultaneously. This flexibility enables BASINS to continue evolving to meet the changing needs of the watershed management community.

Learning from the challenges posed in migrating from one foundational GIS to another, the BASINS development team was able to institute a strict architectural standard. This design drastically simplifies maintenance and minimizes the cost of future enhancements. Moreover, this design standard provides a future migration path for using core GIS functions from other GIS packages or for accommodating future updates to the already-supported GIS packages.

By using open-source GIS tools and non-proprietary data formats, the core of BASINS is now independent of any proprietary GIS platform. As a result BASINS users are not limited by having to purchase any prerequisite software, other than the computer operating system. Just as importantly, BASINS has greater stability and transparency, as the source code for all components is available to developers and end users. With the source code freely available, EPA now has the ability to maintain and/or upgrade core GIS functions as needs and budgets permit, not as dictated by the commercial GIS market.

References

Arnold, J.G., R. Srinivasan, R.S. Muttiah, and J.R. Williams. 1998. Large area hydrologic modeling and assessment part I: model development. Journal of American Water Resources Association 34(1):73–89.

Bicknell, B.R., J.C. Imhoff, J.L. Kittle Jr., T.H. Jobes, and A.S. Donigian Jr. 2005. Hydrological Simulation Program – Fortran (HSPF). User's Manual for Release 12.2. U.S. EPA National Exposure Research Laboratory, Athens, GA, in cooperation with U.S. Geological Survey, Water Resources Division, Reston, VA.

Duda, P.B., J.L. Kittle Jr., M.H. Gray, P.R. Hummel, and R.A. Dusenbury. 2001. WinHSPF – An Interactive Windows Interface to HSPF: User's Manual. U.S. EPA Office of Water, Washington, DC.

Gala, S., G. Cothren, and A. Hannoura. 2006. Modeling of Nonpoint Sources in the Tickfaw River Watershed. ASABE International Conference 2006, April 8–12, 2006. ASABE Publication Number 701P0406.

Hummel, P.R., J.L. Kittle, P.B. Duda, A. Patwardhan. 2003. Calibration of a Watershed Model for Metropolitan Atlanta. WEF TMDL 2003, November 16–19, 2003. Chicago, IL. WEF Specialty Conference Proceedings on CD-ROM.

Kittle, J.L. Jr., A.M. Lumb, P.R. Hummel, P.B. Duda, and M.H. Gray. 1998. A Tool for the Generation and Analysis of Model Simulation Scenarios for Watersheds (GenScn). Water-Resources Investigation Report 98-4134. U.S. Geological Survey, Reston, VA, 152p.

Semmens, D.J., S.N. Miller, M. Hernandez, I.S. Burns, W.P. Miller, D.C. Goodrich, and W.G. Kepner. 2004. Automated Geospatial Watershed Assessment (AGWA) – A GIS-Based Hydrologic Modeling Tool: Documentation and User Manual, Version 1.4. U.S. Department of Agriculture, Agricultural Research Service, ARS-1446.

US EPA, 2000. BASINS Case Study 1: Cottonwood Creek Watershed, Idaho County, ID. EPA-823-R-00-024. U.S. Environmental Protection Agency, Office of Water, Washington, DC.

US EPA, 2004b. AQUATOX Release 2 – Modeling Environmental Fate and Ecological Effects in Aquatic Ecosystems. EPA-823-C-04-001. U.S. Environmental Protection Agency, Office of Water, Washington, DC.

US EPA, 2007. Better Assessment Science Integrating point and Nonpoint Sources – BASINS Version 4.0. EPA-823-C-07-001. U.S. Environmental Protection Agency, Office of Water, Washington, DC. Available at: www.epa.gov/waterscience/basins/.

Wen, M., Y. Zhang, and W. Zeng. 2007. HSPF Model of the Streamflow Simulation for the Lower Flint River Watershed. 2007 Georgia Water Resources Conference, March 27–29, 2007. Athens, Georgia. Proceedings of the 2007 Georgia Water Resources Conference.

Chapter 19
A Decision Support System for the Management of the Sacca di Goro (Italy)

Chiara Mocenni, Marco Casini, Simone Paoletti, Gianmarco Giordani, Pierluigi Viaroli and José-Manuel Zaldívar Comenges

Abstract This chapter addresses some results concerning the development of a Decision Support System (DSS) for the management of Southern European lagoons, and is particularly focused on the application of the proposed DSS to the management of clam farming in the Sacca di Goro lagoon (Italy). After having established a reference framework for the study and management of coastal lagoons, a general model-based decision support structure is introduced. The development of this tool, obtained within the EU project DITTY, was motivated by the need for a common and flexible framework to ease the integration of the outputs of different models and analyses, as well as to deal with the diversity of socio-economic and environmental characteristics of several application sites. The proposed structure helps integrate and manage in a clear and structured fashion the information provided by different kinds of mathematical and analytical models of a lagoon ecosystem, such as biogeochemical, hydrodynamic, ecological and socio-economic models. Data and information obtained from the models can be used to accomplish the decision task by application of multicriteria analysis. Finally, robustness of the decision with respect to external factors beyond the control of the decision maker is considered in the proposed DSS.

19.1 Introduction

Over the last decades, coastal zones have become an extremely valuable, but scarce, economic resource. This increase in their value is mainly due to the enormous potential of coasts for residential, tourism, commerce, and recreational development. On the other hand, concepts like sustainable use of the natural resources and integrated coastal zone management have been often

C. Mocenni (✉)
Centro per lo Studio dei Sistemi Complessi, Università di Siena, Via Tommaso Pendola 37, 53100 Siena, Italy
e-mail: mocenni@dii.unisi.it

A. Marcomini et al. (eds.), *Decision Support Systems for Risk-Based Management of Contaminated Sites*, DOI 10.1007/978-0-387-09722-0_19,
© Springer Science+Business Media, LLC 2009

disregarded. Overcrowding, degradation of water quality, loss of natural habitat, resource exhaustion, conflicting use of resources, multiple and uncoordinated ecosystem modifications undertaken with only limited sectorial objectives in mind, are some of the current issues associated with coastal areas, and contribute to the decrease of the economic potential of these systems. For these reasons, the prevention of further damage and the introduction of sustainable development concepts are being recognized worldwide as fundamental issues in processes for coastal zone regional planning and management.

In Europe, both the European Commission and the individual governments have been investing a considerable amount of financial resources in research projects aiming at analysing and solving the problems related to coastal environments as well as at restoration measures. Since these systems are subject to various kinds of anthropogenic pressures, which are often sources of conflicts among the different users, it is extremely difficult to balance the economic interests with the safeguarding of the unique features of the ecosystems. In this respect, it is now widely recognised that integrated management is the key to the sustainable, equitable and efficient development of European water resources. This means that decisions need to be taken in the light of not only environmental considerations, but also their economic, social, and political impacts. It also requires the active participation of stakeholders in the decision making process. However, the real problem is how to translate these concepts in an efficient and practical approach to achieve these aims.

19.1.1 The IWRM Approach

The European Commission (2000) has adopted the concept of Integrated Water Resource Management (IWRM) as an integral part of the Water Framework Directive (WFD), which was issued in October 2000. Common key features of the IWRM approach and similar approaches that have been undertaken in North America, Australia, Africa and Asia, are the following (Letcher and Giupponi, 2005):

- Interactions between biophysical, ecological and socioeconomic drivers, processes and impacts should be considered through an integrated system approach.
- A balance should be achieved between economic and environmental objectives. This requires assessment of social and economic impacts from policies, and not only biophysical impacts.
- Environmental performance and improvement should be measured with respect to "good condition" targets.
- Public participation and collaboration should be key features of management and decision-making at the catchment level.

In other words, the approach recognises that the impact of management decisions is not restricted to the water resource itself, but inevitably affects a range of stakeholders with interests in the area. These impacts must be identified and evaluated under different scenarios. To make a balanced and fair judgement, a planner must be able to evaluate the effects of a decision based on a wide range of factors. Since many of the impacts may conflict, an integrated policy requires all the types of benefits and drawbacks to be taken into account and evaluated.

A second essential element of integrated management is that it must actively involve the community in any decisions that are to be made, being open to the scrutiny of those who will be affected. A failure achieving this, will ultimately lead to a poor implementation. Involvement does not mean consulting the community only after decisions have been made. Rather a representative cross-section of organisations and individuals should be actively involved in the decision making process itself. This gives the opportunity for people with different points of view to express their opinion. Conflicts may arise, but part of the integrated approach is to incorporate techniques of achieving consensus. In this way, conflict resolution is more likely to be achieved, the whole process becomes more transparent, and the final decision is more likely to be widely accepted due to a sense of ownership of the process and its results.

Although the principles are straightforward, successful implementation of IWRM is anything but straightforward. Linking together all the factors affected by a particular decision in a quantitative way is a complex process. Some factors may be environmental, others social or economic, each having different measurement units. Some factors may be qualitative in nature, with no recognised quantitative measurements. In addition, there are multiple feedbacks and nonlinear relationships between factors which may produce a cascade of direct and indirect effects which may be difficult to understand. Therefore, a water resource planner needs to consider the effects of a range of plausible actions on a large number of interconnected factors, and from this analysis implement strategies that are equitable, sustainable and efficient. This process needs to be carried out while, at the same time, ensuring that effective stakeholder participation in the process is achieved. The main problem is to find a series of techniques and tools that may help achieving all these goals.

From the above discussion, it is clear that many advances in research are required for the success of the WFD and similar policies (Letcher and Giupponi, 2005):

- Modeling needs to be undertaken in a more integrated way.
- Methods for evaluating economic and social impacts of policies need to be developed and implemented.
- Scenario-based approaches need to be developed to allow testing of potential policies and management changes before these are implemented.
- Improved participation methods need to be developed, and their use fully understood.

We believe that through integration of modeling approaches, system analysis and management tools like the DPSIR framework, decision support systems and multicriteria analysis, there already exists the potential for the development of integrated systems able to achieve successful results.

19.1.2 The DITTY Project

As a consequence of their location between land and open sea, coastal lagoons are characterized by large fluctuations in physical and chemical conditions. Their equilibrium is strongly influenced by the quality of inland waters flowing into them, rich in organic and mineral nutrients derived from urban, agricultural and/or industrial effluents and domestic sewages. Moreover, Southern European lagoons are subject to strong anthropogenic pressures from activities such as tourism and intensive shellfish/fish farming, as well as uncoordinated land-use and lagoon structural modifications. All these factors are responsible for important ecosystem alterations characterised by eutrophic or dystrophic conditions, including algal blooms, oxygen depletion and hydrogen sulphide production. Additional problems arise from cost erosion, subsidence and effects related to extreme meteorological events.

The objective of the European Union policy for the environment is to contribute to preservation, protection and improvement of the quality of the environment through a prudent and rational utilisation of the natural resources, based on the principle that preventive actions should be taken, where necessary, to ensure "good water status" (as described in the Water Framework Directive). Since there are different conditions and needs in Member States, which require specific solutions, this diversity should be taken into account in the planning and execution of measures for ensuring protection and sustainable use of water in the framework of coastal zones. Decisions should be taken as close as possible to the locations where water is affected or used. Priority should be given to actions within the responsibility of Member States by adjusting measures to regional and local conditions.

The EU funded DITTY project[1] (*Development of an Information Technology Tool for the Management of European Southern Lagoons under the influence of river-basin runoff*, contract EVK3-CT-2002-00084) aimed at developing the scientific and operational bases for a sustainable and rational utilisation of the available resources in Southern European lagoons by taking into account all the relevant impacts from agricultural, urban and economic activities affecting the aquatic environment, and by developing information technology tools tailored to these types of ecosystems. Reliability of the developed information tools was pursued by putting particular emphasis on model validation and benchmarking, as well as on a detailed socio-economic assessment of different management options through the close involvement of economists and stakeholders.

---

[1] Web site: http://www.dittyproject.org/

One of the outputs of the project was the prototype of a decision support system for the management of coastal lagoons. By integrating different kinds of mathematical and analytical models, this tool is able to simulate and analyze the effects on the ecosystem of multiple factors (e.g., lagoon fluid dynamics, river runoff influence, nutrients cycles, shellfish farming, and macro-algal blooms), as well as the economical implications of different management options. The tool has provided the local authorities with suitable techniques for assessing the effectiveness of actions designed to achieve a good environment quality status, as described, e.g., in the Water Framework Directive.

Implementations and tests of the decision support system were accomplished in five lagoon sites: Ria Formosa (Portugal), Mar Menor (Spain), Etang de Thau (France), Sacca di Goro (Italy), and Gera (Greece). The choice of these sites was mainly dictated by the accessibility of long-term data sets and by good working contacts with local and regional authorities ensuring the exchange of data and information.

The DITTY project built on, extended and integrated the efforts by the European Commission funded research initiatives to investigate the coastal zones and the interactions between land and ocean, being part of the ELOISE[2] (European Land-Ocean Interaction Studies) cluster of research studies.

The objectives of the DITTY project were multiple. Specifically, it intended to:

- Synthesize the knowledge about coastal lagoon ecosystems (also using prior investigations from ELOISE), and gather data in the form of long-term spatio-temporal time series that support this knowledge.
- Develop integrated modeling techniques for the watershed basin, the coastal lagoon, and the coastal zone, and design and validate benchmarking exercises.
- Develop the basis of common information technology tools: Database Management Systems, Geographic Information Systems, and Decision Support Systems.
- Develop scenario analysis for each site taking into account the end-user priorities.
- Use the above tools for:

  1. Assessing the influence of watershed basin runoff.
  2. Assessing the influence of shellfish farming on ecosystem equilibrium.
  3. Studying the spreading factors of bacteria of sanitary concern.
  4. Developing early warning systems for anoxic crises.
  5. Developing a monitoring strategy that may deal with the time scales involved (short term events, seasonal variability, and long term behavior).

- Assessing methodologies for detecting the impact of climate changes on coastal lagoon ecosystems through long-term simulations.

---

[2] Web site: http://www2.nilu.no/eloise/

- Applying global bio-indicators (e.g., exergy[3] and others) for addressing the issue of a rigorous definition of "good ecological status", which has been introduced in the Water Framework Directive.
- Apply the LOICZ (Land-Ocean Interaction in the Coastal Zone) standard methodology for assessing its efficiency in the case of coastal lagoons and contributing to the flux assessment of the EU Mediterranean basin area.

19.2 The DITTY DSS

The diversity of socio-economic and environmental characteristics of the case studies of the DITTY project required a tool capable of a common approach to different decision cases, and responsive in a range of different cultural, political and organisational contexts, but also flexible enough to adapt to the specific objectives and constraints of a particular decision problem. The efforts were thus primarily directed toward the development of a *general model-based DSS structure* into which all site-specific decision problems could be cast.

The DSS structure includes a mechanism for generating the alternatives to be compared. The core of the DSS is then represented by the mathematical and analytical models (e.g., biogeochemical, hydrodynamic, ecological, socio-economic models, geographic information systems, etc.) developed for each site during the course of the project. These are used to simulate the alternatives, and to provide corresponding system performance indicators related to the decision criteria. Multicriteria analysis is finally applied to evaluate and rank the alternatives on the basis of both the values of the indicators and the interaction with the decision maker. In particular, the Analytic Hierarchy Process (AHP) approach (Saaty, 1980) has been implemented. The AHP is a tool for decision making that is suitable in situations where multiple and conflicting criteria are present. However, other multicriteria analysis tools, such as reference point (Wierzbicki, 1998) or ELECTRE (Roy, 1991) methods, can be used.

It is apparent that the characteristics of the DITTY DSS make it a very flexible structure into which decision problems of different types and at different levels of complexity can be easily cast. Indeed, flexibility of the DSS structure was a key issue for the designers, since the project case studies pursued different objectives, and were carried out in different geographical and socio-economic contexts, so that only a flexible tool could be capable of facing these challenges.

19.2.1 Terminology and Logic

The first fundamental step when developing a successful DSS is the decision problem statement, together with a detailed understanding of the actual

---

[3] The thermodynamic definition of *exergy* is the amount of work that a system can perform by being brought into equilibrium with its environment.

decision making processes, the actors and the stakeholders (and thus the intended audience and users), their responsibilities, interests, legal and regulatory framework, as well as the institutional structure and the political and socioeconomic framework within which the decisions are made. This step includes understanding the current system status, and identifying the driving forces and pressures which could prevent the objectives from being achieved. The links between the pressures exerted by society and their environmental impact have to be understood. In practice this means identifying the origins of pressure (driving forces) represented by social, demographic and economic developments, modeling the effects caused by the pressures on the system state, evaluating the impacts, and finally developing responses aimed to prevent, compensate, or mitigate the negative outcomes of state changes.

It is assumed that the decision problem is structured and presented in terms of:

- The *control options*, i.e. the alternative actions, strategies, or policies that can be undertaken to affect the system[4] behavior.
- The *criteria* on which the system performance led by each control option is evaluated.
- The *objectives*, i.e. the type of optimisation to be performed on each criterion.
- The *constraints*, establishing bounds for the criteria in order to make the evaluated alternative acceptable or feasible.

The aim of the DITTY DSS is to support the choice of a control option that both meets the constraints and optimises the objectives. In this respect, the control option definition and design is of central importance. The control options are described by value assignments of the *controllable variables*. For instance, if local authorities are asked to grant new farming concessions for aquaculture, and have to decide the amount of such concessions, the allocated farming area is the controllable variable.

On the other hand, the *uncontrollable variables* describe external factors that are not subject to choice, but do affect the system performance. Their role in the DSS can only be viewed in terms of sensitivity and robustness of the final decision. Typical uncontrollable variables are the weather conditions for the biogeochemical models of the lagoons, and the prices and market data for the economic models.

The criteria are expressed by means of *indicators*, that are used to evaluate the system performance under alternative control options. Note that within the DITTY project, several environmental, social and economic indicators were developed and tested (e.g., Viaroli et al., 2004; 2003; Verdesca et al., 2006). The *objectives* correspond to indicators whose value has to be either maximized or

---

[4] We distinguish the *system*, i.e. the part of the real world (environment, people, activities, etc.) that is the object of the decision maker's interests and actions, and the *decision support system*, i.e. the tool for supporting the decisions.

minimized. The *constraints* impose a maximum and/or minimum value to the indicators. They may correspond to thresholds defined on the basis of regulations and/or experience, and allow elimination of unacceptable alternatives. Additional variables that do not correspond to criteria, but the decision maker might want to constrain, are referred to as *internal variables*. Note that also the controllable and uncontrollable variables could be bounded in order to reflect both physical and practical constraints.

With the above definitions, the basic DITTY DSS logic is simple. A set of alternative control options for the system are generated by changing the values of the controllable variables. Each control option leads to a corresponding system performance, which is expressed by indicators. Performances of the different evaluated control options are finally analysed and compared to arrive at the final preference ranking of the alternatives, and the eventual choice of a preferred alternative as the solution of the decision process.

Note that multiple and conflicting criteria typically require the direct or indirect introduction of *weights* defining the relative importance of different criteria in contributing to the objective function.

19.2.2 Structure of the Decision Support System

Once the decision problem has been identified, and structured in terms of control actions, criteria, objectives, and constraints, the main elements of a decision include the design of promising, feasible alternatives and the subsequent selection of a (possibly optimal) solution from the set of alternatives thus generated.

The main function of a DSS is therefore to design and generate alternatives, and to provide the tools for their selection, given the decision maker's criteria, objectives, and constraints. In model-based decision support systems, mathematical and analytical models provide the main functionality by making it possible to simulate the system behavior under different future conditions (Power, 1997).

Following the basic DSS logic described above, the scheme of the proposed model-based DSS is shown in Fig. 19.1, where the models play a key role between the control option generation and the performance analysis and comparison (multicriteria analysis). The different component blocks of the DSS will be described in detail in the following. It is stressed here that the proposed DSS may answer both "what-if" and "how-to" questions, since simulation models perform scenario analysis, while optimisation/satisfaction is addressed in the multicriteria analysis section. In addition, a feedback mechanism makes it possible to adapt the set of evaluated alternatives in order to improve the decision maker's satisfaction.

A preliminary version of the DSS was presented by Casini et al. (2005) who devoted a particular emphasis at developing an optimisation model for resource

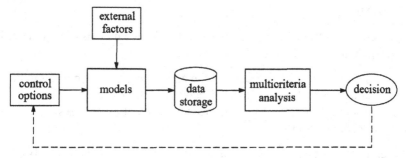

Fig. 19.1 Block scheme of the proposed model-based DSS

allocation in coastal lagoon areas characterized by shellfish farming and agriculture. A complete version of the DSS is described in (Agnetis et al., 2006) and (Casini et al., 2007).

19.2.2.1 Control Options

The block "Control options" provides the alternative (pre-existing or generated on demand) control options by assigning different values to the controllable variables. A discrete approach is assumed, where a finite (possibly very large) set of alternatives is considered. Assuming that p controllable variables are considered, and n different alternatives are generated, the p-dimensional vector V_i contains the values assigned to the controllable variables in the ith alternative, $i = 1, \ldots, n$.

The generation mechanism is not specified, since it may depend on the application, but in general a DSS makes it easy to design and generate alternatives. Note that the discrete approach does not guarantee optimality, so that the smaller the set to choose from, the less likely it will contain a good (in some sense optimal) solution. On the other hand, for highly complex systems it may be the only possible approach, which implies that one should always attempt to generate the largest possible number of alternatives.

19.2.2.2 External Factors

This block provides values for the uncontrollable variables describing the external factors that cannot be controlled or manipulated by the decision maker, but do affect the system performance (e.g., the weather conditions, the water inflow, etc., for the lagoon models; the prices and the market data for the economic models). Uncontrollable variables represent one type of uncertainty affecting the decision process. Inadequate values assigned to them could invalidate the results of the study. Hence, their role in the DSS can be viewed in terms of sensitivity and robustness of the final decision (see the subsequent remark in Section 19.2.2.3).

19.2.2.3 Models

This block represents a suitable interconnection of the models used to describe the system behavior. The use of models is twofold:

- Make simulations and predictions of, e.g., the physical, chemical and biological, as well as the economic and social variables of the system.
- To compute the performance indicators for a quantitative assessment of the evaluated control option.

When the ith control option is considered, $i = 1, \ldots, n$, the block "Models" receives as inputs both the controllable variables characterizing that particular control option, and the uncontrollable variables. The block output is an m-dimensional vector I_i (where m is the number of criteria) containing the values of the system performance indicators under the ith control option. Possible constraints imposed on the indicators, as well as on the internal variables, are checked during the simulation. If one or more constraints are violated, the considered alternative is discarded as infeasible.

The internal structure of the block "Models" is application specific, depending primarily on available models. Figure 19.2 shows a possible first-stage disaggregation of the block for a lagoon ecosystem and its watershed.

Remark. For a given control option, the values of the performance indicators are clearly affected by the uncontrollable variables. Hence, in order to perform a fair evaluation of different alternatives, the system performance must be compared under the same external conditions. In addition, in order to make the DSS more robust with respect to varying external conditions, it is strongly recommended to evaluate and compare the alternative options for several value assignments of the uncontrollable variables, and then to consider either an average or a worst-case ranking, as described in Section 19.3.3.2.

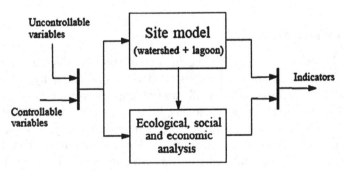

Fig. 19.2 Disaggregation of the block "Models" in Fig. 19.1 for a lagoon ecosystem and its watershed

19.2.2.4 Data Storage

For fixed external conditions, the values of the performance indicators corresponding to the n evaluated control options are stored in the $n \times m$ matrix

$$\mathbf{I} = [\mathbf{I}_1 \; \ldots \; \mathbf{I}_n]^T. \tag{19.1}$$

that is shown for clarity in Fig. 19.3.

19.2.2.5 Multicriteria Analysis

If only the jth criterion is considered ($j = 1, \ldots, m$), the best control option can be simply selected by taking the optimum over the jth column of the matrix $\mathbf{I}$ (see Fig. 19.3). However, when all the m criteria are considered, it is likely that the optimum within each column is not achieved by the same control option. In this case the selection of the best alternative might be neither direct nor intuitive. This justifies the need for multicriteria analysis tools (see the subsequent Section 19.2.3).

19.2.2.6 Decision

This block represents the actual decision made by the decision maker (single or group) based on the results of the multicriteria analysis.

A *feedback* mechanism allows the decision maker also to extend or adapt the set of control options according to his/her/their own preferences and the multicriteria analysis. Indeed, concrete solutions which have been formulated and analysed typically bring a deeper insight and understanding of what the problem actually is, and how it could be better solved. In addition, in some cases the alternative options are not readily available, and have to be discovered.

19.2.3 Multicriteria Analysis

The availability of numerous algorithms to solve multiple-criteria decision problems developed during recent decades (see, e.g., Figueira et al., 2005) characterises both the flexibility and ambiguity of the multicriteria analysis approach. Since its origin, multicriteria analysis has evolved into a number of divergent schools of thinking, each emphasising a different attitude regarding the way decision making can be supported. The methods differ in the type of information they request, the methodology used, the sensitivity tools they offer,

	Indicator #1	...	Indicator #m
Control option #1			
...			
Control option #n			

Fig. 19.3 Matrix I of the performance indicators

and their mathematical properties. Indeed, practical applications of the multi-criteria approach are hindered by the ambiguity of choosing one particular method among all those available. Each method may potentially lead to different rankings, and the choice of a method is subjective and dependent on the decision maker's predisposition. Although several attempts have been made to facilitate the selection of the "best" method, no agreed set of criteria allows the appropriate multicriteria analysis method to be selected in specific decision situations. Indeed, in any multicriteria analysis, the final step of ranking and selection of the alternatives is not a scientific, but a political exercise of negotiation and trade-off among necessarily subjective values and believes, which again makes the direct involvement of end-users mandatory to reach meaningful and practical results.

The multicriteria analysis tool used in the DSS application illustrated in Section 19.3, is the Analytic Hierarchy Process (AHP). The AHP is a tool for making decisions in situations where multiple and conflicting criteria are present, and both qualitative and quantitative aspects of a decision need to be considered (Saaty, 1980). The AHP provides a ranking of the evaluated alternatives by means of simple matrix calculations and pairwise comparisons. In the following, before reviewing the AHP approach, a general framework for multicriteria analysis is presented.

19.2.3.1 A General Framework for Multicriteria Analysis

In this section, a decision setting is considered where the choice among n alternatives $\mathscr{A}_1, \ldots, \mathscr{A}_n$ is to be made. The objective of multicriteria analysis is to select the alternative that achieves the best (or most satisfactory) trade-off performance with respect to m, typically contrasting, decision criteria $\mathscr{C}_1, \ldots, \mathscr{C}_m$.

To this aim, the direct or indirect introduction of criteria weights $w_1, \ldots, w_m$ is usually required. Weights define the different importance that the decision maker gives to the criteria in contributing to the overall evaluation. In the following,

$$w = [w_1 \ldots w_m]^T \tag{19.2}$$

will represent the vector of criteria weights.

It is assumed that the performance of the ith alternative $\mathscr{A}_i$ with respect to the jth criterion $\mathscr{C}_j$ is expressed by the attribute (or indicator) I_{ij}. Thus, to the decision purposes, the alternative A_i is completely described by the attribute vector

$$I_i = [I_{i1} \ldots I_{im}]^T \tag{19.3}$$

and each alternative can then be viewed as a point in the criteria hyperspace.

Since the attributes are typically incommensurable, i.e. they have different units that cannot be readily compared, the attribute vector I_i needs to be converted into a score vector s_i by application of a suitable scaling function f, i.e.

$$s_i = f(I_i) \tag{19.4}$$

where s_i is an m-dimensional vector, namely

$$s_i = [s_{i1} \dots s_{im}]^T \tag{19.5}$$

such that s_{ij} can be interpreted as the score assigned to the ith alternative with respect to the jth decision criterion. A global score v_i for the ith alternative can be now obtained as the weighted sum

$$v_i = w^T s_i = \sum_{j=1}^{m} w_j s_{ij}. \tag{19.6}$$

The mechanism for the computation of the global scores is illustrated in Fig. 19.4. Finally, the ranking of the global scores $v_1, \dots, v_n$ provides the multicriteria ranking of the alternatives.

Many multicriteria approaches can be cast into this framework. They only differ in the way the user is asked to provide the weights (e.g., directly or indirectly), and the scaling of the attributes is performed (the function f can be explicitly or implicitly defined). The AHP approach (Saaty, 1980), reviewed in the following, provides a method for weighting the criteria and scaling the attributes based on decision maker's pairwise comparisons.

Remark. It is worthwhile to stress that the criteria (or attribute) set is not given a priori. Its selection and definition is one of the most critical steps in the decision making process. It can be easily shown that adding or deleting criteria from consideration is a very powerful way to influence decisions. On the other hand, a primary concern is to make sure the criteria are relevant, i.e. they describe aspects of the decision problem that are indeed meaningful and relevant to all stakeholders and actors.

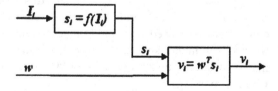

Fig. 19.4 Computation of global scores in a general multicriteria analysis framework

19.2.3.2 The Analytic Hierarchy Process

The *Analytic Hierarchy Process* (AHP), introduced by Thomas Saaty (1980), is a tool for dealing with complex decision making, that may aid the decision maker to set priorities and make the best decision. By reducing complex decisions to a series of pairwise comparisons, and then synthesizing the results, the AHP helps capture both subjective and objective aspects of a decision. In addition, the AHP incorporates a useful technique for checking the consistency of the decision maker's evaluations, thus reducing the bias in the decision making process.

The AHP considers a set of evaluation criteria, and a set of alternative options among which the best decision is to be made. It is important to recall that, since some of the criteria could be contrasting, it is not true in general that the best option is the one optimising each single criterion, rather the one achieving the most suitable trade-off among the different criteria.

The AHP generates a weight for each evaluation criterion according to the decision maker's pairwise comparisons of the criteria. The higher the weight, the more important the corresponding criterion. Next, for a fixed criterion, the AHP assigns a score to each option according to the decision maker's pairwise comparisons of the options based on that criterion. The higher the score, the better the performance of the option with respect to the considered criterion. Finally, the AHP combines the criteria weights and the options scores, thus determining a global score for each option, and a consequent ranking. As previously described, the global score for a given option is a weighted sum of the scores it obtained with respect to all the single criteria.

The AHP is a flexible and powerful tool because the scores, and therefore the final ranking, are obtained on the basis of the pairwise relative evaluations of both the criteria and the options provided by the user. The computations made by the AHP are always guided by the decision maker's experience, and the AHP can thus be considered as a tool that is able to translate the evaluations (both qualitative and quantitative) made by the decision maker into a multicriteria ranking. In addition, the AHP is simple because there is no need of building a complex expert system with the decision maker's knowledge embedded in it.

On the other hand, the AHP may require a large number of evaluations by the user, especially for problems with many criteria and options. Although every single evaluation is very simple, since it only requires the decision maker to express how two options or criteria compare to each other, the load of the evaluation task may become unreasonable. In fact the number of pairwise comparisons grows quadratically with the number of criteria and options. For instance, when comparing 10 alternatives on 4 criteria, $4 \cdot 3/2 = 6$ comparisons are requested to build the weight vector, and $4 \cdot (10 \cdot 9/2) = 180$ pairwise comparisons are needed to build the score matrix. However, in order to reduce the decision maker's workload, the AHP can be completely or partially automated by specifying suitable thresholds for automatically deciding some pairwise comparisons.

19.3 DSS Application to the Sacca di Goro (Italy)

The Sacca di Goro lagoon is located in the southern part of the Po River delta, in the province of Ferrara. It is approximately triangular in shape with a surface area of 26 Km^2 and an average depth of 1.5 m, and is connected to the sea by a mouth about 1.5 Km wide. The tidal amplitude is approximately 80 cm. The watershed of Burana-Volano is a lowland, flat basin, which is in some parts below the sea level. The watershed on the northern and eastern side is bordered by a branch of the Po river entering the Adriatic sea, and covers an area of about 3000 Km^2.

About 80% of the watershed is dedicated to agriculture, whilst the lagoon is one of the most important aquacultural systems in Italy. About 13 Km^2 of the aquatic surface (50% of the total) are exploited for farming of the Manila clam (*Tapes philippinarum*). Fishery and shellfish farming provide work, directly or indirectly, to approximately 5000 people.

Due to the large supply of nutrients, organic matter and sediments coming from the agricultural catchment, the limited water circulation characterised by little water exchange with the sea, and the intensive shellfish production, the water quality is a major problem in the Sacca di Goro. Indeed, from 1987 the Sacca di Goro has experienced an abnormal proliferation of macroalgae (*Ulva sp*). The decomposition of *Ulva* in summer causes oxygen depletion and anoxic crises. The interested reader is referred to (Viaroli et al., 2006).

As described in Table 19.1, in the Sacca di Goro lagoon the problem that the local administration had to face consisted in deciding the optimum amount of the area to be allocated for clam farming. Indicators to evaluate different allocation options include both monetary and environmental indicators. The decision variable for this problem corresponds to the number of hectares to allocate for clam farming, while the exogenous inputs are mainly represented by climate conditions and input fluxes into the lagoon, that cannot be controlled, but do affect the system status. A constraint for this problem is given by the minimum value of the farming area, since the current assignments to fishermen cannot be withdrawn.

However, being well known that shellfish farming activities are responsible for important ecosystem disruption (Zaldívar et al., 2003b), it was expected that increasing the farming area would result into a worsening of the ecosystem health.

Table 19.1 Decision problem description for the Sacca di Goro lagoon

Problem	Global objective	Criteria	Controllable variables	Uncontrollable variables	Constraints
Requests for new aquaculture concessions	Sustainable development	(i) Revenue (ii) Environmental/ economic ratio (iii) Water quality	Total area for aquaculture concessions	Climate conditions, inputs from watershed	Minimum and maximum aquaculture area

Hence, the aim of the DSS was to help the local authorities find a suitable trade-off between the socio-economic interests and the environment preservation.

19.3.1 Decision Problem Definition

The considered decision problem is summarized in Table 19.2 following the lines described in the previous sections.

The only possible action available to end users is to decide the amount of concessions (expressed in hectares) to grant to clam farmers. Minimum and maximum allocable aquaculture area constraints are present: 1300 ha corresponds to the current allocated area for aquaculture, while 1450 ha is the maximum extension that is regarded as feasible. Three criteria are considered to evaluate the performance of different control options. These criteria are related to the following indicators:

- The *Net Present Value* (NPV), to express the aquaculture revenue. This indicator is chosen to take into account only the economic aspect of the problem. Clearly, end-users aim at maximizing the aquaculture revenue to improve the social well-being.
- The *Transitional Water Quality Index* (TWQI), to express a pure environmental criterion related to water quality. Local administrators look at preserving the water quality in order to guarantee a sustainable development of the area. The TWQI was introduced in (Giordani et al., submitted) and is based on a multi-objective value function that integrates several environmental indicators: dissolved oxygen (%SAT), chlorophyll-a (mg/m^3), macroalgae coverage (% of surface area), phanerogams coverage (% of surface area), dissolved inorganic phosphorus (mmol/m^3) in water column, and dissolved inorganic nitrogen (mmol/m^3) in water column. The TWQI is computed as a weighted sum of the quality indexes determined by applying a single value function to each environmental indicator.
- The ratio of the *Wasted Exergy* (WE) to the *Net Present Value* (NPV) for the aquaculture economic sector (denoted by WE/NPV), to express a mixed environmental and economic criterion (Verdesca et al., 2006). This value can be interpreted as the amount of ecosystem "health" lost per unit of net present value. Indeed, the thermodynamic definition of *exergy* is the amount of work that a system can perform by being brought into equilibrium with its environment. Exergy attempts to account for the actual free energy of the

Table 19.2 Decision problem definition for the Sacca di Goro lagoon

Control actions	Controllable variables	Constraints	
Grant new farming concessions	Aquaculture area [ha]	min: 1300 max: 1450	
Criteria	Indicators	Objectives	Constraints
Aquaculture revenue	NPV [MEuro]	Maximize	–
Environmental vs. economic balance	WE/NPV [MJ/Euro]	Minimize	–
Water quality	TWQI [%]	Maximize	–

biomass by including the free energy that is stored in the information embodied in the biomass structure (genes). This makes it possible to use the exergy as a goal function (Bendoricchio and Jørgensen, 1997), as it measures the distance of an ecosystem from the equilibrium, i.e. from the death. The higher is the biodiversity, the larger is the value of exergy of the ecosystem, and hence the better is its health status. It follows that the smaller is the ratio WE/NPV, the more sustainable are the production activities for the environment. Thus, the objective is to minimize the WE/NPV indicator, since a smaller WE/NPV ratio corresponds to a more efficient use of the lagoon ecosystem.

19.3.2 Site Model

For site simulations, the simple zero dimensional (0D) biogeochemical model of the Sacca di Goro lagoon proposed in (Zaldívar et al., 2003a) is used. It considers the nutrient cycles in the water column as well as in the sediments. The nutrient cycles, phytoplankton, zooplankton and macro-algae dynamics, as well as shellfish farming, are modelled. The dynamics of oxygen is also simulated in order to predict anoxic crises in the lagoon. Input fluxes from the watershed are considered, as well as water exchanges with the sea. The aim of this simple model is to capture the dynamics of interest, and, at the same time, to limit the sensitivity to parameter variations and disturbances. Another advantage of a simple model is the possibility of directly incorporating it into the DSS, since the computation time of a complete simulation is relatively short. Hence, it is possible to simulate several scenarios accounting for different conditions in a reasonable time.

For modeling the spatial distribution of variables, the model considers external inputs from the watershed, although it does not take into account explicitly the geography of the area under study (at present it is not coupled with a GIS database). Note that a detailed 3D model of the Sacca di Goro is available (Marinov et al., 2006), but it requires a long computation time (few days) to complete a single simulation run, whereas the 0D model requires only few seconds.

19.3.2.1 External Factors

The *input fluxes* from the watershed into the lagoon, and the *weather conditions* are considered as external variables. In more details, as input fluxes from the watershed the model considers the daily input flow of each channel and river (Canal Bianco-Romanina, Bonello, Giralda, Po di Volano, Po di Goro) together with the nutrients loads (total and soluble reactive phosphorus, reactive silica, ammonium, nitrites and nitrates) for each input flux. The weather conditions are defined by specifying, for each day, variables like the wind speed, the global solar radiation, the temperature, and the rain precipitation.

Some reference years were obtained by specifying input fluxes and weather conditions for each day of the year. In particular, three benchmark years were considered for testing the DSS, namely a dry year (D), a normal year (N), and a wet year (W). These reference years were chosen from the period 1995–2005, for which input fluxes and weather conditions data are available.

19.3.3 DSS Results

In what follows, two different types of results are presented, concerning the application of both the DSS and the Analytic Hierarchy Process (AHP). The proposed results help show different aspects of the decision process. For a clear visualization of the results, only $n = 7$ control options are evaluated and compared, obtained by varying the aquaculture area from 1300 to 1450 ha with steps of 25 ha. Clearly, more detailed simulations can be performed, if needed. Indeed, a possible use of the DSS is first to consider a rough set of options ranging from a minimum aquaculture area (corresponding to the policy of maintaining the current situation) to a maximum area (the maximum allocable aquaculture area) and, in a second phase, to refine the search in the most promising zone.

In particular, the presented results were obtained by varying the criteria weights in order to show the ability of the DSS to model the preferences and the objectives of the end-users. Moreover, robustness issues are illustrated by showing results related to the variations of external factors. The aim is to select a control option which is robust with respect to varying climate conditions. This robustness analysis is carried out on a two- and three-year time horizon, respectively.

19.3.3.1 Varying the Criteria Weights

In this section, a single objective situation is first discussed. Clearly, when only one objective is considered the problem to select the best control option is trivial, since it suffices to run all the control options of interest and choose the one that optimises the criterion under consideration. Nevertheless, these simple cases are useful to gain some insight on the problem at hand. Second, the influence of the criteria weights in evaluating the best control option under a multicriteria point of view, is discussed.

To evaluate the best control option when only one criterion has to be optimised, the system was simulated for 1 year with seven different values of the aquaculture area. A normal reference year was assumed in all simulations. The plots of the performance indicators (NPV, WE/NPV, and TWQI) are shown in Fig. 19.5. Each plot shows the optimum value of aquaculture area

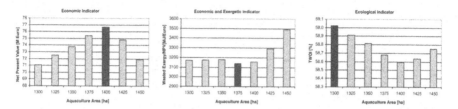

Fig. 19.5 Plot of the performance indicators versus the aquaculture area for 1 year. Normal weather conditions are assumed. The *dark bars* denote the best options

according to the corresponding criterion. A first interesting comment arising from these plots is that the maximum NPV value is not obtained when the aquaculture area takes its maximum value. Indeed, when considering only the economic indicator, 1400 ha of aquaculture area turns out to be the best option. On the other hand, when considering only the ecological indicator (TWQI), the best option is the current situation (minimum area dedicated to aquaculture). Finally when the mixed environmental and economic criterion is considered, the best option settles at 1375 ha, i.e. between the two previous solutions.

Then, the case when all the three criteria (NPW, WE/NPV and TWQI) contribute with different weights to the determination of the overall best control option, is addressed. In particular, two situations are considered; the former privileges environmental issues, while the latter aims at optimising the economic criterion.

Different criteria weights are obtained by the AHP from the following pairwise comparison matrices:

$$A_1 = \begin{bmatrix} 1 & 1/3 & 1/5 \\ 3 & 1 & 1/3 \\ 5 & 3 & 1 \end{bmatrix}, \quad A_2 = \begin{bmatrix} 1 & 3 & 5 \\ 1/3 & 1 & 3 \\ 1/5 & 1/3 & 1 \end{bmatrix}.$$

Note that A_1 privileges the environmental criterion (expressed by the third indicator, LWQI), while A_2 privileges the economic criterion (expressed by the first indicator, NPV). The corresponding AHP scores obtained by running the AHP with the two pairwise comparison matrices A_1 and A_2, are shown in Fig. 19.6. When the decision maker privileges the environmental criterion, the first option, corresponding to the policy of maintaining the current situation, turns out to be the best control option. On the other hand, when the economic criterion is assumed to be more important, the best alternative is setting the aquaculture area to 1400 ha.

By comparing the three plots shown in Fig. 19.5 (single criterion problem), and the AHP scores shown in Fig. 19.6 (multicriteria problem), it is evident that the AHP is actually able to reflect the decision maker's preferences. In fact, the results shown in Fig. 19.6 agree with those of the single criterion analysis, when the criteria weights are suitably selected. Clearly, more complex situations can be devised by setting appropriately the decision maker's preferences.

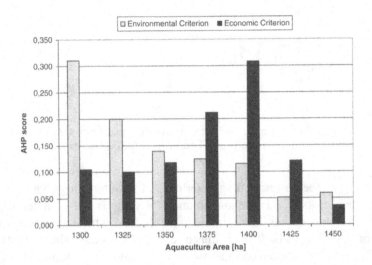

Fig. 19.6 AHP scores obtained by privileging the environmental (*light*) and economic (*dark*) criterion

19.3.3.2 Varying the External Factors

In this section, the DSS results are presented for the case when the variability of the external factors is taken into account. A time horizon of two and three years is considered. The aim is to point out a reliable control option when there is uncertainty in the external factors that may affect the decision.

In both cases (2- and 3-year time horizon) the same methodology was applied. In particular, seven alternative control options corresponding to different values of the aquaculture area were considered. The criteria pairwise comparison matrix A_2(the one privileging the economic criterion) was used in the AHP. For each external condition, obtained by combining three reference years (dry, normal and wet) over the fixed time horizon, the scores associated by the AHP to the different control options are summarized in a table. Each row of the table corresponds to a given configuration of climate conditions, and the highest AHP score is highlighted in bold. The last row of the table reports the mean AHP score for each alternative. The mean AHP score is computed by averaging the entries on the corresponding column, and can be used in order to select the best control option when external factors are uncertain.

In Table 19.3 the system performance led by each alternative on a time horizon of 2 years is evaluated under nine external conditions corresponding to all combinations of dry (D), normal (N), and wet (W) years. For instance, the row "D-D" in Table 19.3 refers to the AHP scores obtained by the seven control options when both the first and the second year are dry. The best control option in this particular weather configuration turns out to be 1375 ha. However, when considering all the combinations of weather conditions, 1325 ha is the most reliable control option in the average, followed by 1350 ha and 1300 ha. If more

Table 19.3 AHP scores under different 2-year external conditions (Dry, Normal, Wet)

	1300	1325	1350	1375	1400	1425	1450
D-D	0.099	0.149	0.255	**0.286**	0.042	0.069	0.100
D-N	0.119	0.108	0.163	0.173	**0.310**	0.096	0.031
D-W	0.311	**0.337**	0.116	0.066	0.046	0.053	0.072
N-D	0.176	0.269	**0.294**	0.102	0.074	0.049	0.037
N-N	0.124	0.096	0.099	0.150	0.186	**0.287**	0.058
N-W	0.106	0.108	0.170	0.119	**0.214**	0.082	0.201
W-N	0.194	0.242	**0.351**	0.037	0.039	0.052	0.086
W-W	0.291	**0.369**	0.119	0.038	0.039	0.061	0.083
Mean	0.190	**0.201**	0.194	0.139	0.109	0.187	0.080

aquaculture area is to be allocated, only 1425 ha seems to be a reasonably reliable solution.

In Table 19.4 the results corresponding to a 3-year time horizon are reported. In this case, 27 external conditions, corresponding to all combinations of dry (D), normal (N), and wet (W) years, are considered.

Table 19.4 AHP scores under different 3-year external conditions (Dry, Normal, Wet)

	1300	1325	1350	1375	1400	1425	1450
D-D-D	0.190	0.227	**0.306**	0.132	0.062	0.045	0.034
D-D-N	0.169	0.211	**0.330**	0.138	0.059	0.061	0.029
D-D-W	0.290	**0.335**	0.152	0.070	0.051	0.065	0.033
D-N-D	0.147	0.149	0.200	**0.327**	0.108	0.031	0.036
D-N-N	0.122	0.116	0.175	**0.341**	0.162	0.033	0.047
D-N-W	0.188	0.080	**0.351**	0.210	0.073	0.104	0.043
D-W-D	0.285	0.028	**0.343**	0.139	0.082	0.036	0.032
D-W-N	0.187	**0.248**	0.125	0.128	0.244	0.040	0.026
D-W-W	0.298	**0.353**	0.082	0.091	0.034	0.048	0.091
N-D-D	0.185	0.261	**0.325**	0.078	0.045	0.041	0.063
N-D-N	0.249	**0.351**	0.087	0.097	0.041	0.056	0.117
N-D-W	**0.456**	0.107	0.141	0.058	0.067	0.111	0.058
N-N-D	0.195	0.218	**0.274**	0.090	0.131	0.047	0.042
N-N-N	0.106	0.100	0.117	0.213	**0.307**	0.118	0.035
N-N-W	0.104	0.090	0.147	**0.204**	0.295	0.098	0.059
N-W-D	**0.248**	0.135	0.176	0.055	0.043	0.230	0.110
N-W-N	0.174	0.127	0.060	0.077	0.112	0.165	**0.281**
N-W-W	0.093	0.086	0.147	0.192	**0.289**	0.153	0.036
W-D-D	0.180	0.106	**0.344**	0.194	0.031	0.047	0.094
W-D-N	0.263	0.187	**0.286**	0.035	0.039	0.061	0.126
W-D-W	0.089	0.060	0.096	**0.358**	0.266	0.091	0.036
W-N-D	0.119	0.125	0.137	0.202	0.047	0.081	**0.285**
W-N-N	0.131	**0.286**	0.257	0.035	0.043	0.076	0.169
W-N-W	0.095	0.092	0.136	0.048	0.074	**0.370**	0.183
W-W-D	**0.423**	0.105	0.130	0.194	0.033	0.038	0.074
W-W-N	0.231	**0.330**	0.220	0.035	0.036	0.050	0.094
W-W-W	**0.445**	0.169	0.102	0.039	0.040	0.074	0.127
Mean	**0.209**	0.173	0.194	0.140	0.104	0.087	0.087

The average AHP scores are shown in the last row of the table, and the best choices fall again between 1300 and 1350 ha of aquaculture area. Note that, when the three-year time horizon is considered, the 1425 ha alternative, which was apparently a rather good solution in the 2-year time horizon analysis, is not anymore a good solution. Note also that, in a couple of weather configurations (namely, N-W-N and W-N-D), 1450 ha (i.e. the maximum allocable aquaculture area) is the optimal solution. Nevertheless, it obtains a low average score, since its performance degrades under all other weather configurations.

By comparing the last rows of Tables 19.3 and 19.4 with Fig. 19.6, it is easily observed that, when robustness issues are taken into account by considering the variability of the external factors, the DSS solution settles at low values of the aquaculture area, even though the economic criterion is privileged. This suggests that the expected economic growth related to increasing the aquaculture area (and thus the clam production) does not compensate the environmental losses.

19.4 Conclusions

It is possible to draw several conclusions from the application of the developed DSS to the Sacca di Goro lagoon.

The models used within the DSS to simulate the biological and socio-economic dynamics are of fundamental importance for the possible applications of the overall DSS. In fact, a fast 0D model allows the users to have a quick-response and easy-to-use DSS to evaluate almost in real-time the effects of possible control decisions. On the other hand, a more detailed model, such as the 3D COHERENS model, would allow the integration of the DSS with a GIS database, thus enabling the users to graphically define the aquaculture area of interest, and to evaluate the effects of different spatial allocations (not only the amount). These two alternatives are not mutually exclusive. In fact, a very promising future research direction is to develop a DSS including a still larger number of models, and use "fast" 0D models for a rough evaluation of the effects of alternative decisions, and then "slow" 3D models for accurately focusing on the most promising alternatives. In this respect, note that the DSS structure proposed in Section 19.2 is general, and scales well with different models as far as the end-users provide all the needed information.

The prototype of the proposed DSS for the Sacca di Goro lagoon was developed in Matlab^{TM} 7.0, and the tests ran on a 3 GHz Pentium 4 processor. The computation time required to run the models for a single scenario was only few seconds, while the overall 3-year time horizon test required only few minutes. Hence, the DSS adopting a 0D simplified biological model can be easily used as an online tool to evaluate different scenarios.

The results obtained by the DSS were validated by end-users during a meeting held in Ferrara (December 12–14, 2005). Moreover, based also on the results provided by the DSS, the Administration of the Province of Ferrara decided to stop the grant of new concessions for clam farming.

It can be assessed from the obtained results that the DITTY DSS represents a valuable contribution whose significance and applicability go beyond the specific objectives of the DITTY project (see also Loubersac et al., 2007, for another DSS application to a DITTY project case study). Although the DSS development was initially targeted to Mediterranean lagoons, in a wider European perspective the proposed DSS structure is in principle applicable to all types of coastal lagoons, and even more generally to "transition water systems" as defined by the Water Framework Directive. In this respect, it is also important to note that several decision support systems developed for specific applications, such as (Carvalho, 2002), (Pallottino et al., 2002) and (Mysiak et al., 2002, 2005), fit the general structure presented here.

Finally, an important feature of the DITTY DSS development is that it drove the entire study. Commonly, in many DSS projects, the DSS is considered at the very end of the project, with the only aim of adding value and justification to whatever analysis went before. More correctly, in the case of the DITTY DSS, previous steps such as data compilation, development of (simplified) models and indicators, scenario analysis, etc., were focused to the requirements of the DSS, to be sure that the information finally available was complete, meaningful, and relevant to the decision problems at hand.

References

Agnetis A, Basosi R, Caballero K, Casini M, Chesi G, Ciaschetti G, Detti P, Federici M, Focardi S, Franchi E, Garulli A, Mocenni C, Paoletti S, Pranzo M, Tiribocchi A, Torsello L, Vercelli A, Verdesca D, Vicino A (2006) Development of a decision support system for the management of Southern European lagoons. (Deliverable of the DITTY Project)

Bendoricchio G, Jørgensen SE (1997) Exergy as goal function of ecosystems dynamics. Ecological Modelling 102(1):5–15

Carvalho A (2002) Simulation tools to evaluate sustainable development in coastal areas. In: Proc. Littoral 2002, The Changing Coast, EUROCOAST/EUCC, Porto, Portugal

Casini M, Mocenni C, Paoletti S, Vicino A (2005) A decision support system for the management of coastal lagoons. In: Proc. 16th IFAC World Congress, Prague, Czech Republic

Casini M, Mocenni C, Paoletti S, Pranzo M (2007) Model-based decision support for integrated management and control of coastal lagoons. In: Proc. of the European Control Conference (ECC'07), Kos, Greece

European Commission (2000) Directive 2000/60/EC of the European Parliament and of the Council of 23 October 2000 establishing a framework for Community action in the field of water policy. O.J. L. 327, pp. 1–73

European Environmental Agency (1999) Environmental indicators: typology and overview. (Technical Report no. 25)

Figueira J, Greco S, Ehrgott M (eds) (2005) Multiple criteria decision analysis: State of the art surveys. International Series in Operations Research and Management Science, vol. 78, Springer-Verlag, Boston, MA

Giordani G, Zaldívar JM, Viaroli P (2008). Simple tools for assessing water quality and trophic status in transitional water ecosystems. Ecological Indicators (submitted).

Letcher RA, Giupponi C (eds) (2005) Special issue on policies and tools for sustainable water management in the European Union. Environmental Modelling and Software 20(2): 93–271

Loubersac L, Do-Chi T, Fiandrino A, Jouan M, Derolez V, Lemsanni A, Rey-Valette H, Mathe S, Page S, Mocenni C, Casini M, Paoletti S, Pranzo M, Valette F, Serais O, Laugier T, Mazouni N, Vincent C, Troussellier M, Aliaume C (2007) Microbial contamination and management scenarios in a Mediterranean coastal lagoon (Etang de Thau, France): Application of a decision support system within the Integrated Coastal Zone Management context. Transitional Waters Bulletin 1(1):107–127

Marinov D, Norro A, Zaldívar JM (2006) Application of COHERENS model for hydrodynamic investigation of Sacca di Goro coastal lagoon (Italian Adriatic Seashore). Ecological Modelling 193(1–2):52–68

Mysiak J, Giupponi C, Fassio A (2002) Decision support for water resource management: An application example of the MULINO DSS. In: Rizzoli AE, Jakeman AJ (eds) Integrated Assessment and Decision Support, vol. 1, pp. 138–143, iEMSs

Mysiak J, Giupponi C, Rosato P (2005) Towards the development of a decision support system for water resource management. Environmental Modelling and Software 20:203–214

Pallottino S, Sechi GM, Zuddas P (2002) A DSS for water resources management under uncertainty. In: Rizzoli AE, Jakeman AJ (eds) Integrated Assessment and Decision Support, vol. 1, pp. 96–101, iEMSs

Power DJ (1997) What is a DSS? The On-Line Executive Journal for Data-Intensive Decision Support 1(3)

Roy B (1991) The outranking approach and the foundations of ELECTRE methods. Theory and Decision 31(1):49–73

Saaty TL (1980) The Analytic Hierarchy Process. McGraw-Hill, New York

Verdesca D, Federici M, Torsello L, Basosi R (2006) Exergy-economic accounting for sea-coastal systems: A novel approach. Ecological Modelling 193(1–2):132–139

Viaroli P, Bartoli M, Giordani G, Azzoni R, Nizzoli D (2003) Short term changes of benthic fluxes during clam harvesting in a coastal lagoon (Sacca di Goro, Po river delta). Chemistry and Ecology 19:189–206

Viaroli P, Bartoli M, Giordani G, Magni P, Welsh DT (2004). Biogeochemical indicators as tools for assessing sediment quality/vulnerability in transitional aquatic ecosystems. Aquatic Conservation: Marine and Freshwater ecosystems 14, S19–S29.

Viaroli P, Giordani G, Bartoli M, Naldi M, Azzoni R, Nizzoli D, Ferrari I, Zaldívar JM, Bencivelli S, Castaldelli G, Fano EA (2006). The Sacca di Goro lagoon and an arm of the Po River. In: P.J. Wangersky (Editor), The Handbook of Environmental Chemistry. Volume H: Estuaries Springer-Verlag GmbH, Berlin, pp. 197–232.

Wierzbicki AP (1998) Reference point methods in vector optimization and decision support. (Interim Report IR-98-017, International Institute for Applied Systems Analysis, Laxenburg, Austria)

Zaldívar JM, Cattaneo E, Plus M, Murray CN, Giordani G, Viaroli P (2003a) Long-term simulation of main biogeochemical events in a coastal lagoon: Sacca di Goro (Northern Adriatic Coast, Italy). Continental Shelf Research 23(17–19):1847–1875

Zaldívar JM, Plus M, Giordani G, Viaroli P (2003b) Modelling the impact of clams in the biogeochemical cycles of a Mediterranean lagoon. In: Proc. 6th International Conference on the Mediterranean Coastal Environment, MEDCOAST'03, pp. 1291–1302

Chapter 20
Decision Support Systems (DSSs) for Inland and Coastal Waters Management – Gaps and Challenges

Elena Semenzin and Glenn W. Suter II

20.1 Introduction

The chapters of this section provided an overview of decision support systems for management of inland and coastal waters. Specifically, Chapters 14 and 15 present a review of DSSs developed and used in US and Europe respectively. These chapters explain the main challenges encountered by decision makers, the requirements of current legislations and how existing DSSs support or could support the decisional processes. These chapters also explain why inland and coastal waters problems are characterized by high complexity. Specifically, aquatic systems are dynamic and thus their environmental condition highly depends on climate factors and external input. Also, they usually include a variety of areas with different environmental features (i.e., physico-chemical and biological characteristics), and they are affected by multiple stressors at local, regional, and global scales. Subsequent chapters describe actively-used DSSs for inland and coastal waters management. Each of those chapters provide information regarding their system's framework and functionality, system structure, decision aspects, stakeholder involvement, applications and ongoing development, if applicable. Finally, this chapter summarizes identified gaps in existing inland and coastal waters management DSSs and the challenges faced by anyone attempting to build and implement a DSS.

20.2 Gaps

As highlighted in Chapters 14 and 15, existing (or still in development) DSSs in both the US and Europe cover many problems encountered in water management and offer potentially appropriate tools for supporting decisional

E. Semenzin (✉)
IDEAS (Interdepartmental Centre for Dynamic Interactions between Economy, Environment and Society), University of Ca' Foscari, San Giobbe 873, 30121, Venice, Italy; Consorzio Venezia Ricerche, Via della Libertà 2-12, Marghera-Venice, Italy
e-mail: semenzin@unive.it

A. Marcomini et al. (eds.), *Decision Support Systems for Risk-Based Management of Contaminated Sites*, DOI 10.1007/978-0-387-09722-0_20, © Springer Science+Business Media, LLC 2009

processes. Available DSSs range from simple to very complex software systems and from generic systems to systems targeted to a specific problem or location. They also include various tools such as Geographic Information System (GIS) and Multi Criteria Decision Analysis (MCDA). However, DSSs are not routinely used by decision makers and thus the main gap to be bridged is between DSSs (or DSSs' developers) and users.

Decision makers are usually reluctant to use DSSs for two reasons.

(1) They perceive DSSs to be black boxes, so they do not trust them and prefer to make decisions by themselves.
(2) They do not want to spend more time and effort on a tool that makes the problem more complex. DSSs are not always quick and easy in their application, but they can help the user to better do their job by addressing issues that they previously ignored or by providing a robust and repeatable structure for the decisional process.

20.3 Challenges

The development of DSSs for environmental management is a new enterprise; for this reason, challenges to bridging the gap between DSSs and users are fundamental. That is, developers need to know what sorts of DSSs are useful in what circumstances and how to best go about developing them.

20.3.1 Effective Presentation of DSSs

The functionalities and capabilities of a DSS, together with its limitations, must be clearly presented to the potential users. DSSs often use complex and technical language and mathematics. What a DSS can do and how it is done must be understandable by different potential users with different degrees of technical expertise. Therefore, more effort should be put in providing user-friendly interfaces, detailed manuals of use, and appropriate training courses. This would help the users to build confidence in the system, to properly apply it, and to correctly interpret and use the results of DSS implementation.

20.3.2 Involvement of Users

As pointed out also in Chapter 13, a group of potential users should always be involved in the development of a DSS in order to assure that their relevant needs are addressed. The system may still be a black box to other users, but they would at least have the assurance that fellow users contributed the content of the box and that the system is reasonable and gives useful results. Moreover, it would allow the DSS to be updated over time according to additional users' needs and expectations.

20.3.3 Flexibility

As defined in Chapter 15, flexibility "is the characteristic of the system to be adaptable, in terms of change of input parameters or addition of new models and functionalities. It is also linked to the possibility to be adaptable to different coasts or basins than those of the case-studies." Adding flexibility to the developed system is a difficult task but it would enhance its use. For example, a DSS adaptable to different coastal areas can be adopted at national level in order to manage the whole coastline using a common approach.

20.3.4 Degree of Automation

Decision support systems range from those that automate the decision making process (e.g., expert systems that diagnose diseases from symptoms) to those that simply provide appropriate information. In general, environmental decisions are too complex to fully automate, but some such as MODELKEY (Chapter 16) provide assumptions, models, and a decision structure that automate much of the assessment process. CADDIS supports decisions concerning the cause of biological impairments observed at a site by providing an inferential structure and some supporting information and tools, but the actual inferences and model development must be performed ad hoc. Finally, BASINS is a set of tools that are easily linked to solve a variety of pollution problems in watersheds. DSS designers must balance the convenience and consistency of a relatively automated system against the transparency and flexibility of systems that provide technical support without directing the decision process. Hence, DSS developers are challenged to either (1) work closely with a specific set of decision makers to provide the appropriate degree of automation of the assessment and decision making processes or (2) develop systems that allow users to choose among optional implementations that determine the degree to which the software automates the process.

20.3.5 Additional Obstacles in Using DSSs

Often DSSs are not used for very practical reasons. One is that commercial DSSs or those that use commercial software may be expensive to license or purchase. Those costs may exclude some users. The adoption of open source software may increase their use (see, for example, Chapter 18).

Another obstacle could be the language in which the DSSs are written. The majority of DSSs is written in English in order to be understandable and applicable worldwide. However in non-English-speaking countries, decision makers and stakeholders very often are not familiar with English and would prefer to use a system written in their own language to avoid errors or misinterpretations.

20.4 Conclusions

Additional suggestions on how to improve DSSs for inland and coastal management and to increase their application appear in the section's introductory chapters (14 and 15). Chapter 14 offers some suggestions to make sophisticated DSSs more useful, particularly by incorporating Multi Criteria Decision Analysis (MCDA). Chapter 15 highlights some characteristics such as ease of use and low cost that would make a DSS more appealing to decision makers. However, the greatest needs are for experience in using these systems, feedback from users, and monitoring of the results of remediation and restoration. That is, do DSSs actually facilitate decisions and lead to improvements in the environment?

Index